physica status solidi (b)

ISSN 0370-1972 · SUBJECT AND AUTHOR INDEX (b) VOLUMES 91 TO 96 · 1979

ISSN 0031-8965 · SUBJECT AND AUTHOR INDEX (a) VOLUMES 51 TO 56 · 1979

physica status solidi (a)

THEORIA CUM PRAXI

AKADEMIE-VERLAG · BERLIN

Classification Scheme

1. Structure of Crystalline Solids
   - 1.1 Perfectly Periodic Structure
   - 1.2 Solid-State Phase Transformations
   - 1.3 Alloys. Metallurgy
   - 1.4 Microstructure (Magnetic Domains See 18; Ferroelectric Domains See 14.4.1)
   - 1.5 Films
   - 1.6 Surfaces
2. Non-Crystalline State
3. Crystal Growth
4. Bonding Properties
5. Mössbauer Spectroscopy
6. Lattice Dynamics. Phonons
7. Acoustic Properties
8. Thermal Properties
9. Diffusion
10. Defect Properties (Irradiation Defects See 11)
    - 10.1 Metals
    - 10.2 Non-Metals
11. Irradiation Effects (X-Ray Diffraction Investigations See 1 and 10)
12. Mechanical Properties (Plastic Deformations See 10)
    - 12.1 Metals
    - 12.2 Non-Metals
13. Electron States
    - 13.1 Band Structure
    - 13.2 Fermi Surfaces
    - 13.3 Surface and Interface States
    - 13.4 Impurity and Defect States
    - 13.5 Elementary Excitations (Phonons See 6)
      - 13.5.1 Excitons
      - 13.5.2 Plasmons
      - 13.5.3 Polarons
      - 13.5.4 Magnons
14. Electrical Properties. Transport Phenomena
    - 14.1 Metals. Semi-Metals
    - 14.2 Superconductivity. Superconducting Materials and Devices
    - 14.3 Semiconductors
      - 14.3.1 Films
      - 14.3.2 Surfaces and Interfaces
      - 14.3.3 Devices. Junctions (Contact Problems See 14.3.4)
      - 14.3.4 High-Field Phenomena, Space-Charge Effects, Inhomogeneities, Injected Carriers (Electroluminescence See 20.3; Junctions See 14.3.3)
    - 14.4 Dielectrics
      - 14.4.1 Ferroelectrics
15. Thermoelectric and Thermomagnetic Properties
16. Photoconductivity. Photovoltaic Effects
17. Emission of Electrons and Ions
    - 17.1 Field Emission Microscope Investigations
18. Magnetic Properties
    - 18.1 Paramagnetic Properties
    - 18.2 Ferromagnetic Properties
      - 18.2.1 Ferromagnetic Films
    - 18.3 Ferrimagnetic Properties
    - 18.4 Antiferromagnetic Properties

(Continued on over three)

physica status solidi (b)

basic research

physica status solidi (a)

applied research

SUBJECT AND AUTHOR INDEX

Volumes 91 to 96, 1979 — Volumes 51 to 56, 1979

ISSN 0370-1972/0031-8965

Edited by

E. BORCHARDT

AKADEMIE-VERLAG · BERLIN

Subscriptions and orders for this Subject and Author Index
or single copies of physica status solidi (b) and physica status solidi (a) should be directed
in the DDR:

to the Postzeitungsvertrieb,
to a book-shop, or
to the Akademie-Verlag, DDR-1080 Berlin, Leipziger Straße 3 – 4;

in the other socialist countries:
to a book-shop for foreign language literature or
to the competent news-distributing agency;

in the BRD and BERLIN (WEST):
to a book-shop or
to the wholesale distributing agency Kunst und Wissen, Erich Bieber, D-7000 Stuttgart 1, Wilhelmstr. 4 – 6;

in AUSTRIA:
to the Globus-Buchvertrieb, A-1201 Wien, Höchstädtplatz 3;

in the other Western European countries:
to Kunst und Wissen Erich Bieber GmbH, CH-8008 Zürich, Dufourstr. 51;

in USA and CANADA
to Verlag Chemie International, Inc., Plaza Centre, Suite E, 1020 N.W. 6th Street, Deerfield Beach, FL 33 441, USA:

in other countries:
to the international book- and journal-selling trade,
to Buchexport, Volkseigener Außenhandelsbetrieb der Deutschen Demokratischen Republik, DDR-7010 Leipzig, Postfach 160, or
to the Akademie-Verlag, DDR-1080 Berlin, Leipziger Straße 3 – 4.

Schriftleiter und verantwortlich für den Inhalt der Zeitschriften:
Professor Dr. Dr. h. c. P. Görlich, DDR-1020 Berlin, Neue Schönhauser Str. 20 bzw. DDR-6900 Jena, Schillbachstr. 24.

Herausgeber und verantwortlich für den Inhalt des Registers:
Dipl.-Phys. E. Borchardt, DDR-1020 Berlin, Neue Schönhauser Straße 20.

Verlag:
Akademie-Verlag, DDR-1080 Berlin, Leipziger Straße 3–4; Fernruf: 2236221 und 2236229; Telex-Nr.: 114420; Bank: Staatsbank der DDR, Berlin, Kto.-Nr.: 6836-26-20712.

Veröffentlicht unter der Lizenznummer 1310 und 1620 des Presseamtes beim Vorsitzenden des Ministerrates der Deutschen Demokratischen Republik.

Gesamtherstellung:
VEB Druckerei „Thomas Müntzer", DDR-5820 Bad Langensalza.

Bestellnummer des Registers 1068/85 R 12.

Printed in the German Democratic Republic.
AN(EDV) 20 831/20 735

Preface

The twelfth index volume contains volumes 91 to 96 of "physica status solidi (b) — basic research" as well as volumes 51 to 56 of "physica status solidi (a) — applied research", i.e. all volumes having appeared in 1979 containing 1515 papers.

The general scheme of the last index volume was maintained. The basis for quoting is the subject classification given for every paper in the respective issue. Nevertheless, before taking over these subject classifications into the index volume they were once more carefully revised and in some cases improved and completed in order to get a most uniform and clear classification for all papers. The average number of quotations per paper in this index volume is nearly the same as in the last one, namely about 3.0.

Again I have to express my thanks to the readers for their interest, and I will always readily take up any critical remarks and proposals for improvements. I am especially obliged to Prof. Dr. P. Görlich, Prof. Dr. E. Gutsche, Dr. H.-J. Hänsch, Dr. H. Lange, Dr. S. Oberländer, and Dipl. Phys. K. Müller for giving me advice and unselfish help. Furthermore I have to thank Mrs. E. Schwan for her exact and reliable secretary's work at the manuscript.

Berlin, January 1980

Eva Borchardt

Classification Scheme

1. Structure of Crystalline Solids
   1.1 Perfectly Periodic Structure
   1.2 Solid-State Phase Transformations
   1.3 Alloys. Metallurgy
   1.4 Microstructure (Magnetic Domains See 18; Ferroelectric Domains See 14.4.1)
   1.5 Films
   1.6 Surfaces
2. Non-Crystalline State
3. Crystal Growth
4. Bonding Properties
5. Mössbauer Spectroscopy
6. Lattice Dynamics. Phonons
7. Acoustic Properties
8. Thermal Properties
9. Diffusion
10. Defect Properties (Irradiation Defects See 11)
    10.1 Metals
    10.2 Non-Metals
11. Irradiation Effects (X-Ray Diffraction Investigations See 1 and 10)
12. Mechanical Properties (Plastic Deformations See 10)
    12.1 Metals
    12.2 Non-Metals
13. Electron States
    13.1 Band Structure
    13.2 Fermi Surfaces
    13.3 Surface and Interface States
    13.4 Impurity and Defect States
    13.5 Elementary Excitations (Phonons See 6)
        13.5.1 Excitons
        13.5.2 Plasmons
        13.5.3 Polarons
        13.5.4 Magnons
14. Electrical Properties. Transport Phenomena
    14.1 Metals. Semi-Metals
    14.2 Superconductivity. Superconducting Materials and Devices
    14.3 Semiconductors
        14.3.1 Films
        14.3.2 Surfaces and Interfaces
        14.3.3 Devices. Junctions (Contact Problems See 14.3.4)
        14.3.4 High-Field Phenomena, Space-Charge Effects, Inhomogeneities, Injected Carriers (Electroluminescence See 20.3; Junctions See 14.3.3)
    14.4 Dielectrics
        14.4.1 Ferroelectrics
15. Thermoelectric and Thermomagnetic Properties
16. Photoconductivity. Photovoltaic Effects
17. Emission of Electrons and Ions
    17.1 Field Emission Microscope Investigations

18. Magnetic Properties
    18.1 Paramagnetic Properties
    18.2 Ferromagnetic Properties
        18.2.1 Ferromagnetic Films
    18.3 Ferrimagnetic Properties
    18.4 Antiferromagnetic Properties

19. Magnetic Resonance

20. Optical Properties
    20.1 Spectra. Optical Constants (X-Ray Spectra See 20)
    20.2 Lasers
    20.3 Luminescence (X-Ray Spectra See 20)

21. Metals. Alloys
    21.1 Transition Metals (Group Sc to Zn) and Their Alloys
        21.1.1 Fe and Fe Alloys
    21.2 Alkali Metals
    21.3 Alkali-Earth Metals
    21.4 Rare-Earth Metals
    21.5 Actinides
    21.6 Noble Metals
    21.7 Semi-Metals (Semiconducting Alloys See 22.7)

22. Semiconductors and Ionic Crystals
    22.1 Elements
        22.1.1 Germanium
        22.1.2 Silicon
        22.1.3 Group VI Elements
    22.2 III–V Compounds
        22.2.1 Arsenides
        22.2.2 Phosphides
        22.2.3 Antimonides
        22.2.4 Mixed Crystals
    22.3 IV–IV, II–V, and III–VI Compounds
        22.3.1 Mixed Crystals
    22.4 II–VI Compounds
        22.4.1 Sulphides
        22.4.2 Selenides
        22.4.3 Tellurides
        22.4.4 Mixed Crystals
    22.5 Halides
        22.5.1 Silver Halides
        22.5.2 Alkali Halides
        22.5.3 Alkali-Earth Halides
        22.5.4 Mixed Crystals
    22.6 Simple Oxides (CdO, ZnO See 22.4)
        22.6.1 Mixed Crystals
    22.7 Semiconducting Intermetallic Compounds
        22.7.1 Mixed Crystals
    22.8 Tertiary and Higher Compounds
        22.8.1 Salts of Oxy-Acids (Carbonates, Phosphates, Silicates Including Mica etc.)
        22.8.2 Spinels, Garnets, Ferrites
    22.9 Organic Semiconductors

23. Solidified Gases

For comments on the use of the classification scheme see p. 6.

Some Comments on the Use of the Subject and Author Index

The classification system[1]) comprises 20 broad sections each devoted to a major field of solid state physics and 3 broad sections devoted to the main groups of substances. A section may contain a number of subsections dealing with aspects of this field or special substance, respectively, which tend to receive particular attention. A paper dealing with this field will either be classified under one or more of the subsections or, if none of these are relevant, it will just appear under the general section heading (i.e. sections and subsections are equivalent in this sense). For example, a paper dealing with problems of paramagnetism would appear under 18.1 but not 18, whereas a paper on diamagnetism would be quoted under 18; a paper dealing with GaP would appear under 22.2.2, not under 22 or 22.2, whereas a paper on AlN would be quoted under 22.2, and a paper on a II–IV or a IV–VI compound would be quoted under 22.

All papers will be quoted under each relevant section. There is made a distinction between one or two principal entries (refering to the main subject matter of the paper) and secondary entries (refering to subsidiary subject matter and substance, respectively). In the principal entries the title of the paper is given in its original language and in English. In addition, every principal entry is marked by * in front of the title of the paper. In the secondary entries the title is only given in English.

Each entry gives the volume, the number of the issue, the initial and final page numbers, and the year in which the paper appeared. A paper of physica status solidi (b) or physica status solidi (a) is marked by (b) or (a) in front of the line, respectively. The letter "K" before the page numbers indicates a short note published in the offset part of the journal. Each entry is followed by references to the other entries. These references give the subject classifications and the pages in the index where the other entries appear. The references to the principal entries are given in italics.

Review articles appear at the head of the list of entries under a given subject and are given the subtitle "Review Articles". These articles are listed in the chronological order of their publication, irrespective of being a paper of physica status solidi (b) or physica status solidi (a). The subject classification for a review article consisting of several parts is based on the article as a whole rather than on the individual parts.

The entries for original papers, short notes, and errata are made together in a single list for each subject and this follows the entries for review articles. These are again ordered according to data of publication, irrespective of being a paper of physica status solidi (b) or physica status solidi (a). Original papers are sometimes published in several parts under the same principal title (in general with special subtitles). The various parts of papers of this type are treated separately and each part is classified according to its particular contents.

In quoting errata the following rule is obeyed. If the erratum refers to a paper quoted in the preceding index, it is listed in the chronological order of its publication, otherwise the erratum is quoted together with the paper to which it refers.

---

[1]) See cover two and three, and pages 4 and 5. For the case of losing the cover by bookbinding we repeat the classification scheme on pages 4 and 5.

In using the subject index attention should be given to the following details:

subject	quoted in	remarks
representations of space groups	1 and 13.1	
Alfven waves, helicon waves	13 and 20	
hot electrons	13 and 14.3.4	
transport phenomena	14 or subsection of 14	if not clearly to be assigned to 8 or 15
galvanomagnetic effects	14 or subsection of 14	not in 18
electro-optical effects	14 and 20 (or subsection of 14 and 20)	
piezoelectricity, electrostriction	14 or subsection of 14	not in 12 or subsection of 12; different electromechanical effects are quoted in 12 or subsection of 12
ionic conduction	14.4	not in 14.3
magneto-optical effects	18 and 20 (or subsection of 18 and 20)	possibly in additional sections, e. g. Faraday effect in 13
ferromagnetic films	18.2.1	not additionally in 1.5
X-ray spectra	20	but not X-ray studies of structure
MIS and MOS structures	the section assigned to the semiconductor	not in the section assigned to the oxide or the metal
MIM structures	the section assigned to the insulator	generally not in the section assigned to the metal
epitaxial layers	the section assigned to the layer	not in the section assigned to the substrate

The author index gives the names of the authors in alphabetical order. Each name is followed by the bibliographic details of the paper and the numbers of the pages on which the paper appears in the subject index. A paper of physica status solidi (b) or physica status solidi (a) is again marked by (b) or (a) in front of the number of the volume, respectively. The principal entry is again given in italics. When several papers have been published by the same author they are arranged according to the dates of publication, irrespective of being a paper of physica status solidi (b) or physica status solidi (a).

Subject Index

1. Structure of Crystalline Solids

Review Articles

Original Papers and Short Notes

P. M. van den Broek
*The Clebsch-Gordan Coefficients of the Crystallographic Space Groups
(b) Vol. **94**, No. 2, p. 487–496 (1979).
See also *13*, p. *139*.

Y. Epelboin and P. Riglet
*Boundary Conditions in the Numerical Integration of Takagi-Taupin Equations. Application to the Laue and Bragg Cases
(a) Vol. **54**, No. 2, p. 547–556 (1979).
See also 22.2.4, p. 442.

H.-G. Neumann and Th. Roy
*Diffraction Investigation on the Structure of SiO_x
(a) Vol. **54**, No. 2, p. K107–K111 (1979).
See also *2*, p. *46*; 22.6, p. 483.

P. Rez
*The Absorption of Bloch Waves in High Energy Electron Diffraction
(a) Vol. **55**, No. 1, p. K79–K82 (1979).

B. Dietrich
*Sensitivity of X-Ray Diffraction Methods for the Measurement of Extended Deformation Fields
(a) Vol. **56**, No. 1, p. 199–206 (1979).
See also *10*, p. *91*.

G. Kleer, H. Schmitt, and A. Klöpperpieper
*Investigation of Stoichiometry Defects in Ammonium Rubidium Sulphate Single Crystals by Electron Probe Microanalysis
(a) Vol. **56**, No. 2, p. K75–K77 (1979).
See also 14.4.1, p. 247; 22.8.1, p. 504.

1.1 Perfectly Periodic Structure

Original Papers and Short Notes

K. Z. Botros, R. Perez, and S. S. Sheinin
*On the Relationship of the Effects of Systematic Reflections in H.C.P. Cobalt to the Critical Voltage
(a) Vol. 51, No. 1, p. 41 – 47 (1979).
See also 21.1, p. 354.

F. Studer et B. Raveau
*Erratum to: Etude par microscopie électronique des phénomènes d'intercroissance dans les siliconiobates de baryum (II): Intercroissances multiples et microdomaines
(Erratum to: Eectron Microscopic Study of Intergrowth Phenomena in Barium Siliconiobates (II):Multiple Intergrowths and Microdomains)
(a) Vol. 51, No. 1, p. 313 (1979).
See also *1.4*, p. *26*; 22.8, p. 491.

J. Przedmojski and B. Pałosz
*X-Ray Investigation of Needle-Like Crystals of $CdIn_2Se_4$
(a) Vol. 51, No. 1, p. K1 – K3 (1979).
See also 22.8, p. 491.

J. Leciejewicz, H. Ptasiewicz-Bąk, and A. Zygmunt
*Magnetic Phase Transition in UPd_2Si_2
(a) Vol. 51, No. 1, p. K71 – K73 (1979).
See also *18.2*, p. *271*; 18.4, p. 286; 22.8, p. 491.

S. S. Sheinin and B. K. Jap
*On the Validity of Assuming Symmetrical Laue Diffraction Conditions in Calculations of Diffraction Contrast
(b) Vol. 91, No. 2, p. 407 – 412 (1979).
See also 10.1, p. 92; 21.6, p. 387.

O. Alstrup
Investigation of Radiation Damage in Ion Implanted Silicon Crystals by Pendellösung Topography
(a) Vol. 51, No. 2, p. 407 – 418 (1979).
See also 10.2, p. 104; *11*, p. *119*; 22.1.2. p. 416.

E. N. Kislovskii and P. V. Petrashen
*Intensity of Laue Diffracted Beams in the General Case of Diffraction in an Elastically Bent Crystal
(a) Vol. 51, No. 2, p. 527 – 531 (1979).
See also *10.2*, p. *104*; 22.1.2, p. 416.

A. Iida and K. Kohra
*Separate Measurements of Dynamical and Kinematical X-Ray Diffractions from Silicon Crystals with a Triple Crystal Diffractometer
(a) Vol. 51, No. 2, p. 533 – 542 (1979).
See also 22.1.2, p. 416.

Th. Karakostas, G. Nouet, and P. Delavignette
*Grain Boundary Analysis in TEM (II): A Modified O-Lattice for CSL and DSC Lattice Determination
(a) Vol. 52, No. 1, p. 65–76 (1979).
See also *1.4*, p. *27*.

C. Jouanin, B. Yangui, C. Cesari, C. Boulesteix, and G. Nihoul-Boutang
*Semi-Quantitative Study of Contrast near Bend Contours of Upper Laue Areas with the Incident Electron Beam Nearly Parallel to a Crystallographic Axis
(a) Vol. 52, No. 1, p. 121–129 (1979).
See also 1.4, p. 27; 22.6, p. 478.

D. van Dyck
*Improved Methods for the High Speed Calculation of Electron Microscopic Structure Images
(a) Vol. 52, No. 1, p. 283–292 (1979).

S. Matthies
*On the Reproducibility of the Orientation Distribution Function of Texture Samples from Pole Figures (Ghost Phenomena)
(b) Vol. 92, No. 2, p. K135–K138 (1979).

H. Berger, B. Kuhrig, G. Oelgart, U. Pietsch, and D. Schikora
*Investigation of Chemical Micro-Inhomogeneities in $Bi_{100-x}Sb_x$ Single Crystals
(a) Vol. 52, No. 2, p. 427–432 (1979).
See also *1.4*, p. *27*; 22.7, p. 488.

S. B. Patil, H. V. Keer, and D. K. Chakrabarty
Structural, Electrical, and Magnetic Properties in the System $Ba_xLa_{1-x}CoO_3$
(a) Vol. 52, No. 2, p. 681–686 (1979).
See also *14*, p. *197*; *18.2*, p. *273*; 22.8, p. 493.

I. H. Ismailzade, A. Yu. Kudzin, and L. Ya. Sadovskaya
X-Ray Diffractional and Optical Investigations of Ferroelectric $SrTeO_3$
(a) Vol. 52, No. 2, p. K105–K109 (1979).
See also *14.4.1*, p. *245*; 22.8.1, p. 499.

J. Arsene, J. Lopitaux, M. Drifford et M. Lenglet
*Mise en évidence de la coordination tétraédrique de l'ion Cr^{3+} dans un chromigallate de lithium par analyse du spectre électronique
(Evidence of Tetrahedral Coordination of Cr^{3+} in a Lithium Chromium Gallate from Electronic Spectra)
(a) Vol. 52, No. 2, p. K111–K114 (1979).
See also *20.1*, p. *310*; 22.8.1, p. 499.

M. Popescu and C. Ghizdeanu
*Cation Distribution in Cobalt Ferrite-Aluminates
(a) Vol. 52, No. 2, p. K169–K172 (1979).
See also *18.3*, p. *283*; 22.8.2, p. 505.

C. MANOLIKAS, J. VAN LANDUYT, and S. AMELINCKX
*Electron Microscopy and Electron Diffraction Study of the Domain Structures, the Dislocation Fine Structure, and the Phase Transformations in β-$MoTe_2$
(a) Vol. **53**, No. 1, p. 327–338 (1979).
See also 1.2, p. 17; 22, p. 399.

A. YA. DZYUBLIK
*Dynamical Effects in Inelastic Scattering of Thermal Neutrons by a Crystal
(b) Vol. **93**, No. 2, p. 515–521 (1979).
See also *6*, p. *69*.

G. B. ABDULLAEV, K. R. ALLAKHVERDIEV, R. KH. NANI, E. YU. SALAEV, and M. M. TAGYEV
*Neutron Diffraction, Infrared, and Raman Scattering Investigations of the Layered GaS_xSe_{1-x} System
(a) Vol. **53**, No. 2, p. 549–555 (1979).
See also 6, p. 70; *20.1*, p. *314*; 22.3.1, p. 447.

O. BRÜMMER, H. R. HÖCHE, and J. NIEBER
*X-Ray Diffraction in the Bragg Case at Bragg Angels of about $\pi/2$
(a) Vol. **53**, No. 2, p. 565–570 (1979).

H. ZOGHEIB, B. PARDO et D. RENARD
*Conditions d'apparition et d'extinction en théorie dynamique, de taches de diffraction électronique interdites cinématiquement
(Conditions of Appearance and Extinction of Kinematically Forbidden Electronic Diffraction Spots in the Dynamical Theory)
(a) Vol. **53**, No. 2, p. 577–587 (1979).
See also 1.4, p. 29; 22.6, p. 481.

A. IIDA
*Applications of X-Ray Triple Crystal Diffractometry to Studies on the Diffusion-Induced Defects in Silicon Crystals
(a) Vol. **54**, No. 2, p. 701–706 (1979).
See also *10.2*, p. *111*; 22.1.2, p. 423.

P. KH. BEZIRGANYAN, KH. R. DRMEYAN, and G. M. ALADZHADZHYAN
*Fine Structure of X-Ray Interferograms (I): A Two-Crystal System with a Non-Diffracting Zone (a Narrow Gap)
(a) Vol. **54**, No. 2, p. 729–733 (1979).

A. RATUSZNA, A. PIETRASZKO, A. CHEŁKOWSKI, and K. ŁUKASZEWICZ
*The Temperature Dependence of Lattice Parameters of $KMeF_3$ and $KMn_{0.9}Me_{0.1}F_3$ Compounds (Me = Mn^{2+}, Co^{2+}, and Ni^{2+})
(a) Vol. **54**, No. 2, p. 739–743 (1979).
See also *1.2*, p. *19*; 22.8, p. 495.

M. KRYSTEK
*Lattice Parameters of $(Ba_xSr_{100-x})SO_4$ Doped with Europium
(a) Vol. **54**, No. 2, p. K133–K135 (1979).
See also 22.8.1, p. 501.

M. J. GORINGE, B. LOBERG, and D. A. SMITH
*On the Determination of Intergranular Orientation Relationships by Kikuchi Electron Diffraction
(a) Vol. 55, No. 2, p. 569–572 (1979).

N. D. ZAKHAROV, W. NEUMANN, R. HILLEBRAND, and V. N. ROZHANSKII
*Analysis of Many-Beam Lattice Images of Y–Al Garnet Crystals
(a) Vol. 55, No. 2, p. 573–582 (1979).
See also 1.4, p. 31; 22.8.2, p. 507.

TH. KARAKOSTAS, G. L. BLERIS, and J. G. ANTONOPOULOS
*Grain Boundary Analysis in TEM (III): Determination of Conditions for CSL and Application in Copper
(a) Vol. 55, No. 2, p. 801–809 (1979).
See also *1.4*, p. *32*; 21.1, p. 366.

M. PIOTROWSKI, H. PTASIEWICZ-BĄK, and A. MURASIK
*The Crystal Structures of HoF_3 and TbF_3 by Neutron Diffraction
(a) Vol. 55, No. 2, p. K163–K166 (1979).
See also 22.5, p. 462.

S. ACHARYA and O. N. SRIVASTAVA
Electronic Bandgap Measurements of SnS_2 Polytypes
(a) Vol. 56, No. 1, p. K1–K4 (1979).
See also *13.1*, p. *154*; 14.3, p. 217; 22, p. 406.

P. V. PETRASHEN and E. N. KISLOVSKII
*Experimental Study of the X-Ray Diffraction in a Bent Crystal with a Large Strain Gradient
(a) Vol. 56, No. 2, p. 663–667 (1979).
See also *10.2*, p. *117*; 22.1.2, p. 426.

A. SIEMS
*Quantitative Untersuchungen an Kikuchi-Strukturen (II): Thermisch diffuse Streuung (Quantitative Investigation of Kikuchi Structures (II): Thermal Diffuse Scattering)
(a) Vol. 56, No. 2, p. 697–709 (1979).
See also 6, p. 77; 22.1.2, p. 426.

1.2 Solid-State Phase Transformations

Review Articles

V. K. Pershin and Vl. K. Pershin
Conditions of the Existence of a Molecular Crystal as a Mesogenous System
(b) Vol. **92**, No. 1, p. 9–22 (1979).
See also 2, p. 40; *6*, p. *63*; 23, p. 515.

J. Fousek and J. Petzelt
*Changes of Refractive Indices of Crystals Induced by Structural Phase Transitions
(a) Vol. **55**, No. 1, p. 11–40 (1979).
See also 12.2, p. 133; *14.4.1*, p. *244*; 20, p. 298; 22.8.1, p. 498.

Original Papers and Short Notes

F. L. Edelman, B. N. Zaitsev, V. Z. Latuta, and A. A. Khoromenko
*The Structure of Silicon Nitride Films (II): Non-Stoichiometric Silicon Nitride
(a) Vol. **51**, No. 1, p. 49–56 (1979).
See also *1.5*, p. *34*; 22, p. 393.

W. Baden, P. C. Schmidt, and Alarich Weiss
The Intermetallic System $LiCd_{1-x}Tl_x$. X-Ray Investigations and Measurements of the Knight Shift of ^{205}Tl and ^{113}Cd
(a) Vol. **51**, No. 1, p. 183–190 (1979).
See also *1.3*, p. *22*; 19, p. 290; 21, p. 339.

Yu. A. Vodakov, E. N. Mokhov, A. D. Roenkov, and D. T. Saidbekov
Effect of Crystallographic Orientation on the Polytype Stabilization and Transformation of Silicon Carbide
(a) Vol. **51**, No. 1, p. 209–215 (1979).
See also *1.5*, p. *34*; *3*, p. *52*; 22.3, p. 444.

R. Hrabański, P. B. Sczaniecki, and J. Stankowski
*Phase Transition Effects in the EPR Spectrum of Mn^{2+} in Magnesium Fluosilicate
(a) Vol. **51**, No. 1, p. 243–247 (1979).
See also *19*, p. *290*; 22.8, p. 491.

S. K. Pabi
*Computer Simulation of the Two-Phase Diffusion-Controlled Dissolution in the Planar Finite Multilayer Couples
(a) Vol. **51**, No. 1, p. 281–289 (1979).
See also *9*, p. *85*.

I. O. Bashkin, E. G. Ponyatovskii, and M. E. Kost
*On the Possibility of Hydrogen Ordering in Rare-Earth Hydrides
(b) Vol. **91**, No. 2, p. 401–405 (1979).
See also 14.1, p. 199; 14.3, p. 210; 21.4, p. 382.

J. ALBERS
A Model Theory of Ferroelectric and Other Structural Phase Transitions of First and Second Order
(b) Vol. **91**, No. 2, p. 509–520 (1979).
See also *14.4.1*, p. *244*.

F. L. EDELMAN
*The Structure of Silicon Nitride Films (III): Oxidation of Silicon Nitride – Defects
(a) Vol. **51**, No. 2, p. 375–381 (1979).
See also *1.5*, p. *34*; 22, p. 395.

A. LE LANN and A. DUBERTRET
*A Development of Kronberg's Model for $\{10\bar{1}2\}$ Twins in H.C.P. Metals. Extension to $\{11\bar{2}0\}$ Twins
(a) Vol. **51**, No. 2, p. 497–507 (1979).
See also 1.4, p. 26; 21, p. 340.

Z. CZAPLA, T. LIS, and L. SOBCZYK
*Ferroelectric Properties of NH_4HSeO_4 Crystals
(a) Vol. **51**, No. 2, p. 609–612 (1979).
See also *14.4.1*, p. *245*; 22.8.1, p. 499.

E. V. KAPITANOV and E. N. YAKOVLEV
*Mössbauer Investigation of SnO under Hydrostatic Pressure up to 41 kbar
(a) Vol. **51**, No. 2, p. 641–644 (1979).
See also *5*, p. *59*; 22.6, p. 478.

D. COLAITIS, P. DELAVIGNETTE, D. VAN DYCK, and S. AMELINCKX
*Long Period Superstructures Related to Partial Ordering in $Ni_{3\pm x}Te_2$ as Studied by Means of Electron Diffraction
(a) Vol. **51**, No. 2, p. 657–672 (1979).
See also 1.4, p. 26; 21, p. 340.

H. SHIMIZU, H. YAMADA, N. YASUDA, and S. FUJIMOTO
*Effect of Hydrostatic Pressure on the Ferroelectric Phase Transition in NH_4IO_3
(a) Vol. **51**, No. 2, p. K121–K124 (1979).
See also *14.4.1*, p. *245*; 22.8.1, p. 499.

G. SORGE and A. ALMEIDA
Elastic Compliances and Ultrasonic Attenuation in $(NH_4)_2BeF_4$ in the Temperature Range from 170 to 190 K
(a) Vol. **51**, No. 2, p. K181–K184 (1979).
See also *14.4.1*, p. *245*; 22.8, p. 492.

T. SOMA
*Equation of State under High Pressure for Metallic Si and Ge
(b) Vol. **92**, No. 1, p. K51–K53 (1979).
See also 4, p. 55; 12.1, p. 129; 22.1.1, p. 410; 22.1.2, p. 417.

Y. Mlik, A. Daoud, and M. Couzi
*Raman Spectra and Structural Phase Transitions in Chain Compounds $(CH_3)_4NMnCl_3$ and $(CH_3)_4NCdCl_3$
(a) Vol. **52**, No. 1, p. 175–188 (1979).
See also 6, p. 66; 20.1, p. 308; 22.9, p. 510.

R. C. Ecob, J. V. Bee, and B. Ralph
*The Structure of the β-Phase in Dilute Copper–Titanium Alloys
(a) Vol. **52**, No. 1, p. 201–210 (1979).
See also *1.4*, p. *27*; 21.1, p. 357.

A. V. Gaisanyuk, K. I. Kugel, and V. A. Petrov
*Some Peculiarities of the Hall Constant in Disordered Materials: the ZrC–C System as an Example
(a) Vol. **52**, No. 1, p. K81–K83 (1979).
See also *14*, p. *197*; 22.3, p. 444.

E. Dynowska and J. Stankowski
*X-Ray Study of the Structural Phase Transition in Nickel Hexammine Perchlorate
(a) Vol. **52**, No. 2, p. 381–386 (1979).
See also 22.8, p. 493.

M. Peyrard and R. Perret
FIR Spectroscopy Study of the First-Order Phase Transition of $(CH_3CH_2NH_3)_2CdCl_4$
(a) Vol. **52**, No. 2, p. 521–528 (1979).
See also *6*, p. *67*; *20.1*, p. *310*; 22.9, p. 510.

M. E. Gurevich, A. I. Mitsek, and A. F. Zhuravlev
*Anomalies in Thermal Expansion and Heat Capacity of Ferromagnets
(b) Vol. **93**, No. 1, p. 127–134 (1979).
See also 8, p. 81; 12.1, p. 129; 18.2, p. 274; 21.1, p. 358.

B. G. S. Doman
An Improved Molecular Field Approach to the Triangular Ising Antiferromagnet
(b) Vol. **93**, No. 1, p. 147–151 (1979).
See also *18.2*, p. *274*; *18.4*, p. *287*.

C. Malinowska-Adamska and L. Wojtczak
Influence of Lattice Constant Fluctuations on the Thermal Properties of Crystals
(b) Vol. **93**, No. 1, p. K5–K8 (1979).
See also *8*, p. *81*.

M. Stock and W. Dultz
*Raman Spectroscopic Studies of the Crystalline Phases of NaCN and KCN at Pressures up to 55 kbar
(a) Vol. **53**, No. 1, p. 237–241 (1979).
See also 20.1, p. 312; 22, p. 399.

C. Manolikas, J. van Landuyt, and S. Amelinckx
Electron Microscopy and Electron Diffraction Study of the Domain Structures, the Dislocation Fine Structure, and the Phase Transformations in β-$MoTe_2$
(a) Vol. **53**, No. 1, p. 327–338 (1979).
See also *1.1*, p. *13*; 22, p. 399.

E. V. KAPITANOV and E. N. YAKOVLEV
*Mössbauer Study of Phase Transitions under High Hydrostatic Pressures (II): The Phase Transition of SnTe
(a) Vol. 54, No. 1, p. 139–143 (1979).
See also *5*, p. *61*; 22.7, p. 489.

E. BŁASZCZAK and B. WESTWAŃSKI
*Phase Transitions in Two-Sublattice Biquadratic Systems ($S = 3/2$)
(b) Vol. 94, No. 2, p. 405–411 (1979).
See also 18.3, p. 284.

S. NAKASHIMA, A. AZIZA, M. LE POSTOLLEC, and M. BALKANSKI
Raman and Infrared Spectra in Organic Layer Crystals of Squaric Acid
(b) Vol. 94, No. 2, p. 529–539 (1979).
See also 4, p. 57; *20.1*, p. *316*; 22.9, p. 512.

V. I. ZUBOV
*On the Theory of Phase Transition between a Strongly Anharmonic Crystal and Gas
(b) Vol. 94, No. 2, p. K105–K109 (1979).
See also 23, p. 515.

M. GUYMONT, D. GRATIAS et A. BISSON
*Fluctuations prétransitionnelles dans AuCu
(Pre-Transition Fluctuations in AuCu)
(a) Vol. 54, No. 2, p. 573–576 (1979).
See also 1.4, p. 29; 21.1, p. 363; 21.6, p. 388.

A. RATUSZNA, A. PIETRASZKO, A. CHEŁKOWSKI, and K. ŁUKASZEWICZ
*The Temperature Dependence of Lattice Parameters of $KMeF_3$ and $KMn_{0.9}Me_{0.1}F_3$ Compounds (Me = Mn^{2+}, Co^{2+}, and Ni^{2+})
(a) Vol. 54, No. 2, p. 739–743 (1979).
See also *1.1*, p. *13*; 22.8, p. 495.

A. K. SINGH, C. SURYANARAYANA, and O. N. SRIVASTAVA
*Electron Microscopic Studies of Phase Transformations in NiSe Thin Films
(a) Vol. 54, No. 2, p. K103–K105 (1979).
See also *1.5*, p. *35*; 21.1, p. 363.

M. RAZEGHI, J. P. BUISSON, and B. HOULIER
*EPR Investigation of Gd^{3+} and Eu^{2+} in the α- and β-Phase of Lead Phosphate
(b) Vol. 95, No. 1, p. 283–289 (1979).
See also *19*, p. *295*; 22.8.1, p. 501.

M. VAN SANDE, J. VAN LANDUYT, and S. AMELINCKX
*Direct Imaging of the Structure and Structural Defects of Rhombohedral γ-Brasses
(a) Vol. 55, No. 1, p. 41–49 (1979).
See also 1.3, p. 24; *1.4*, p. *30*; 21, p. 348.

O. BOSTANJOGLO, W. GIESE, and R. LIEDTKE
Effect of Codeposited Gases on the Crystallization of Amorphous FeNi Films
(a) Vol. 55, No. 1, p. K7–K9 (1979).
See also *1.5*, p. *36*; *2*, p. *47*; 21.1, p. 365; 21.1.1, p. 375.

F. Okuyama
*Oriented Structure Transformation from W_2C to W as Revealed by Field Electron and Transmission Electron Microscopy
(a) Vol. **56**, No. 1, p. 285–292 (1979).
See also 17.1, p. 260; 21, p. 352; 22, p. 406.

M. Hidaka, I. G. Wood, and B. J. Garrard
*Structural Phase Transitions in $CsVF_4$ above Room Temperature
(a) Vol. **56**, No. 1, p. 349–354 (1979).
See also 22.5.4, p. 476.

R. Blinc, M. Burgar, J. Slak, V. Rutar, and F. Milia
*^{13}C NMR Study of the Incommensurate Phase Transition in $[N(CH_3)_4]_2$ $ZnCl_4$
(a) Vol. **56**, No. 1, p. K65–K69 (1979).
See also 14.4.1, p. 247; *19*, p. *297*; 22.9, p. 514.

B. Lorenz and D. Ihle
*Electron–Phonon versus Coulomb Interaction Effects at the Verwey Transition of Fe_3O_4
(b) Vol. **96**, No. 2, p. 659–669 (1979).
See also 4, p. 58; 6, p. 76; 13, p. 142; 22.6, p. 485.

J. Pebler, K. Schmidt und G. Weiser
*Magnetische Suszeptibilitäts- und elektrische Leitfähigkeitsmessungen an $V_{1-x}Fe_xO_{2-x}F_x$ im Konzentrationsbereich $0 < x < 0{,}20$
(Magnetic Susceptibility and Electrical Conductivity Measurements of $V_{1-x}Fe_xO_{2-x}F_x$ in the Concentration Range $0 < x < 0.20$)
(a) Vol. **56**, No. 2, p. 457–465 (1979).
See also 13.4, p. 180; 14.3, p. 218; 18.1, p. 269; 22.6.1, p. 487.

P. E. Tomaszewski and A. Pietraszko
*Thermal Expansion of Lithium Ammonium Sulphate
(a) Vol. **56**, No. 2, p. 467–472 (1979).
See also 14.4.1, p. 247; 22.8.1, p. 503.

Y. Tanji, Y. Nakagawa, Y. Saito, K. Nishimura, and K. Nakatsuka
Anomalous Thermodynamic Properties of Iron–Nickel (F.C.C.) Alloys
(a) Vol. **56**, No. 2, p. 513–519 (1979).
See also *8*, p. *84*; 21.1, p. 368; 21.1.1, p. 377.

K. Zellama, J. F. Morhange, P. Germain, and J. C. Bourgoin
Growth Rate of Crystallization in Amorphous Germanium Produced by Ion Implantation. A Raman Spectroscopy Study
(a) Vol. **56**, No. 2, p. 717–723 (1979).
See also *2*, p. *50*; 6, p. 77; *11*, p. *126*; 20.1, p. 324; 22.1.1, p. 413.

M. Kalliomäki, V. Meisalo, and A. Laisaar
*High Pressure Transformations in Cuprous Oxide
(a) Vol. **56**, No. 2, p. K127–K131 (1979).
See also 22.6, p. 486.

1.3 Alloys. Metallurgy

Review Article

Original Papers and Short Notes

M. Ohta, M. Yamada, T. Kanadani, and A. Sakakibara

*Effects of the Fluctuation of Solute Concentration in Al–10 wt% Zn Alloy on Aging at Low Temperature

(a) Vol. **52**, No. 1, p. K21 – K24 (1979).

See also *14.1*, p. *200*; 21, p. 341; 21.1, p. 357.

R. Mosseri, C. Sella, and J. Dixmier

*X-Ray Diffraction Study of the Effect of Hydrogen Atoms on the Si–Si Atomic Short-Range Order in Amorphous Silicon

(a) Vol. **52**, No. 2, p. 475 – 479 (1979).

See also 2, p. 42; 22.1.2, p. 418.

J. de Groot, P. M. Bronsveld, and J. Th. M. de Hosson

*Superlattice Dislocations in Cu_2NiZn

(a) Vol. **52**, No. 2, p. 635 – 645 (1979).

See also *10.1*, p. *95*; 21.1, p. 358.

P. N. Koul, T. K. Hatwar, M. N. Ghatikar, and B. D. Padalia

Chemical Shifts of the L_3-Absorption Edge of Pt in the Rare Earth–Platinum Intermetallic Compounds

(b) Vol. **93**, No. 1, p. 223 – 228 (1979).

See also *20.1*, p. *311*; 21.4, p. 383.

D. van Dyck, D. Colaïtis, P. Delavignette, and S. Amelinckx

*On the Diffraction Features of Two-Layer Systems. Application to $Ni_{3\pm x}Te_2$

(a) Vol. **53**, No. 1, p. 105 – 112 (1979).

See also 21, p. 343.

F. Lukeš, E. Schmidt, J. Humlíček, M. K. Kekoua, and E. Khoutsichvili

*Determination of the Composition of GeSi Alloy by Ellipsometry

(a) Vol. **53**, No. 1, p. 321 – 325 (1979).

See also 20.1, p. 312; 22.3, p. 445.

M. J. Besnus, J. M. Bouton, and R. Clad

*Crystal Structure and Magnetic Properties of $Y_6(Fe_{1-x}Al_x)_{23}$ Compounds

(a) Vol. **53**, No. 1, p. 351 – 354 (1979).

See also *18.2*, p. *274*; 21.1.1, p. 373; 21.4, p. 383.

M. Fumeron, J. P. Guillot, A. Dauger et J. Caisso

*Etude par diffusion des rayons X, de la forme des zones de Guinier-Preston et de leur effet de taille dans un alliage Al–6,8% at. Zn

(X-Ray Scattering Study of the Shape of Gunier-Preston Zones and Their Size Effect in Al–6.8 at% Zn Alloy)

(a) Vol. **53**, No. 2, p. 605 – 610 (1979).

See also 10.1, p. 96; 12.1, p. 130; 21.1, p. 360.

H. R. Anand

Mössbauer Study of Iron Precipitates in LiF and NaF

(a) Vol. **53**, No. 2, p. 671 – 679 (1979).

See also *5*, p. *61*; 22.5.2, p. 468; 22.5.4, p. 475.

I. C. Baianu, K. Rubinson, and J. Patterson
The Observation of Structural Relaxation in FeNiPB Glass by Ferromagnetic Resonance and X-Ray Scattering
(a) Vol. **53**, No. 2, p. K133—K138 (1979).
See also *2*, p. *44*; *18.2*, p. *276*; 19, p. 293; 21.1, p. 360; 21.1.1, p. 374.

S. F. Dubinin, S. G. Teplouchov, and S. K. Sidorov
*Atomic Structure of the Invar Fe-Ni Alloys
(a) Vol. **54**, No. 1, p. 239—244 (1979).
See also 18.2, p. 277; 21.1, p. 362; 21.1.1, p. 374.

C. Templier et J. Delafond
*Etude par diffusion centrale des rayons X de la démixtion d'un alliage Al-Zn 22 at%
(Study of Segregation of an Al-22 at% Zn Alloy by Central X-Ray Diffraction)
(a) Vol. **54**, No. 1, p K17—K21 (1979).
See also 21, p. 346; 21.1, p. 362.

M. van Sande, G. van Tendeloo, S. Amelinckx, and P. Airo
*High Resolution Electron Microscopic and Electron Diffraction Study of the Au-Mg System (I): The Monoclinic $Au_{15}Mg_4$ Phase
(a) Vol. **54**, No. 2, p. 499—512 (1979).
See also *1.4*, p. *29*; 21.3, p. 381; 21.6, p. 388.

M. van Sande, J. van Landuyt, and S. Amelinckx
Direct Imaging of the Structure and Structural Defects of Rhombohedral γ-Brasses
(a) Vol. **55**, No. 1, p. 41—49 (1979).
See also *1.2*, p. *19*; *1.4*, p. *30*; 21, p. 348.

L. van Goethem, R. de Ridder, D. van Dyck, and S. Amelinckx
Study of the Ordering of Octahedral Intercalates in Transition Metal Dichalcogenides
(a) Vol. **55**, No. 1, p. 67—77 (1979).
See also *1.4*, p. *30*; 22, p. 404.

J. M. Yellup and B. A. Parker
The Determination of Compositions in Non-Homogeneous Ferromagnetic Materials by Curie Temperature Measurement
(a) Vol. **55**, No. 1, p. 137—145 (1979).
See also *18.2*, p. *278*; 21.1, p. 364.

B. K. Samantaray and G. D. Nigam
*A Simple Relation for the Additivity of the Debye Characteristic Temperature of Binary Alloys
(a) Vol. **55**, No. 1, p. K23—K26 (1979).
See also *6*, p. *73*; 8, p. 83; 21, p. 349.

A. Laupheimer and M. Wilkens
*TEM Studies of Displacement Cascades in Ion-Bombarded Cu_3Pd and Fe_3Al
(a) Vol. **55**, No. 2, p. 497—507 (1979).
See also *11*, p. *124*; 21.1, p. 365; 21.1.1, p. 376.

P. Bartuška, R. Kroggel, H. Löffler, M. Simerská, V. Syneček, and G. Wendrock
*On the Room Temperature Stability of Metastable Phases Formed at 200 °C in Al-Zn Alloys
(a) Vol. 55, No. 2, p. 673–679 (1979).
See also *10.1*, p. *99*; 21, p. 350; 21.1, p. 365.

G. van Tendeloo, M. van Sande, and S. Amelinckx
*Imaging Hexagonal Based Superstructures. A New Stable Orthorhombic Structure in the Au–Mn System
(a) Vol. 55, No. 2, p. 681–689 (1979).
See also 1.4, p. 31; 22.6, p. 389.

V. A. Goltsov, N. I. Timofeev, and I. Yu. Machikina
*The Phenomenon of Controllable Hydrogen Phase "Naklep" of Metals and Alloys
(a) Vol. 55, No. 2, p. 691–698 (1979).
See also *1.2*, p. *20*; 10.1, p. 99; 21, p. 351.

G. Inden
*Determination of the Interchange Energies $W^{(2)}_{CoSi}$, $W^{(1)}_{CoSi}$ for B.C.C. Solid Solutions from High Temperature Neutron Diffraction on Ternary Fe–Co–Si Crystals
(a) Vol. 56, No. 1, p. 177–186 (1979).
See also 1.2, p. 20; 21.1.1, p. 376.

J. U. Madsen and R. M. J. Cotterill
*On the Size Statistics of Bernal Holes
(a) Vol. 56, No. 1, p. K5–K10 (1979).
See also *2*, p. *49*.

W. O. Meyer and G. Inden
Quantitative Determination of Ordered Configurations in B.C.C. Fe–Co–Si Solid Solutions by Mössbauer Spectroscopy
(a) Vol. 56, No. 2, p. 481–486 (1979).
See also *5*, p. *62*; 21.1.1, p. 377.

1.4 Microstructure (Magnetic Domains See 18; Ferroelectric Domains See 14.4.1)

Original Papers and Short Notes

S. Hashimoto and H. Iwasaki
*A Microdomain Model Analysis for Diffuse X-Ray Scattering from Quenched CuPt Alloy
(a) Vol. **51**, No. 2, p 673–682 (1979).
See also 21.1, p. 356; 21.6, p. 387.

R. A. Varin
*The Validity of the CSL Model of Grain Boundary Structure for Describing Grain Boundary Behaviour
(a) Vol. **51**, No. 2, p. K189–K193 (1979).
See also 21, p. 340; 21.1, p. 356.

Th. Karakostas, G. Nouet, and P. Delavignette
*Grain Boundary Analysis in TEM (II): A Modified O-Lattice for CSL and DSC Lattice Determination
(a) Vol. **52**, No. 1, p. 65–76 (1979).
See also *1.1*, p. *12*.

C. Jouanin, B. Yangui, C. Cesari, C. Boulesteix, and G. Nihoul-Boutang
*Semi-Quantitative Study of Contrast near Bend Contours of Upper Laue Areas with the Incident Electron Beam Nearly Parallel to a Crystallographic Axis
(a) Vol. **52**, No. 1, p. 121–129 (1979).
See also *1.1*, p. *12*; 22.6, p. 478.

J. Gastaldi and C. Jourdan
*Observation of the Annealing Twin Growth by Synchrotron Radiation X-Ray Topography
(a) Vol. **52**, No. 1, p. 139–148 (1979).
See also *10.1*, p. *93*; 21, p. 341.

R. C. Ecob, J. V. Bee, and B. Ralph
*The Structure of the β-Phase in Dilute Copper–Titanium Alloys
(a) Vol. **52**, No. 1, p. 201–210 (1979).
See also *1.2*, p. *17*; 21.1, p. 357.

R. A. Varin
*Spreading of Extrinsic Grain Boundary Dislocations in Austenitic Steel
(a) Vol. **52**, No. 1, p. 347–356 (1979).
See also 10.1, p. 94; 21, p. 341.

H. Berger, B. Kuhrig, G. Oelgart, U. Pietsch, and D. Schikora
*Investigation of Chemical Micro-Inhomogeneities in $Bi_{100-x}Sb_x$ Single Crystals
(a) Vol. **52**, No. 2, p. 427–432 (1979).
See also *1.1*, p. *12*; 22.7, p. 488.

R. Z. Valiev, O. A. Kaibyshev, and Sh. Kh. Khannanov
Grain Boundaries during Superplastic Deformation
(a) Vol. **52**, No 2, p. 447–453 (1979).
See also *10.1*, p. *94*; 21, p. 342.

J. G. Erlings and F. W. Schapink
*Electron Diffraction from Periodic ⟨111⟩ Twist Grain Boundary Structures
(a) Vol. **52**, No. 2, p. 529–539 (1979).
See also 10.1, p. 95; 21.6, p. 387.

O. V. Presnyakova, V. I. Zaitsev, and N. A. Doroshenko
*Aluminium Polygonization under High Pressure
(a) Vol. **52**, No. 2, p. 627–633 (1979).
See also *10.1*, p. *95*; 21, p. 343.

C. S. Pande
*Transmission Electron Microscopy of Radiation Damage in Nb_3Sn
(a) Vol. **52**, No. 2, p. 687–696 (1979).
See also *11*, p. *121*; 14.2, p. 205.

V. Teodorescu, L. C. Nistor, and S. V. Nistor
*Electron Microscopy Study of Pure and Doped Synthetically Grown CaF_2 Crystals
(a) Vol. **52**, No. 2, p. 711–717 (1979).
See also 10.2, p. 106; 22.5.3, p. 473.

A. González Arias
*On the Theory of Broadening in X-Ray Diffraction Peaks Caused by Substitutional Disorder
(b) Vol. **93**, No. 1, p. 305–314 (1979).

U. Knauer
*Distribution of the Iron Dopant in Barium Titanate Ceramic, Determined by the Scanning Transmission Electron Microscope
(a) Vol. **53**, No. 1, p. 207–210 (1979).
See also 10.2, p. 108; 22.8.1, p. 500.

J. D. Stephenson, T. Tuomi, V. Kelhä, and M. Tilli
*Quasi-Lifetime Topography of Magnetic Domain Movements in Grain Oriented (100) [001] Fe–3 wt% Si Polycrystals Using White Synchrotron Radiation (II): The Influence of Small Tensile Stress
(a) Vol. **53**, No. 1, p. 271–276 (1979).
Erratum
(a) Vol. **55**, No. 1, p. 339 (1979).
See also 12.1, p. 130; *18.2*, p. *274*; 21.1.1, p. 373.

J. Gastaldi and C. Jourdan
Erratum to: New Possibilities for Recrystallization Study by X-Ray Synchrotron Radiation Topography
(a) Vol. **53**, No. 1, p. 409 (1979).
See also *3*, p. *52*; 21, p. 344.

M. J. Marcinkowski
*Unifying Concepts in the Theories of Eleasticity, Plasticity, and Dislocations
(b) Vol. **93**, No. 2, p. 703–711 (1979).
See also 10, p. 89.

D. Colaitis, D. van Dyck, P. Delavignette, and S. Amelinckx
*Direct Structure Imaging of Long Period $Ni_{3\pm x}Te_2$ Superstructures
(a) Vol. **53**, No. 2, p. 423–431 (1979).
See also 1.2, p. 18; 21.1, p. 359.

P. Köhler, W. Möhling, and M. Pasemann
*Microdefects in Silicon Different from Swirls
(a) Vol. **53**, No. 2, p. 509–517 (1979).
Erratum
(a) Vol. **56**, No. 2, p. 769 (1979).
See also 10.2, p. 109; 22.1.2, p. 420.

T. TUOMI, J. D. STEPHENSON, M. TILLI, and V. KELHÄ
*Application of Synchrotron Topography to the Study of Magnetic Domain Movements in (100) Fe–3 wt% Si Produced by a Mechanical Stress
(a) Vol. **53**, No. 2, p. 571–576 (1979).
See also 12.1, p. 130; *18.2*, p. *275*; 21.1.1, p. 374.

H. ZOGHEIB, B. PARDO and D. RENARD
Conditions of Appearance and Extinction of Kinematically Forbidden Electronic Diffraction Spots in the Dynamical Theory
(a) Vol. **53**, No. 2, p. 577–587 (1979).
See also *1.1*, p. *13*; 22.6, p. 481.

F. N. CHUKHOVSKII and A. M. ARUSTAMYAN
*A Dynamical Analytical Theory of the Dislocation Image Contrast in Electron Transmission Microscopy
(a) Vol. **54**, No. 1, p. 45–54 (1979).
See also *10*, p. *89*.

M. PASEMANN and P. WERNER
*Identification of Small Defects in Silicon
(a) Vol. **54**, No. 1, p. 179–188 (1979).
See also *10.2*, p. *109*; 22.1.2, p. 421.

F. MONTHEILLET and J. M. HAUDIN
*Coherent Precipitation near Dislocations. A Theoretical Analysis
(a) Vol. **54**, No. 1, p. 271–280 (1979).
See also *10.2*, p. *110*; 22.8.2, p. 506.

M. VAN SANDE, G. VAN TENDELOO, S. AMELINCKX, and P. AIRO
*High Resolution Electron Microscopic and Electron Diffraction Study of the Au–Mg System (I): The Monoclinic $Au_{15}Mg_4$ Phase
(a) Vol. **54**, No. 2, p. 499–512 (1979).
See also *1.3*, p. *24*; 21.3, p. 381; 21.6, p. 388.

R. RANGANATHAN and S. DATTAGUPTA
Theory of Ultrasonic Scattering by Grains in Polycrystals
(a) Vol. **54**, No. 2, p. 537–546 (1979).
See also *7*, p. *79*; 21, p. 347.

M. GUYMONT, D. GRATIAS, and A. BISSON
Pre-Transition Fluctuations in AuCu
(a) Vol. **54**, No. 2, p. 573–576 (1979).
See also *1.2*, p. *19*; 21.1, p. 363; 21.6, p. 388.

K. HEINEMANN, M. J. YACAMÁN, and H. POPPA
*A New Approach for Simultaneous Orientation and Three-Dimensional Shape Determination of Small Metal Particles
(a) Vol. **54**, No. 2, p. 675–680 (1979).
See also *1.5*, p. *35*; 21.6, p. 388.

F. Appel, U. Messerschmidt, and M. Kuna
Crack Propagation in MgO during In-Situ Deformation in the High-Voltage Electron Microscope
(a) Vol. **55**, No. 2, p. 529–536 (1979).
See also *10.2*, p. *114*; 22.6, p. 484.

V. E. Cosslett
*Penetration and Resolution in STEM and CTEM in Amorphous and Polycrystalline Materials
(a) Vol. **55**, No. 2, p. 545–548 (1979).

B. Jouffrey
*Methodical Aspects of HVEM in In-Situ Experiments in the Field of Condensed Matter
(a) Vol. **55**, No. 2, p. 549–562 (1979).
See also 21, p. 350; 22, p. 405.

G. Thomas and K. H. Westmacott
*The Role of Environmental Conditions in In-Situ Experiments in the High-Voltage Electron Microscope
(a) Vol. **55**, No. 2, p. 563–568 (1979).
See also 11, p. 124; 21, p. 350; 21.1.1, p. 376.

N. D. Zakharov, W. Neumann, R. Hillebrand, and V. N. Rozhanskii
*Analysis of Many-Beam Lattice Images of Y-Al Garnet Crystals
(a) Vol. **55**, No. 2, p. 573–582 (1979).
See also *1.1*, p. *14*; 22.8.2, p. 507.

N. Pashov, S. Simon, M. Kalitzova, M. G. Michailov, and A. Djakov
*Alignment of Defects in Ge Implanted with Te^+ Ions
(a) Vol. **55**, No. 2, p. 599–602 (1979).
See also *11*, p. *124*; 22.1.1, p. 412.

H. Blumtritt, R. Gleichmann, J. Heydenreich, and H. Johansen
*Combined Scanning (EBIC) and Transmission Electron Microscopic Investigations of Dislocations in Semiconductors
(a) Vol. **55**, No. 2, p. 611–620 (1979).
See also *10.2*, p. *114*; 14.3.3, p. 230; 22.1.2, p. 424.

G. van Tendeloo, M. van Sande, and S. Amelinckx
Imaging Hexagonal Based Superstructures. A New Stable Orthorhombic Structure in the Au–Mn System
(a) Vol. **55**, No. 2, p. 681–689 (1979).
See also *1.3*, p. *25*; 21.6, p. 389.

C. Manolikas, J. van Landuyt, R. de Ridder, and S. Amelinckx
*Electron Microscopic Study of the Domain Structure and of the Transition State in $Cu_{0.5}In_{2.5}Se_4$
(a) Vol. **55**, No. 2, p. 709–722 (1979).
See also 10.2, p. 115; 22.8, p. 496.

J. B. Bilde-Sørensen
*TEM Method for Identification of Domains in Materials Which Undergo a Pm3m → R3c or R$\bar{3}$c Phase Transition
(a) Vol. **55**, No. 2, p. 747–750 (1979).
See also *1.2*, p. *20*; 22.8.1, p. 503.

R. Schindler, J. E. Clemans, and R. W. Balluffi
*On Grain Boundary Dislocations in Plane Matching Grain Boundaries
(a) Vol. **56**, No. 2, p. 749—761 (1979).
See also *10.1*, p. *102*; 21.6, p. 390.

1.5 Films

Original Papers and Short Notes

KAMNA AGGARWAL and R. G. MENDIRATTA
*Mössbauer Investigations on Thin Films of $FeSi_2$
(a) Vol. **53**, No. 1, p. K95–K98 (1979).
See also *5*, p. *60*; 21.1.1, p. 373.

E. M. TRUKHANOV, S. I. STENIN, and A. G. NOSKOV
*Dislocations and Stresses in a Crystal with an Island Film
(a) Vol. **53**, No. 2, p. 433–440 (1979).
See also *10.2*, p. *109*; 22.1.1, p. 411.

Ż. R. ŻYTKIEWICZ
*Lattice Matching in $Ga_xIn_{1-x}As_yP_{1-y}$–GaAs and $Ga_xIn_{1-x}As_yP_{1-y}$–InP Structures
(a) Vol. **53**, No. 2, p. K165–K168 (1979).
See also *3*, p. *52*; 22.2.4, p. 442.

L. S. KHAZAN, L. A. MATVEEVA, G. N. SEMENOVA, and YU. A. TKHORIK
*Account of Dislocation Structure at Stress Measurements in Heteroepitaxial Systems (I): Theory
(a) Vol. **54**, No. 2, p. 447–456 (1979).
See also *10*, p. *90*.

A. V. DVURECHENSKII, L. P. POTAPOVA, and V. V. KALININ
*Efficiency of Defect Build-up in Ion-Implanted Polycrystalline Silicon Films
(a) Vol. **54**, No. 2, p. 457–462 (1979).
See also 10.2, p. 111; *11*, p. *123*; 19, p. 294; 22.1.2, p. 422.

L. N. ALEKSANDROV, A. S. LUTOVICH, and E. D. BELORUSETS
*The Mechanism of Silicon Epitaxial Layer Growth from Ion-Molecular Beams
(a) Vol. **54**, No. 2, p. 463–469 (1979).
See also *3*, p. *52*; 22.1.2, p. 422.

K. HEINEMANN, M. J. YACAMÁN, and H. POPPA
*A New Approach for Simultaneous Orientation and Three-Dimensional Shape Determination of Small Metal Particles
(a) Vol. **54**, No. 2, p. 675–680 (1979).
See also *1.4*, p. *29*; 21.6, p. 388.

M. YANO, T. TAKASE, and M. KIMATA
*Heteroepitaxial InSb Films Grown by Molecular Beam Epitaxy
(a) Vol. **54**, No. 2, p. 707–713 (1979).
See also *3*, p. *53*; 14.3.1, p. 220; 22.2.3, p. 440.

A. K. SINGH, C. SURYANARAYANA, and O. N. SRIVASTAVA
*Electron Microscopic Studies of Phase Transformations in NiSe Thin Films
(a) Vol. **54**, No. 2, p. K103–K105 (1979).
See also *1.2*, p. *19*; 21.1, p. 363.

L. KÜCHLER and H.-J. HUNGER
*Quantitative Electron Probe Microanalysis of Noble Gases in Sputtered Layers of Semiconductors
(a) Vol. **54**, No. 2, p. K141–K144 (1979).
See also *10.2*, p. *112*.

L. S. Khazan, L. A. Matveeva, G. N. Semenova, and Yu. A. Tkhorik
*Account of Dislocation Structure at Stress Measurements in Heteroepitaxial Systems (II): Experiment
(a) Vol. 56, No. 1, p. 47–54 (1979).
See also *10*, p. *90*.

K. Narayandas, M. Radhakrishnan, and C. Balasubramanian
Dependence of Defect Density and Activation Energy on Deposition Rates in Silver Films
(a) Vol. 56, No. 1, p. 195–198 (1979).
See also *10.1*, p. *100*; *14.1*, p. *202*; 21.6, p. 389.

J. Toušková, J. Toušek, E. Klier, and R. Kužel
Preparation and Basic Electrical Properties of CdTe Thick Films
(a) Vol. 56, No. 1, p. 315–322 (1979).
See also *14.3.1*, p. *221*; 22.4.3, p. 456.

D. Valentovič, J. Červenák, S. Luby, M. L. Aldea, and T. Boţilă
*Some Non-Equilibrium Phenomena in Sputtered CdTe Thin Films
(a) Vol. 56, No. 1, p. 341–347 (1979).
See also 13.4, p. 179; 16, p. 254; 22. 4.3, p. 457.

M. Wendt
*Detection of Thin Surface Films by Electron Probe Microanalysis in Dependence on the Substrate
(a) Vol. 56, No. 1, p. K21–K23 (1979).
See also 21.1, p. 367.

G. Marchal, Ph. Mangin, M. Piecuch, and Chr. Janot
*Some Comments about Crystallization of Fe–Si Amorphous Films
(a) Vol. 56, No. 1, p. K29–K31 (1979).
See also 2, p. 49; *5*, p. *62*; 21.1.1, p. 377.

E. M. Trukhanov, S. I. Stenin, and A. G. Noskov
*Dislocation Interactions in Germanium Covered with Dielectric Films
(a) Vol. 56, No. 2, p. 443–450 (1979).
See also 2, p. 50; *10.2*, p. *117*; 22, p. 407; 22.1.1, p. 413; 22.6, p. 485.

O. Bostanjoglo, U. Heinecke, and R. Liedtke
Possible Stabilization of Amorphous FeNi Films by Conduction Electrons
(a) Vol. 56, No. 2, p. 569–572 (1979).
See also *2*, p. *50*; 21.1, p. 368; 21.1.1, p. 377.

W. Andrä and H. Danan
*Columnar Microstructure and Magnetic Anisotropy in Thin Films
(a) Vol. 56, No. 2, p. K145–K148 (1979).
See also *18*, p. *264*.

1.6 Surfaces

Original Papers and Short Notes

L. E. Murr and O. T. Inal
The Role of Substrate Dislocations and Grain Boundaries in the Nucleation and Growth of Thin Electrochemical Overgrowths. TEM and FIM Studies
(a) Vol. **51**, No. 2, p. 345–358 (1979).
See also *1.4*, p. *26*; 1.5, p. 34; 21.1, p. 355.

B. N. Oshcherin
*On Surface Energy and Compressibility of $A^2B^4C^5_2$ Semiconducting Compounds
(a) Vol. **51**, No. 2, p. K175–K179 (1979).
See also *4*, p. *55*; 12.2, p. 133; 22.8, p. 492.

U. Berg, R. Fuhrmann, and O. Brümmer
*Angular-Dependent X-Ray Photoemission Study on the Surfaces of GaAs (111) and ($\bar{1}\bar{1}\bar{1}$) Single Crystals
(a) Vol. **53**, No. 1, p. K1–K4 (1979).
See also *17*, p. *256*; 22.2.1, p. 433.

H. Pfeiffer
*Basic Considerations for an Analytic Theory of Two-Component Crystal Growth
(b) Vol. **93**, No. 2, p. K149–K153 (1979).
See also *3*, p. *52*.

H. Pfeiffer
*On the Analytic Treatment of the Growth of Rough Interfaces
(b) Vol. **94**, No. 1, p. K11–K14 (1979).
See also *3*, p. *52*.

N. Toyoda, M. Mihara, and T. Hara
*Impurity Effect on the Formation of Terraces in GaAs LPE Growth
(a) Vol. **54**, No. 1, p. 225–230 (1979).
See also 10.2, p. 110; 22.2.1, p. 434.

J. Koshy and C. C. Ogbonna
*Etching of Synthetic Barium Fluoride Single Crystals
(a) Vol. **54**, No. 1, p. K11–K12 (1979).
See also 22.5.3, p. 473.

H. Pfeiffer
*A Simple Elementary Rate Analysis of Growth Parameters of Isolated Steps on Vicinal Faces
(b) Vol. **94**, No. 2, p. K115–K118 (1979).
See also *3*, p. *52*.

V. Gotovac and B. Lovreček
*The Temperature and Compositional Dependence of the Electrical Conductivity of Sintered Aluminium Oxide with Metal Excess
(a) Vol. **54**, No. 2, p. 745–754 (1979).
See also *14.4*, p. *242*; 15, p. 250; 22.6, p. 482.

Yu. A. Baikov, Yu. V. Zelenev, and W. Haubenreisser
*A Fluctuation Theory of Normal Crystal Growth for Metallic One-Component Systems. Two Cases of Modelling the Melt Growth Interface Motion
(a) Vol. 55, No. 1, p. 123–135 (1979).
See also *3*, p. *53*.

Th. Klupsch
*On Pair Distribution Functions and Free Energy Functionals of the Crystal–Liquid Phase Boundary
(b) Vol. 95, No. 2, p. K147–K150 (1979).
See also *13.3*, p. *161*.

M. Krohn, G. Gerth, and H. Stenzel
*Conditions of Step Decoration on Crystal Surfaces
(a) Vol. 55, No. 2, p. 375–383 (1979).
See also 22.5.2, p. 471.

M. M. Dupuis, G. Grange, and B. Mutaftschiev
*Observation of NaCl–LiCl High-Temperature Phases by Gold Decoration on the Solid–Melt Interface of Czochralski Grown Crystals
(a) Vol. 55, No. 2, p. 385–394 (1979).
See also *3*, p. *53*; 22.5.4, p. 476.

L. Malicskó and H. Stenzel
*On the Gold Decoration of KCl:Pb Crystal Surfaces
(a) Vol. 55, No. 2, p. 395–401 (1979).
See also 22.5.2, p. 471.

P. Bennema, J. P. van der Eerden, W. J. P. van Enckevort, B. van der Hoek, and K. Tsukamoto
Observation and Interpretation of Spirals for Spiral Growth from the Liquid State
(a) Vol. 55, No. 2, p. 403–413 (1979).
See also *3*, p. *53*; 22.5.2, p. 471; 22.8.1, p. 503.

W. Arabczyk, H.-J. Müssig, and F. Storbeck
*Sulphur Segregation on the (111) Iron Surface Studied by LEED, AES, and WFC
(a) Vol. 55, No. 2, p. 437–446 (1979).
See also 21.1.1, p. 375.

Ch. Kleint
*Stationary Stochastic Adparticle Flip-Flop as a Precursor to Surface Diffusion
(a) Vol. 55, No. 2, p. 447–456 (1979).
See also *9*, p. *88*; 17.1, p. 260; 22.9, p. 513.

G. V. Spivak and E. I. Rau
Measurements of Microfields on Semiconductors
(a) Vol. 55, No. 2, p. 589–597 (1979).
See also 14.3.2 p. 224; *17*, p. *258*.

R. Ducros, M. Housley, and G. Piquard
*Influence of Structural Defects, Steps, and Kinks on Oxygen Adsorption on Rhenium Single Crystals
(a) Vol. 56, No. 1, p. 187–193 (1979).
See also 10.1, p. 100; 21, p. 352.

2. Non-Crystalline State

Review Article

V. K. Pershin and Vl. K. Pershin
Conditions of the Existence of a Molecular Crystal as a Mesogenous System
(b) Vol. 92, No. 1, p. 9–22 (1979).
See also 1.2, p. 15; *6*, p. *63*; 23, p. 515.

Original Papers and Short Notes

Y. Waseda and W. A. Miller
*The Structure of Liquid Mercury over a Wide Range of Temperature (–30 to 290 °C)
(b) Vol. 91, No. 1, p. 141–146 (1979).
See also 14.1, p. 199; 15, p. 249; 21, p. 338.

D. Ploumbidis and T. Sutter
Temperature and Concentration Dependence of the Knight Shift of ^{115}In in Liquid In–Hg Alloys
(b) Vol. 91, No. 1, p. 185–188 (1979).
See also *19*, p. *289*; 21, p. 338.

T. Satow, O. Uemura, and M. Sanzyō
The Electrical Conductivity of Bismuth Chalcogenides in the Liquid State
(b) Vol. 91, No. 1, p. 325–330 (1979).
See also *14.3*, p. *209*; 22, p. 393.

R. Yamamoto, H. Matsuoka, and M. Doyama
*A Three-Dimensional Computer Simulation for the Tensile Deformation of Amorphous Iron
(a) Vol. 51, No. 1, p. 163–172 (1979).
See also *10.1*, p. *92*; 21.1.1, p. 370.

J. Schoenes, E. Kaldis, W. Thöni, and P. Wachter
Optical, Magnetic and Magnetooptical Properties of the Europium Silicate Glass $Eu_{0.14}Si_{0.31}O_{0.55}$
(a) Vol. 51, No. 1, p. 173–181 (1979).
See also *18*, p. *261*; *20.1*, p. *305*; 22.6, p. 477.

N. Saito, H. Fujiyasu, and S. Yamada
*Doping Effects of Sb and Ga on Evaporated Amorphous Ge
(a) Vol. 51, No. 1, p. 235–242 (1979).
See also *14.3.1*, p. *219*; 15, p. 249; 22.1.1, p. 410.

S. Tanigawa, K. Hinode, R. Nagai, K. Kanbe, M. Doyama, and N. Shiotani
*Positron Annihilation in Amorphous Alloys
(a) Vol. 51, No. 1, p. 249–255 (1979).
See also 21, p. 339.

I. I. Fishchuk and V. N. Rudko
*On Theory of the Hall and Faraday Effects in Disordered Systems of Two-Level Atoms
(b) Vol. 91, No. 2, p. K179–K184 (1979).
See also *14.3*, p. *210*; 22, p. 394.

T. Ayano, H. Ohno, K. Kawamura, and K. Furukawa
*Structural Analysis of Amorphous $Pd_{0.8}Si_{0.2}$ Alloy by the Dense Random Packing Model of Differently Sized Compressible Spheres
(a) Vol. 51, No. 2, p. 325–332 (1979).
See also 21, p. 340.

B. Reimer and H. Bässler
Fast Hole Transport in Polyvinylcarbazole
(a) Vol. 51, No. 2, p. 445–451 (1979).
See also *16*, p. *252*; 22.9, p. 510.

H. S. Chen and Y. Waseda
*Structure of Glassy Zr–Cu and Nb–Ni Alloys
(a) Vol. 51, No. 2, p. 593–599 (1979).
See also 21.1, p. 355.

I. Ardelean and W. Mieskes
*Diffuse Reflectance Spectra of $x Fe_2O_3 \cdot (1 - x)$ [3 $B_2O_3 \cdot PbO$] Glasses
(a) Vol. 51, No. 2, p. K129–K132 (1979).
See also *20.1*, p. *306*; 22.6.1, p. 487.

E. A. Lebedev, E. Nebauer, and P. Süptitz
*Broadening of Electrodiffusion Profiles at High Electric Fields in Glassy As_2Se_3:Cu
(a) Vol. 51, No. 2, p. K207–K211 (1979).
See also *9*, p. *85*; 14.3.4, p. 232; 22, p. 395.

H. Wieder, M. Cardona, and C. R. Guarnieri
*Vibrational Spectrum of Hydrogenated Amorphous Si–C Films
(b) Vol. 92, No. 1, p. 99–112 (1979).
See also 4, p. 55; *20.1*, p. *307*; 22.3, p. 444.

K. Handrich
*Rotons in Amorphous Ferromagnets
(b) Vol. 92, No. 1, p. K29–K32 (1979).
See also *18.2*, p. *272*.

Y. L. Rao and C. Mande
K X-Ray Absorption Discontinuity in Crystalline and Amorphous Selenium
(b) Vol. 92, No. 1, p. K61–K64 (1979).
See also *20*, p. *299*; 22.1.3, p. 428.

K. Hermann and S. Kobe
*Thermodynamical Behaviour of Finite Amorphous Ising Chains with Antiferromagnetic Interaction
(b) Vol. 92, No. 1, p. K69–K72 (1979).
See also *18.4*, p. *286*; 22.8.1, p. 499.

D. Hafner and H. Hoffmann
Microscopic and Macroscopic Inhomogeneity of Magnetization and Anisotropy in Amorphous Rare Earth/Transition Metal Films
(a) Vol. 52, No. 2, p. 549–558 (1979).
See also *18.2.1*, p. *281*; *20.1*, p. *310*; 21.1, p. 357; 21.4, p. 383.

J. Novotný and Z. Spurný
Absorption Shift in Aluminophosphate Glasses
(a) Vol. 52, No. 2, p. 603–608 (1979).
See also 13.4, p. 168; *20.1*, p. *310*; 22.8, p. 493.

V. G. Zhdanov, B. T. Kolomiets, V. M. Lyubin, and V. K. Malinovskii
*Photoinduced Optical Anisotropy in Chalcogenide Vitreous Semiconducting Films
(a) Vol. 52, No. 2, p. 621–626 (1979).
See also *20.1*, p. *310*; 22, p. 397.

H. Fiedler and H.-R. Beilich
*Influence of the Substrate Temperature on the Dielectric Relaxation of Amorphous Selenium Layers in the Glass Transition Range
(a) Vol. 52, No. 2, p. K157–K159 (1979).
See also *14.4*, p. *239*; 22.1.3, p. 428.

W. Paul and B. Fromm
Some Remarks on the AC Behaviour of Thin Sputtered Amorphous GeSe Films
(a) Vol. 52, No. 2, p. K209–K212 (1979).
See also *14.3.1*, p. *219*; 22.7, p. 489.

M. Grünewald and P. Thomas
Percolation and Effective Medium Theory of Conductivity and Thermopower in the Hopping Transport Regime
(b) Vol. 93, No. 1, p. K17–K20 (1979).
See also *14*, p. *197*; *15*, p. *250*; 22, p. 398.

G. Jones and R. A. Collins
Threshold and Memory Switching in Amorphous Selenium Thin Films
(a) Vol. 53, No. 1, p. 339–350 (1979).
See also *14.3.1*, p. *220*; *14.3.3*, p. *227*; 22.1.3, p. 428.

Kh. I. Amirkhanov, Ya. B. Magomedov, Sh. M. Ismailov, and N. L. Kramynina
*The Thermal Conductivity of Semiconductors of the System $As_2(Se_{1-x}Te_x)_3$ Depending on Temperature and Pressure
(a) Vol. 53, No. 1, p. 361–366 (1979).
See also 6, p. 68; *8*, p. *81*; 22, p. 399.

H. Fiedler and H.-R. Beilich
*Dielectric Relaxation in Amorphous Selenium Layers in the Glass Transition Range and Ageing
(a) Vol. 53, No. 1, p. K9–K10 (1979).
See also *14.4*, p. *240*; 22.1.3, p. 428.

P. Doubrava and M. Závětová
*Photo-Induced Changes in the Infrared Vibrational Spectra of As–Se Films
(a) Vol. 53, No. 2, p. K211–K213 (1979).
See also *20.1*, p. *314*; 22, p. 400.

M. Grünewald and P. Thomas
A Hopping Model for Activated Charge Transport in Amorphous Silicon
(b) Vol. 94, No. 1, p. 125–133 (1979).
See also *14.3*, p. *213*; 15, p. 250; 22.1.2, p. 420.

M. Grünewald, P. Thomas, and D. Würtz
*Simple Calculation of the Hopping Conductivity for an Exponential Band Tail
(b) Vol. 94, No. 1, p. K1–K4 (1979).
See also *14.3*, p. *214*; 22, p. 401.

M. Giehler and W. Pilz
Major Difference of the Vibrational Spectra and Structure of $(As_3S_3)_{1-x}(GeS_2)_x$ and $As_2S_3Ge_x$ Glasses
(b) Vol. 94, No. 1, p K81–K84 (1979).
See also *6*, p. *71*; *20.1*, p. *315*; 22, p. 402.

V. K. Saxena
*Erratum to: Constant Coupling Approximation for the Curie Temperature of Ferromagnets with Radom Anisotropy
(b) Vol. 94, No. 1, p. K103 (1979).
See also *18.2*, p. *276*.

M. G. Grimaldi, P. Baeri, S. U. Campisano, G. Foti, and E. Rimini
*Laser Induced As Profile Broadening in Amorphous Silicon Layers
(a) Vol. 54, No. 1, p. 55–59 (1979).
See also 9, p. 87; *11*. p. *122*; 22.1.2, p. 421.

V. I. Arkhipov, M. S. Iovu, A. I. Rudenko, and S. D. Shutov
*An Analysis of the Dispersive Charge Transport in Vitreous 0.55 As_2S_3:0.45 Sb_2S_3
(a) Vol. 54, No. 1, p. 67–77 (1979).
See also *14.3*, p. *214*; 22, p. 402.

K. Rozwadowska
*Preparation and DC Conductivity of the Low-Temperature Glassy Carbon
(a) Vol. 54, No. 1, p. 93–97 (1979).
See also *14.3.1*, p. *220*; 22.9, p. 511.

J. Hajtó and P. J. S. Ewen
*Natural Optical Activity and Related Phenomena in As_2S_3 Glasses
(a) Vol. 54, No. 1, p. 385–390 (1979).
See also *20.1*, p. *316*; 22, p. 402.

D. Weaire and F. Wooten
*Interpretation of Structural Anomalies in Hydrogenated Amorphous Silicon
(a) Vol. 54, No. 1, p. K41–K43 (1979).
See also 22.1.2, p. 421.

F. HARBACH
Disorder-Induced Raman Scattering in Non-Crystalline Films of Alkaline-Earth Halides
(b) Vol. 95, No. 1, p. 195–202 (1979).
See also *6*, p. *73*; *20.1*, p. *318*; 22.5.3, p. 474.

I. P. ZVYAGIN
*On the Hopping Mechanism of Dispersive Transport
(b) Vol. 95, No. 1, p. 227–235 (1979).
See also 13.4, p. 175; *14.3*, p. *214*.

S. UEDA and T. SHIMIZU
*NMR Studies of the Effect of Alloying Cu, Ag, Tl, I, and Ge into Amorphous As_2Se_3
(b) Vol. 95, No. 1, p. 279–282 (1979).
See also *19*, p. *295*; 22, p. 404.

L. KUBLER, A. JAÉGLÉ, and J. J. KOULMANN
The Effect of Oxygen on the Properties of Evaporated Amorphous Silicon. ESR Studies
(b) Vol. 95, No. 1, p. 307–315 (1979).
See also *13.4*, p. *175*; *19*, p. *295*; 22.1.2, p. 423.

Y. P. JOSHI
Scattering of Phonons by Density Fluctuations and Thermal Conductivity of Amorphous Solids
(b) Vol. 95, No. 1, p. 317–324 (1979).
See also *6*, p. *73*; *8*, p. *82*; 22.8, p. 496.

G. RÖPKE and V. CHRISTOPH
Derivation of DC Hopping Conductivity Formula Using the Correlation Function Method
(b) Vol. 95, No. 1, p. K15–K20 (1979).
See also *14.3*, p. *215*; 22, p. 404.

A. FREUDENHAMMER and D. DEUTGES
On the $1/T$-Behaviour of the Specific Heat of a RKKY Ising Spin Glass
(b) Vol. 95, No. 1, p. K55–K59 (1979).
See also *8*, p. *83*; 18, p. 263; 21, p. 348.

J. F. GRACZYK
*Structure of Glow Discharge Amorphous Silicon
(a) Vol. 55, No. 1, p. 231–242 (1979).
See also *1.5*, p. *36*; 22.1.2, p. 423.

O. BOSTANJOGLO, W. GIESE, and R. LIEDTKE
*Effect of Codeposited Gases on the Crystallization of Amorphous FeNi Films
(a) Vol. 55, No. 1, p. K7–K9 (1979).
See also 1.2, p. 19; *1.5*, p. *36*; 21.1, p. 365; 21.1.1, p. 375.

J. G. KLIAVA and J. J. PURĀNS
*Fine Structure Parameter Distributions for Mn^{2+} in Glasses
(a) Vol. 55, No. 1, p. K59–K62 (1979).
See also *19*, p. *296*; 22.8.1, p. 502.

J. U. Madsen and R. M. J. Cotterill
*On the Size Statistics of Bernal Holes
(a) Vol. 56, No. 1, p. K5–K10 (1979).
See also *1.3*, p. *25*.

L. N. Skuja and A. R. Silin
*Optical Properties and Energetic Structure of Non-Bridging Oxygen Centers in Vitreous SiO_2
(a) Vol. 56, No. 1, p. K11–K13 (1979).
See also *20.1*, p. *322*; 20.3, p. 336; 22.6, p. 485.

J. Purāns, J. Kliava, and I. Millere
*Computer Simulation of Mn^{2+} EPR Spectra in Glasses with $g_{ef} = 4.29$
(a) Vol. 56, No. 1, p. K25–K28 (1979).
See also *19*, p. *297*; 22.6, p. 485.

G. Marchal, Ph. Mangin, M. Piecuch, and Chr. Janot
*Some Comments about Crystallization of Fe–Si Amorphous Films
(a) Vol. 56, No. 1, p. K29–K31 (1979).
See also *1.5*, p. *37*; *5*, p. *62*; 21.1.1, p. 377.

C. A. Hogarth and M. Y. Nadeem
*The Effect of Electroforming on the Alternating Current Behaviour of Thin Amorphous Insulating Films
(a) Vol. 56, No. 1, p. K37–K40 (1979).
See also *14.3.1*, p. *221*; 22.6.1, p. 487.

M. Suzuki, K. Maruyama, M. Kumeda, and T. Shimizu
*Properties of a-$Si_{100-x}Al_x$Prepared by High Pressure RF Sputtering
(a) Vol. 56, No. 1, p. K49–K52 (1979).
See also *14.3.1*, p. *221*; 19, p. 297; 22.1.2, p. 426.

V. B. Sulimov
*Balance Equation for Non-Equilibrium Electrons in the Mobility Gap of a Disordered Semiconductor in the Presence of Light
(b) Vol. 96, No. 2, p. 507–515 (1979).
See also 13.4, p. 179; *14.3*, p. *217*; 22, p. 406.

A. Swaroop and D. M. Esterling
Interatomic Potentials from Experimental Phonon Spectra (II): General Results
(b) Vol. 96, No. 2, p. 691–696 (1979).
See also *4*, p. *58*; 6, p. 76; 12.1, p. 131; 21, p. 353; 21.1, p. 368; 21.2, p. 380; 21.6, p. 390.

J. Szczyrbowski
*On the Shape of the Absorption Tail in Amorphous Solids
(b) Vol. 96, No. 2, p. 769–774 (1979).
See also 13.1, p. 154; *20.1*, p. *323*.

V. K. Saxena
*A Variational Cluster Method for Amorphous Ferromagnets
(b) Vol. 96, No. 2, p. K125–K129 (1979).
See also *18.2*, p. *280*.

J. SCHNEIDER, K. ZAVETA, A. HANDSTEIN, F. ZOUNOVA, and V. NEKVASIL
*Temperature Dependence of the Magnetization in Amorphous $(Fe_{0.04}Ni_{0.96})_{80}P_{10}B_{10}$
(a) Vol. **56**, No. 2, p. K141—K144 (1979).
See also *18*, p. *264*; 22.8, p. 497.

3. Crystal Growth

Original Papers and Short Notes

Yu. A. Vodakov, E. N. Mokhov, A. D. Roenkov, and D. T. Saidbekov
*Effect of Crystallographic Orientation on the Polytype Stabilization and Transformation of Silicon Carbide
(a) Vol. **51**, No. 1, p. 209–215 (1979).
See also 1.2, p. 15; *1.5*, p. *34*; 22.3, p. 444.

J. Daunay, J. Baillou, J. Daunay et P. Bugnet
*Application des figures d'attaque à l'étude des mâcles de croissance dans le ZnSe monocristallin
(Application of Etch Figures to the Study of Growth Twins in Single-Crystalline ZnSe)
(a) Vol. **51**, No. 1, p. K5–K8 (1979).
See also 10.2, p. 103; 22.4.2, p. 453.

J. Šimšová, I. Tomáš, P. Görnert, M. Nevřiva, and M. Maryško
*Preparation and Properties of Co-Substituted (001) YIG Films with Rectangular Bubble Domains
(a) Vol. **53**, No. 1, p. 297–301 (1979).
See also 18.2.1, p. 281; 22.8.2, p. 506.

J. Gastaldi and C. Jourdan
*Erratum to: New Possibilities for Recrystallization Study by X-Ray Synchroton Radiation Topography
(a) Vol. **53**, No. 1, p. 409 (1979).
See also 1.4, p. 28; 21, p. 344.

H. Pfeiffer
*Basic Considerations for an Analytic Theory of Two-Component Crystal Growth
(b) Vol. **93**, No. 2, p. K149–K153 (1979).
See also *1.6*, p. *38*.

Z. R. Żytkiewicz
*Lattice Matching in $Ga_xIn_{1-x}As_yP_{1-y}$–GaAs and $Ga_xIn_{1-x}As_yP_{1-y}$–InP Structures
(a) Vol. **53**, No. 2, p. K165–K168 (1979).
See also *1.5*, p. *35*; 22.2.4, p. 442.

H. Pfeiffer
*On the Analytic Treatment of the Growth of Rough Interfaces
(b) Vol. **94**, No. 1, p. K11–K14 (1979).
See also *1.6*, p. *38*.

H. Pfeiffer
*A Simple Elementary Rate Analysis of Growth Parameters of Isolated Steps on Vicinal Faces
(b) Vol. **94**, No. 2, p. K115–K118 (1979).
See also *1.6*, p. *38*.

L. N. Aleksandrov, A. S. Lutovich, and E. D. Belorusets
*The Mechanism of Silicon Epitaxial Layer Growth from Ion-Molecular Beams
(a) Vol. **54**, No. 2, p. 463–469 (1979).
See also *1.5*, p. *35*; 22.1.2, p. 422.

P. Revesz and J. W. Mayer
*Regrowth Behaviour of Ge Implanted ⟨100⟩ Si
(a) Vol. **54**, No. 2, p. 513–516 (1979).
See also *11*, p. *123*; 22.1.2, p. 422.

O. A. Troitskii, V. I. Spitsyn, M. M. Moiseenko, and A. D. Tsiganov
*Fractionation of Stable Isotopes and Impurities during Zone Recrystallization of Metals and Preparation of Single Crystals Rich in Individual Isotopes
(a) Vol. **54**, No. 2, p. 651–654 (1979).
See also 9, p. 87; 21, p. 347; 21.1, p. 363.

M. Yano, T. Takase, and M. Kimata
*Heteroepitaxial InSb Films Grown by Molecular Beam Epitaxy
(a) Vol. **54**, No. 2, p. 707–713 (1979).
See also *1.5*, p. *35*; 14.3.1, p. 220; 22.2.3, p. 440.

Yu. A. Baikov, Yu. V. Zelenev, and W. Haubenreisser
*A Fluctuation Theory of Normal Crystal Growth for Metallic One-Component Systems. Two Cases of Modelling the Melt Growth Interface Motion
(a) Vol. **55**, No. 1, p. 123–135 (1979).
See also *1.6*, p. *38*.

K. Fischer, D. Linzen, and E. Sinn
*On the Structure of Oxidic High Temperature Solutions Used for the Growth of LPE Garnet Films
(a) Vol. **55**, No. 1, p. K75–K77 (1979).
See also 22.8.2, p. 507.

V. L. Laska, V. P. Ledovskoi, N. A. Potsar, E. Seidowski, and A. Kh. Chizhov
*Quasi-Closed Volume as the Method of Preparing the Complex Superconducting Films
(a) Vol. **55**, No. 1, p. K107–K108 (1979).
See also *14.2*, p. *207*; 21, p. 349.

G. Honjo, K. Takayanagi, K. Kobayashi, and K. Yagi
On Cluster Mobilities in Nucleation and Growth Processes of Epitaxial Thin Films
(a) Vol. **55**, No. 2, p. 353–367 (1979).
See also *1.5*, p. *36*; 21.6, p. 389; 22, p. 404.

D. Kashchiev
On the Saturation Cluster Density in Deposition of Thin Films from Vapours
(a) Vol. **55**, No. 2, p. 369–374 (1979).
See also *1.5*, p. *36*; 10.2, p. 114; 21.6, p. 389; 22.5.2, p. 471.

M. M. Dupuis, G. Grange, and B. Mutaftschiev
*Observation of NaCl–LiCl High-Temperature Phases by Gold Decoration on the Solid–Melt Interface of Czochralski Grown Crystals
(a) Vol. **55**, No. 2, p. 385–394 (1979).
See also *1.6*, p. *39*; 22.5.4, p. 476.

P. Bennema, J. P. van der Eerden, W. J. P. van Enckevort, B. van der Hoek, and K. Tsukamoto
*Observation and Interpretation of Spirals for Spiral Growth from the Liquid State
(a) Vol. **55**, No. 2, p. 403–413 (1979).
See also 1.6, p. 39; 22.5.2, p. 471; 22.8.1, p. 503.

4. Bonding Properties

Review Article

A. A. Katsnelson, V. M. Silonov, and Farid A. Khwaja
*Electronic Theory of Short-Range Order in Alloys Using the Pseudopotential Approximation and Its Comparison with Experiments
(b) Vol. **91**, No. 1, p. 11–33 (1979).
See also 1.3, p. 22; *13*, p. *136*; 13.1, p. 143; 21, p. 338.

Original Papers and Short Notes

K. N. Khanna and P. K. Sharma
*Binding Energy, Compressibility, Interionic Potential, and Transition Temperature of Simple Metals Obtained by the Pseudopotential Method
(b) Vol. **91**, No. 1, p. 251–254 (1979).
See also *12.1*, p. *129*; 14.2, p. 204; 21.2, p. 379.

J. Shanker, A. K. G. Lashkari, and V. P. Gupta
*Effect of Three-Body Interactions on the Electronic Polarizabilities and Photoelastic Behaviour of Alkali Halides
(b) Vol. **91**, No. 1, p. 263–268 (1979).
See also *12.2*, p. *133*; 14.4, p. 238; 22.5.2, p. 465.

M. S. Blanter and A. G. Khachaturyan
Stress-Induced Interaction and Ordering in B.C.C. Solutions of V, Nb, Ta, and α-Fe
(a) Vol. **51**, No. 1, p. 291–301 (1979).
See also *1.3*, p. *22*; 21.1, p. 354; 21.1.1, p. 370.

B. Perrin
*Cauchy Relations Revisited
(b) Vol. **91**, No. 2, p. K115–K120 (1979).
See also *12*, p. *127*.

A. Ya. Altshul, V. K. Bashenov, and A. G. Petukhov
*Relation between Elastic Constants from Keating's Valence-Force Field
(b) Vol. **91**, No. 2, p. K121–K123 (1979).
See also *12.2*, p. *133*; 22, p. 394.

B. N. Oshcherin
*On Surface Energy and Compressibility of $A^2B^4C^5_2$ Semiconducting Compounds
(a) Vol. **51**, No. 2, p. K175–K179 (1979).
See also *1.6*, p. *38*; 12.2, p. 133; 22.8, p. 492.

H. Wieder, M. Cardona, and C. R. Guarnieri
Vibrational Spectrum of Hydrogenated Amorphous Si–C Films
(b) Vol. **92**, No. 1, p. 99–112 (1979).
See also *2*, p. *41*; *20.1*, p. *307*; 22.3, p. 444.

T. Soma
Equation of State under High Pressure for Metallic Si and Ge
(b) Vol. **92**, No. 1, p. K51–K53 (1979).
See also *1.2*, p. *16*; 12.1, p. 129; 22.1.1, p. 410; 22.1.2, p. 417.

S. Lehri and M. P. Verma
Phonon Dispersion in Rare Gas Solids
(b) Vol. **92**, No. 2, p. 363–370 (1979).
See also *6*, p. *66*; 12.2, p. 133; **23**, p. 515.

R. P. Singh
*A Lattice Dynamical Study of Potassium Halide Solids
(b) Vol. **92**, No. 2, p. 557–563 (1979).
See also 6, p. 67; 8, p. 81; 22.5.2, p. 466.

C. Blaauw, G. Stroink, W. Leiper, and M. Zentilli
Crystal-Field Properties of Fe in Brucite $Mg(OH)_2$
(b) Vol. **92**, No. 2, p. 639–643 (1979).
See also *5*, p. *60*; 22.8, p. 492.

D. K. Bowen and V. Vitek
The Construction of Central-Force Interatomic Potentials from Whisker Deformation Data
(b) Vol. **92**, No. 2, p. 645–652 (1979).
See also *10.1*, p. *94*; *12.1*, p. *129*; 21.1.1, p. 372.

K. Hübner
*Frequency- and Wavenumber-Dependent Dielectric Function in Terms of Bond Parameters
(b) Vol. **92**, No. 2, p. K73–K76 (1979).
See also *13.1*, p. *146*; 14.4, p. 239; 22.6, p. 479.

K. Hübner
*Chemical Bond and Related Properties of SiO_2 (VI): Electronic Structure of SiO_x
(a) Vol. **52**, No. 2, p. 541–547 (1979).
See also *13.1*, p. *147*; 14.4, p. 239; 22.6, p. 479.

J. Shanker and V. C. Jain
*Analysis of the Crystal Binding of Alkali Hydrides
(a) Vol. **52**, No. 2, p. 675–680 (1979).
See also 12.2, p. 134; 22, p. 397.

L. Dagens
*Electronic States of the Dilute Transition Metal–Simple Metal Alloy According to the Resonant Model Potential Method
(b) Vol. **93**, No. 1, p. 279–284 (1979).
See also *13*, p. *138*; 21.1, p. 358.

G. Knuyt and L. M. Stals
The Na Vacancy Formation Energy Value from Phonon Data
(b) Vol. **93**, No. 1, p. K9–K11 (1979).
See also 6, p. 68; *10.1*, p. *95*; 21.2, p. 379.

K. Hübner and R. Engelke
*Statistical Investigations of the Structure of SiO_x
(a) Vol. **53**, No. 1, p. K79–K83 (1979).
See also *2*, p. *44*; 22.6, p. 480.

V. K. Bashenov, I. K. Doicho, and A. G. Petukhov
*Selection between Various Local Pseudopotentials for Silicon
(b) Vol. 93, No. 2, p. K131—K134 (1979).
See also *13.1*, p. *149*; 22.1.2, p. 420.

A. P. Pathak
"Simplicity Considerations" in Interatomic Potential for Channeling Calculations
(b) Vol. 93, No. 2, p. K181—K185 (1979).
See also *11*, p. *122*; 13, p. 139.

C. P. Mallett and P. Weinberger
A Preliminary Study of the Electronic Structure of Uranium Chalcogenides and Pnictides Using a Relativistic Single-Site Scattering Model
(b) Vol. 94, No. 1, p. 257—261 (1979).
See also *13.1*, p. *150*; 22, p. 401.

S. Nakashima, A. Aziza, M. le Postollec, and M. Balkanski
*Raman and Infrared Spectra in Organic Layer Crystals of Squaric Acid
(b) Vol. 94, No. 2, p. 529—539 (1979).
See also 1.2, p. 19; *20.1*, p. *316*; 22.9, p. 512.

G. Dräger, O. Brümmer, and J. Bonitz
*Polarized K-Emission Valence Bands, Structure, and Bonding of Some Hexagonal Close-Packed Metals
(b) Vol. 94, No. 2, p. K111—K114 (1979).
See also *13.1*, p. *151*; 20, p. 301; 21.1, p. 362; 21.3, p. 381.

M. Ciftan and V. Ruck
*The Adsorptive Chemostress Effect
(b) Vol. 95, No. 1, p. 237—242 (1979).
See also 12, p. 128.

T. Soma
*The Electronic Theory of Si–Ge Solid Solutions
(b) Vol. 95, No. 2, p. 427—431 (1979).
See also 13, p. 141; 22.3, p. 445.

O. K. Alekseeva, V. A. Onishchuk, V. P. Shantarovich, I. Ya. Dekhtyar, and V. I. Shevchenko
*Influence of Absorbed Hydrogen on Positron Lifetime Spectra in Pd, Pd–Ag, Pd–Cu
(b) Vol. 95, No. 2, p. K135—K139 (1979).
See also *10.1*, p. *99*; 21, p. 350.

R. M. Nayak and B. D. Padalia
*Chemical Shifts of the K-Absorption Edge of Mn in Some Mixed Oxidic Spinels
(b) Vol. 96, No. 1, p. 259—262 (1979).
See also *20*, p. *302*; 22.8.2, p. 507.

D. M. Esterling and A. Swaroop
*Interatomic Potentials from Experimental Phonon Spectra (I): Prototypes
(b) Vol. 96, No. 1, p. 401—411 (1979).
See also 2, p. 48; 6, p. 75; 12.1, p. 131; 21.1, p. 366; 21.1.1, p. 376; 21.2, p. 379.

5. Mössbauer Spectroscopy

Original Papers and Short Notes

P. P. SERGIN, F. S. NASREDINOV, and A. SH. BAKHTIYAROV
*Study of the Charge Exchange Process of Iron Impurity Atoms in GaAs aud GaP by Mössbauer Spectroscopy
(b) Vol. **91**, No. 1, p. 35–41 (1979).
See also 22.2.1, p. 432; 22.2.2, p. 437.

T. ZEMČÍK
*Influence of Order on Mössbauer Spectra of ^{119}Sn in Ni_3Fe
(a) Vol. **51**, No. 1, p. K13–K15 (1979).
See also 21.1, p. 354; 21.1.1, p. 370.

M. F. DE JESUS FILHO, J. P. SANCHEZ, and J. M. FRIEDT
*^{121}Sb Mössbauer Spectroscopy in Some Mixed-Valence Antimony Hexachlorides
(b) Vol. **91**, No. 2, p. 443–446 (1979).
See also 22.5.4, p. 475.

C. BLAAUW, C. G. WHITE, and W. LEIPER
*Hyperfine Interactions of ^{57}Fe in $Fe_{1+x}Sb$
(b) Vol. **91**, No. 2, p. K129–K132 (1979).
See also 21.1.1, p. 371.

G. WEYER, A. KETTSCHAU, G. GREBE, W. SCHRÖTER, and W. BERGHOLZ
*Comparison of Mössbauer Spectra of ^{57}Co in Silicon for High-Temperature Ion-Implanted and for Diffused Samples
(a) Vol. **51**, No. 2, p. 459–466 (1979).
See also 22.1.2, p. 416.

E. V. KAPITANOV and E. N. YAKOVLEV
*Mössbauer Investigation of SnO under Hydrostatic Pressure up to 41 kbar
(a) Vol. **51**, No. 2, p. 641–644 (1979).
See also *1.2*, p. *16*; 22.6, p. 478.

A. R. MKRTCHYAN, G. A. ARUTYUNYAN, A. R. ARAKELYAN, and R. G. GABRIELYAN
*Modulation of Mössbauer Radiation by Coherent Ultrasonic Excitation in Crystals
(b) Vol. **92**, No. 1, p. 23–29 (1979).

L. CSER and L. J. NASZÓDI
Relationship between Electron Density Distribution and Atomic Order in Iron–Aluminium and Iron–Silicon Alloys
(b) Vol. 92, No. 1, p. K55–K56 (1979).
See also *1.3*, p. *22*; 21.1.1, p. 371.

M. VAN ROSSUM, J. DE BRUYN, I. DÉZSI, G. LANGOUCHE, and R. COUSSEMENT
*On the Quadrupole Interaction of Xe Implanted in Diamond
(a) Vol. **52**, No. 1, p. K31–K34 (1979).
See also 11, p. 120; 22.1, p. 408.

Sh. Sh. Bashkirov and V. N. Lebedev
*Isomer Shift and Quadrupolar Splitting of the Mössbauer Spectrum of Ferroelectrics
(b) Vol. **92**, No. 2, p. 397–402 (1979).
See also 14.4.1, p. 245; 22.8.1, p. 499.

D. Schroeer, Ch. S. Kuo, and R. L. Lambe
*Isomer Shift of Eu^{2+} in Fluorites
(b) Vol. **92**, No. 2, p. 565–572 (1979).
See also 10.2, p. 106; 22.5.3, p. 473.

C. Blaauw, G. Stroink, W. Leiper, and M. Zentilli
*Crystal-Field Properties of Fe in Brucite $Mg(OH)_2$
(b) Vol. **92**, No. 2, p. 639–643 (1979).
See also 4, p. 56; 22.8, p. 492.

M. F. Vereshchak, A. K. Zhetbaev, and Sh. Sh. Ibragimov
*A Mössbauer Study of the Effect of γ-Rays and Neutron Irradiation on Iron Oxalates
(a) Vol. **52**, No. 2, p. 415–419 (1979).
See also *11*, p. *121*; 22.8, p. 493.

A. A. Bahgat
*A New Method for Quantitative Analysis of the Mössbauer Effect
(a) Vol. **52**, No. 2, p. K217–K220 (1979).
See also 21.1.1, p. 372.

Sh. Sh. Bashkirov, A. L. Belyanin, and E. K. Sadykov
*The Scattering of γ-Rays in the Regime of Double Resonance
(b) Vol. **93**, No. 1, p. 437–442 (1979).
See also *19*, p. *292*.

V. G. Gavrilyuk, V. G. Prokopenko, and O. N. Razumov
*Mössbauer Study of the Effect of Alloying Elements on the Carbide Phase in Cold-Worked Steel
(a) Vol. **53**, No. 1, p. 147–154 (1979).
See also 10.1, p. 95; 21.1.1, p. 373.

Kamna Aggarwal and R. G. Mendiratta
*Mössbauer Investigations on Thin Films of $FeSi_2$
(a) Vol. **53**, No. 1, p. K95–K98 (1979).
See also *1.5*, p. *35*; 21.1.1, p. 373.

A. V. Kolpakov, E. N. Ovchinnikova, and R. N. Kuzmin
*Reflections of a New Type in the Mössbauer Diffraction Pattern
(b) Vol. **93**, No. 2, p. 511–514 (1979).
See also 22.8, p. 494.

G. Langouche, I. Dézsi, M. van Rossum, J. de Bruyn, and R. Coussement
*Observation of a Strong Dose Dependence in ^{57}Co Implantations in Si and Ge
(b) Vol. **93**, No. 2, p. K107–K110 (1979).
See also *11*, p. *122*; 22.1.1, p. 411; 22.1.2, p. 420.

E. V. Kapitanov and E. N. Yakovlev
*Mössbauer Study of Phase Transitions under High Hydrostatic Pressures (I): The Phase Transition of Mg_2Sn
(a) Vol. **53**, No. 2, p. 473 – 480 (1979).
See also *1.2*, p. *18*; 22.7, p. 489.

K. H. J. Buschow and A. M. van der Kraan
Magnetic Properties of Amorphous $Th_{1-x}Fe_x$ Alloys
(a) Vol. **53**, No. 2, p. 665 – 669 (1979).
See also 2, p. 44; *18.1*, p. *267*; *18.2*, p. *276*; 21.1.1, p. 374; 21.4, p. 383.

H. R. Anand
*Mössbauer Study of Iron Precipitates in LiF and NaF
(a) Vol. **53**, No. 2, p. 671 – 679 (1979).
See also 1.3, p. 23; 22.5.2, p. 468; 22.5.4, p. 475.

J. Suwalski, O. Schneeweiss, T. Zemčík, and Z. Tucholski
*Investigation of Prefered Substitution of Co Atoms in Fe–Si–Al Sendust Type Alloys
(a) Vol. **53**, No. 2, p. K195 – K198 (1979).
See also 21.1.1, p. 374.

E. V. Kapitanov and E. N. Yakovlev
*Mössbauer Study of Phase Transitions under High Hydrostatic Pressures (II): The Phase Transition of SnTe
(a) Vol. **54**, No. 1, p. 139 – 143 (1979).
See also *1.2*, p. *19*; 22.7, p. 489.

P. M. Rao, A. Gérard, and F. Grandjean
*A Mössbauer Study of the Effects of the Substitution of Fe by Cr in $SrFe_{12}O_{19}$
(a) Vol. **54**, No. 2, p. 529 – 536 (1979).
See also 18.1, p. 268; 22.8.2, p. 506.

A. Amamou
*Mössbauer Effect and Short-Range Order in Fe–P–B and Fe–P–C Amorphous Alloys
(a) Vol. **54**, No. 2, p. 565 – 572 (1979).
See also *2*, p. *46*; 18.2, p. 277; 21.1.1, p. 375.

K. K. P. Srivastava and H. C. Singh
*Effect of Second-Order Spin–Orbit Coupling $\lambda' (\boldsymbol{L} \cdot \boldsymbol{S})^2$ on Electronic Spin Relaxation of Paramagnetic Fe^{2+} in $CaCO_3$ and $CdCO_3$
(b) Vol. **95**, No. 1, p. 259 – 262 (1979).
See also 18.1, p. 268; 22.8, p. 496.

A. K. Tewari, S. K. Roy, and D. L. Bhattacharya
*Resonance Energy Shift for Mössbauer Impurity Atoms Trapped by Vacancies
(b) Vol. **95**, No. 2, p. 607 – 614 (1979).
See also 10.1, p. 98; 21, p. 350.

J. K. Srivastava and B. W. Dale
*Effect of Annealing Temperature on Cross Relaxation in the Cr^{3+}-Doped Magnetically Dilute Paramagnetic Al_2O_3–Fe_2O_3 System
(b) Vol. **96**, No. 1, p. K47 – K50 (1979).
See also 13.4, p. 178; 22.6.1, p. 487.

J. PEBLER and K. SCHMIDT
*Röntgen-Pulver- und Mössbauer-spektroskopische Untersuchungen an $V_{1-x}Fe_xO_{2-x}F_x$ im Konzentrationsbereich $0 < x < 0{,}20$
(X-Ray Diffraction and Mössbauer Measurements of $V_{1-x}Fe_xO_{2-x}F_x$ in the Concentration Range $0 < x < 0.20$)
(a) Vol. **56**, No. 1, p. 75–85 (1979).
See also *1.2*, p. *20*; 22.6.1, p. 487.

L. TAKÁCS
Electronic Structure of Transition Metal–Metalloid Crystalline and Glassy Alloys
(a) Vol. **56**, No. 1, p. 371–377 (1979).
See also *2*, p. *48*; *13.1*, p. *154*; 18.2, p. 279; 21.1, p. 367; 21.1.1, p. 377.

G. MARCHAL, PH. MANGIN, M. PIECUCH, and CHR. JANOT
*Some Comments about Crystallization of Fe–Si Amorphous Films
(a) Vol. **56**, No. 1, p. K29–K31 (1979).
See also *1.5*, p. *37*; 2, p. 49; 21.1.1, p. 377.

E. JAPA and K. KROP
Hyperfine Fields in Fe–Si Alloys
(b) Vol. **96**, No. 2, p. K65–K68 (1979).
See also *18.2*, p. *280*; 21.1.1, p. 377.

B. D. SAWICKA, J. A. SAWICKI, J. STANEK, T. TYLISCZAK, and J. KOWALSKI
*Temperature-Dependent Conversion Electron Mössbauer Measurements of ^{57}Fe Implanted in Silicon and Germanium
(a) Vol. **56**, No. 2, p. 451–456 (1979).
See also *2*, p. *50*; 22.1.1, p. 413; 22.1.2, p. 426.

W. O. MEYER and G. INDEN
*Quantitative Determination of Ordered Configurations in B.C.C. Fe–Co–Si Solid Solutions by Mössbauer Spectroscopy
(a) Vol. **56**, No. 2, p. 481–486 (1979).
See also 1.3, p. 25; 21.1.1, p. 377.

M. MÜLLNER, J. MAETZ, and H. JEX
*Thermal Diffuse Scattering from Pyrolytic Graphite Investigated with Mössbauer γ-Ray Diffraction
(a) Vol. **56**, No. 2, p. 541–544 (1979).
See also 22.1, p. 409.

6. Lattice Dynamics. Phonons

Review Articles

V. K. Pershin and Vl. K. Pershin
*Conditions of the Existence of a Molecular Crystal as a Mesogenous System
(b) Vol. **92**, No. 1, p. 9–22 (1979).
See also 1.2, p. 15; 2, p. 40; 23, p. 515.

R. H. Enns and S. S. Rangnekar
The Three-Wave Interaction in Nonlinear Optics
(b) Vol. **94**, No. 1, p. 9–28 (1979).
See also 20.1, p. 304, *20.2*, p. *325*.

Original Papers and Short Notes

D. Laplaze
*Second-Order Raman Spectra of LiH
(b) Vol. **91**, No. 1, p. 59–69 (1979).
See also *20.1*, p. *304*; 22.5, p. 460.

G. G. Nilsson
*Phonon Frequencies in Off-Symmetry Directions of NaCl Determined by Neutron Inelastic Scattering
(b) Vol. **91**, No. 1, p. 83–86 (1979).
See also 22.5.2, p. 465.

Th. F. Nonnenmacher and R. Wunderle
*Influence of the Size Effect and Surface Roughness on Phonon Transport in Bounded Systems
(b) Vol. **91**, No. 1, p. 147–153 (1979).
See also *8*, p. *80*.

E. Jahne, W. Pilz, M. Giehler, and L. Hildisch
*Non-Isodisplacement of P Atoms in Long-Wavelength Optical Phonons in $In_{1-x}Ga_xP$
(b) Vol. **91**, No. 1, p. 155–165 (1979).
See also *20.1*, p. *304*; 22.2.4, p. 442.

A. J. C. Sampaio and R. Luzzi
*Resonant Magneto-Raman Scattering in Polar Semiconductors
(b) Vol. **91**, No. 1, p. 305–310 (1979).
See also *20.1*, p. *304*; 22.2.3, p. 439.

N. A. Bakhyshov, N. M. Gasanly, B. M. Yavadov, V. I. Tagirov, and Sh. M. Efendiev
*Mixed One- and Two-Mode Behaviour of Optical Phonons in $TlGaS_{2x}Se_{2(1-x)}$ and $TlInS_{2x}Se_{2(1-x)}$ Layer Solid Solutions
(b) Vol. **91**, No. 1, p. K1–K3 (1979).
See also *20.1*, p. *304*; 22.8, p. 491.

L. N. OVANDER and N. S. TYU
Theory of Raman Scattering on Polar Phonons
(b) Vol. **91**, No. 2, p. 763—772 (1979).
See also *13.5*, p. *181*; *20*, p. *299*.

A. SZYSZKOWSKI
Polarized IR Reflection Spectra of Crystals of Derivatives of Pyridine-Tetracyanoquinodimethane. An Evidence for Electron–Vibration Coupling
(b) Vol. **91**, No. 2, p. K95—K98 (1979).
See also *20.1*, p. *306*; 22.9, p. 509.

V. K. BASHENOV, I. BAUMANN, and D. I. MARVAKOV
Elastic Properties of Layer Gallium Selenide
(b) Vol. **91**, No. 2, p. K125—K128 (1979).
See also *12.2*, p. *133*; 22.3, p. 444.

K. S. DUBEY and H. A. KASSIM
*Lattice Thermal Resistivity Due to the Bound State Electrons
(b) Vol. **91**, No. 2, p. K149—K154 (1979).
See also *8*, p. *80*; 22.1.1, p. 410.

H. N. SPECTOR
*Electromagnetic Generation of Ultrasound in Semiconductors
(b) Vol. **91**, No. 2, p. K161—K163 (1979).
See also *7*, p. *78*; 22, p. 394.

T. O. GEGECHKORI and Z. S. KACHLISHVILI
Influence of the Capture of Charge Carriers upon the *I–U* Curves of Hot Electron Semiconductors with Account of Phonon Heating
(a) Vol. **51**, No. 2, p. 333—336 (1979).
See also *13*, p. *137*; *14.3.4*, p. *232*; 22, p. 395.

H. VOGT and G. NEUMANN
*Observation of Infrared Active and Silent Modes in Cubic Crystals by Hyper-Raman Scattering
(b) Vol. **92**, No. 1, p. 57—63 (1979).
See also *20.1*, p. *306*; 22.5.2, p. 465; 22.8.1, p. 499.

A. STAHL
*Boundary Conditions for Polarization Waves in Spatially Dispersive Dielectrics
(b) Vol. **92**, No. 1, p. 113—122 (1979).
See also 13.5.1, p. 183; *20.1*, p. *307*.

R. ENDERLEIN, K. PEUKER, and F. BECHSTEDT
*General Theory of Light Scattering in Solids
(b) Vol. **92**, No. 1, p. 149—158 (1979).
See also *20*, p. *299*; 22, p. 395.

V. V. HIZNYAKOV and A. V. SHERMAN
*On the Theory of Resonant Secondary Radiation of Excitons Strongly Interacting with Phonons
(b) Vol. **92**, No. 1, p. 177—183 (1979).
See also *13.5.1*, p. *184*; 20.1, p. 307; 20.3, p. 328; 22, p. 395.

M. Hoclet, P. Plumelle, M. Vandevyver, R. Triboulet, and Y. Marfaing
*Lattice Dynamics of Cd-Rich $Cd_{1-x}Hg_xTe$ Alloys
(b) Vol. **92**, No. 2, p. 545—549 (1979).
See also *20.1*, p. *309*; 22.4.4, p. 458.

R. P. Singh
A Lattice Dynamical Study of Potassium Halide Solids
(b) Vol. **92**, No. 2, p. 557—563 (1979).
See also *4*, p. *56*; 8, p. 81; 22.5.2, p. 466.

S. Singh and G. S. Verma
*Resonance Scattering of Phonons in Inhomogeneously Strained n-Ge
(b) Vol. **92**, No. 2, p. 627—631 (1979).
See also *8*, p. *81*; 22.1.1, p. 411.

K. R. Allakhverdiev, E. Yu. Salaev, N. Yu. Safarov, L. K. Vodopyanov, and L. Golubev
*Light Scattering in $TlInSe_2$ Single Crystals
(b) Vol. **92**, No. 2, p. K91—K93 (1979).
See also *20.1*, p. *309*; 22.8, p. 492.

S. Guha and S. Ghosh
*Longitudinal Phonon–Plasmon Interaction in Inhomogeneous Semiconductors
(b) Vol. **92**, No. 2, p. K95—K97 (1979).
See also *13.5.2*, p. *191*; 22, p. 397.

N. M. Gasanly, A. F. Goncharov, B. M. Dzhavadov, N. N. Melnik, V. I. Tagirov, and E. A. Vinogradov
*Optical Phonons in Layer $TlInSe_2$ Single Crystals
(b) Vol. **92**, No. 2, p. K139—K142 (1979).
See also *20.1*, p. *310*; 22.3.1, p. 447.

M. Peyrard and R. Perret
*FIR Spectroscopy Study of the First-Order Phase Transition of $(CH_3CH_2NH_3)_2CdCl_4$
(a) Vol. **52**, No. 2, p. 521—528 (1979).
See also 1.2, p. 17; *20.1*, p. *310*; 22.9, p. 510.

A. Montaner, M. Galtier, C. Benoit, and H. Bill
*Optical Constants of Sodium Sulphide
(a) Vol. **52**, No. 2, p. 597—601 (1979).
See also *20.1*, p. *310*; 22. p. 397.

L. V. Chebotarev
*Surface Waves in Low-Dimensional Crystals
(b) Vol. **93**, No. 1, p. 71—77 (1979).
See also 7, p. 78.

K. Sato, M. Kobayashi, S. Iwabuchi, and I. Yokota
*TO Phonon–Helicon Coupling in Polar Semiconductors
(b) Vol. **93**, No. 1, p. 79—87 (1979).
See also *13.5*, p. *181*; 14.3, p. 212; 22.2.3, p. 440.

A. Ya. Dzyublik
*Dynamical Effects in Inelastic Scattering of Thermal Neutrons by a Crystal
(b) Vol. **93**, No. 2, p. 515–521 (1979).
See also *1.1*, p. *13*.

I. S. Gorban, V. A. Gorynya, V. I. Lugovoi, N. P. Krasnolob, G. I. Salivon, and I. I. Tychina
*One- and Two-Phonon Raman Scattering in Ternary Phosphides $ZnSiP_2$, $ZnGeP_2$, $CdSiP_2$
(b) Vol. **93**, No. 2, p. 531–538 (1979).
See also *20.1*, p. *313*; 22.8, p. 494.

H. Schmidt and E. Mann
*Theory of Anisotropic Electron–Phonon Interaction in Copper (I): Electron–Phonon Matrix Elements
(b) Vol. **93**, No. 2, p. 561–573 (1979).
See also *13.1*, p. *148*; 14.1, p. 200; 21.1, p. 359.

M. Giehler and W. Pilz
*Infrared Reflection and Raman Scattering Studies of the Vibrational Properties and Structure of $As_2S_3Ge_x$ Glasses
(b) Vol. **93**, No. 2, p. 641–651 (1979).
See also *20.1*, p. *313*; 22, p. 400.

J. Prechtl, J. Kalus, E. Lüscher, L. Pintschovius, and R. Ghosh
*Measurement of Mode Grüneisen Parameters in Si
(b) Vol. **93**, No. 2, p. 653–659 (1979).
See also 22.1.2, p. 420.

K. M. Kesharwani
*Localized Phonons Due to Substitutional Impurities in Dilute Copper Alloys
(b) Vol. **93**, No. 2, p. 691–701 (1979).
See also 10.1, p. 96; 21.1, p. 359.

F. Dochy
*Theoretical Calculation of the Electron–Phonon Coupling Parameters of the F-Centre in Alkali Halides
(b) Vol. **93**, No. 2, p. 743–753 (1979).
See also 12.2, p. 134; *13.4*, p. *170*; 22.5.2, p. 468.

H. N. Gupta and R. S. Padhyaya
*A Critical Study of Dynamical Properties of Ionic Crystals
(b) Vol. **93**, No. 2, p. 781–791 (1979).
See also 22.5.2, p. 468.

V. Ramamurthy and S. B. Rajendraprasad
*On the Homology of Lattice Vibrations of Rare-Gas Solids
(b) Vol. **93**, No. 2, p. 793–797 (1979).
See also 21.2, p. 379; 23, p. 515.

R. S. Turtelli and A. R. B. de Castro
*LO Phonon Raman Lineshape in Highly Excited GaAs
(b) Vol. **93**, No. 2, p. 811–815 (1979).
See also *20.1*, p. *313*; 20.3, p. 331; 22.2.1, p. 433.

K. R. ALLAKHVERDIEV, A. N. ABBASOV, T. R. MEKHTIEV, and R. KH. NANI
*Far-Infrared Reflection Spectra of $CdInGaS_4$ Single Crystals
(b) Vol. 94, No. 1, p. K31 – K34 (1979).
See also *20.1*, p. *315*; 22.8, p. 495.

R. P. GOYAL and M. K. SINGHAL
*Dispersion of Phonons in Transition Metal Oxides (FeO)
(b) Vol. 94, No. 1, p. K51 – K52 (1979).
See also 22.6, p. 481.

P. BUSSEMER
A Simple Model for the Explanation of Urbach's Rule
(b) Vol. 94, No. 1, p. K77 – K79 (1979).
See also *20.1*, p. *315*.

M. GIEHLER and W. PILZ
*Major Difference of the Vibrational Spectra and Structure of $(As_3S_3)_{1-x}(GeS_2)_x$ and $As_2S_3Ge_x$ Glasses
(b) Vol. 94, No. 1, p. K81 – K84 (1979).
See also 2, p. 45; *20.1*, p. *315*; 22, p. 402.

J. WESZKA, I. MIOTKOWSKI, K. KLOC, and W. ŻDANOWICZ
*Raman Scattering in CdP_2 Single Crystals
(b) Vol. 94, No. 1, p. K97 – K99 (1979).
See also *20.1*, p. *315*; 22.3, p. 445.

G. E. ALBERGA and J. T. DEVREESE
*Erratum to: Mobility of Electrons in n-Type InSb at 77K in the Avalanche Regime: Theory and Experiment
(b) Vol. 94, No. 1, p. K101 (1979).
See also *14.3.4*, p. *234*; 22.2.3, p. 440.

R. M. LUCHYTSKII, V. S. MANZHARA, and P. M. STARYK
Investigation of n-$Pb_{1-x}Sn_xSe$ Absorption Spectra
(b) Vol. 94, No. 2, p. 381 – 386 (1979).
See also 13.1, p. 151; *20.1*, p. *316*; 22, p. 402.

N. E. MASSA, S. S. MITRA, and J. F. VETELINO
*Volume Dependent Lattice Dynamics of Solid Argon. A Pseudo-Harmonic Approach
(b) Vol. 94, No. 2, p. 387 – 396 (1979).
See also 8, p. 82; 23, p. 515.

A. P. LOGHINOV
*Local and Resonance Vibrations in Defect Polymer Solids
(b) Vol. 94, No. 2, p. 397 – 403 (1979).
See also 10.2, p. 110; 13.4, p. 173; 22.9, p. 512.

W. SCHMID
Determination of Thermodynamical Parameters of the EHD in Silicon from Luminescence Data
(b) Vol. 94, No. 2, p. 413 – 419 (1979).
See also *13.5.1*, p. *187*; *20.3*, p. *333*; 22.1.2, p. 422.

P. Kleinert
*Cluster Bethe Lattice Treatment for Phonons in Pseudobinary Alloy Systems
(b) Vol. **95**, No. 1, p. 163–170 (1979).
See also *20.1*, p. *318*; 22.2.4, p. 442.

H. Finkenrath, G. Franz, and N. Uhle
*Reststrahlen Spectra of $PbSe_{1-x}Te_x$
(b) Vol. **95**, No. 1, p. 179–184 (1979).
See also *20.1*, p. *318*; 22, p. 404.

F. Harbach
*Disorder-Induced Raman Scattering in Non-Crystalline Films of Alkaline-Earth Halides
(b) Vol. **95**, No. 1, p. 195–202 (1979).
See also 2, p. 47; *20.1*, p. *318*; 22.5.3, p. 474.

A. Singh and K. H. Shrivastava
*Optical-Acoustic Two-Phonon Relaxation in Spin Systems
(b) Vol. **95**, No. 1, p. 273–277 (1979).
See also 18, p. 263.

Y. P. Joshi
*Scattering of Phonons by Density Fluctuations and Thermal Conductivity of Amorphous Solids
(b) Vol. **95**, No. 1, p. 317–324 (1979).
See also 2, p. 47; *8*, p. *82*; 22.8, p. 496.

N. M. Gasanly, B. M. Dzhavadov, V. I. Tagirov, and E. A. Vinogradov
*Long-Wave Lattice Vibrations of $TlInS_{2x}Se_{2(1-x)}$ and $TlGaS_{2x}Se_{2(1-x)}$ Layer Solid Solutions
(b) Vol. **95**, No. 1, p. K27–K30 (1979).
See also *20.1*, p. *319*; 22.8, p. 496.

N. M. Gasanly, N. F. Gakhramanov, B. M. Dzhavadov, V. I. Tagirov, and E. A. Vinogradov
*Infrared Spectra of the Layer Compound InS
(b) Vol. **95**, No. 1, p. K89–K92 (1979).
See also *20.1*, p. *319*; 22.3, p. 445.

H.-G. Unruh, W. Eller, and G. Kirf
Spectroscopic and Dielectric Investigations of K_2SeO_4
(a) Vol. **55**, No. 1, p. 173–180 (1979).
See also *14.4.1*, p. *246*; 20.1, p. 319; 22.8.1, p. 502.

B. K. Samantaray and G. D. Nigam
*A Simple Relation for the Additivity of the Debye Characteristic Temperature of Binary Alloys
(a) Vol. **55**, No. 1, p. K23–K26 (1979).
See also *1.3*, p. *24*; 8, p. 83; 21, p. 349.

T. Suski and W. Müller-Lierheim
*Anharmonic Effects in Ferroelectric $Pb_5Ge_3O_{11}$
(b) Vol. **95**, No. 2, p. 349–357 (1979).
See also 8, p. 83; *14.4.1*, p. *246*; 22.8.1, p. 502.

M. Godlewska, J. Ściesiński, and E. Ściesińska
*Spectroscopic Study of Phase Polymorphism and Vibrational Assignments for 2,2-Dinitropropane
(a) Vol. 55, No. 2, p. 723–734 (1979).
See also 1.2, p. 20; *20.1*, p. *320*. 22.9, p. 513.

P. D. Pathak and N. P. Shah
*Debye Temperatures of Silver and Aluminium at High Temperatures. Some New Correlations
(a) Vol. 55, No. 2, p. K159–K162 (1979).
See also *8*, p. *83*; 21, p. 351; 21.6, p. 389.

D. V. Kazakovtsev and Y. B. Levinson
*Phonon Diffusion with Frequency Down-Conversion
(b) Vol. 96, No. 1, p. 117–127 (1979).

V. Z. Polinger and G. I. Bersuker
*Multimode Jahn-Teller Effect for an E-Term with Strong Vibronic Coupling (II): Band Shapes of the Infrared and Raman Spectra
(b) Vol. 96, No. 1, p. 153–161 (1979).
See also *13.4*, p. *178*; 20.1, p. 321.

K. R. Allakhverdiev, S. S. Babaev, E. Yu. Salaev, and M. M. Tagyev
*Angular Behaviour of the Polar Optical Phonons in $A^{III}B^{VI}$ Layered Semiconductors
(b) Vol. 96, No. 1, p. 177–182 (1979).
See also *20.1*, p. *321*; 22.3, p. 446.

M. S. Kushwaha
*Debye-Waller Factors for Rare Gas Solids
(b) Vol. 96, No. 1, p. 301–305 (1979).
See also 23, p. 516.

V. M. Fomin
Non-Linear Optical Properties of Band Charge Carriers Due to Scattering by Lattice Defects
(b) Vol. 96, No. 1, p. 389–399 (1979).
See also 13.4, p. 178; *20.2*, p. *326*; 22, p. 405.

D. M. Esterling and A. Swaroop
Interatomic Potentials from Experimental Phonon Spectra (I): Prototypes
(b) Vol. 96, No. 1, p. 401–411 (1979).
See also 2, p. 48; *4*, p. *57*; 12.1, p. 131; 21.1, p. 366; 21.1.1, p. 376; 21.2, p. 379.

E. Borchi, S. de Gennaro, G. Pelosi, and A. Rettori
*The Lattice Thermal Conductivity of Aluminium at Low Temperatures
(b) Vol. 96, No. 1, p. 425–428 (1979).
See also *8*, p. *83*; 21, p. 351.

E. L. Albuquerque
*Surface Modes in Piezoelectric Crystals
(b) Vol. 96, No. 1, p. 475–482 (1979).
See also 7, p. 79.

K. Tempelhoff, F. Spiegelberg, R. Gleichmann, and D. Wruck
Precipitation of Oxygen in Dislocation-Free Silicon
(a) Vol. **56**, No. 1, p. 213 – 223 (1979).
See also *1.4*, p. *32*; *10.2*, p. *116*; 20.1, p. 322; 22.1.2, p. 425.

R. Claus
*On the Lattice Dynamical Background of Optical Activity in Crystals
(b) Vol. **96**, No. 2, p. 545 – 554 (1979).
See also *13.5*, p. *182*; *20*, p. *302*.

S. L. Chaplot and V. C. Sahni
*Lattice Dynamics of $(NH_4)_2CuCl_4 \cdot 2\,H_2O$
(b) Vol. **96**, No. 2, p. 575 – 580 (1979).
See also 22.8.1, p. 503.

D. Vergnat-Grandjean, P. Vergnat, J.-F. Brice, and R. Leveque
*Infrared Spectra of Calcium Hydride Fluoride $CaF_{2-x}H_x$
(b) Vol. **96**, No. 2, p. 611 – 616 (1979).
See also *20.1*, p. *322*; 22.5.4, p. 476.

Th. F. Nonnenmacher and P. E. Zweifel
A Boltzmann Equation for Phonons and Electrons
(b) Vol. **96**, No. 2, p. 653 – 658 (1979).
See also 8, p. 83; *14*, p. *198*.

B. Lorenz and D. Ihle
Electron–Phonon versus Coulomb Interaction Effects at the Verwey Transition of Fe_3O_4
(b) Vol. **96**, No. 2, p. 659 – 669 (1979).
See also *1.2*, p. *21*; 4, p. 58; 13, p. 142; 22.6, p. 485.

A. Swaroop and D. M. Esterling
Interatomic Potentials from Experimental Phonon Spectra (II): General Results
(b) Vol. **96**, No. 2, p. 691 – 696 (1979).
See also 2, p. 49; *4*, p. *58*; 12.1, p. 131; 21, p. 353; 21.1, p. 368; 21.2, p. 380; 21.6, p. 390.

C. Malinowska-Adamska and L. Wojtczak
*Pseudoharmonic Effect of Phonons in Electrical Resistivity
(b) Vol. **96**, No. 2, p. 697 – 702 (1979).
See also *14.1*, p. *203*; 21.1, p. 368; 21.6, p. 390.

G. K. Roy and A. N. Das
*Mode Damping by Electrostrictive Interaction in a Uniaxial Ferroelectric
(b) Vol. **96**, No. 2, p. 727 – 734 (1979).
See also *14.4.1*, p. *247*.

A. Zwick and M. A. Renucci
*Long-Wavelength Optical Phonons in $ZrSe_3$
(b) Vol. **96**, No. 2, p. 757 – 767 (1979).
See also *20.1*, p. *323*; 22, p. 407.

V. G. Vaks, N. E. Zein, S. P. Kravchuk, and A. V. Trefilov
*Spectral Density of Phonons in Alkali Metals
(b) Vol. **96**, No. 2, p. 857–865 (1979).
See also 21.2, p. 380.

I. Savatinova, E. Anachkova, J. Rohleder, and B. Jakubovski
*C≡N Vibrations in Potassium Ferrocyanide Trihydrate
(b) Vol. **96**, No. 2, p. K97–K101 (1979).
See also 14.4.1, p. 247; *20.1*, p. *323*; 22.8, p. 497.

H. H. Boghossian and K. S. Dubey
*Phonon Conductivity Correction Term and Electron-Phonon Scattering Relaxation Rate: Application to Doped and Undoped Ge
(b) Vol. **96**, No. 2, p. K107–K112 (1979).
See also *8*, p. *84*; 22.1.1, p. 413.

J. Y. Harbec, C. Deville Cavellin, and S. Jandl
*Raman Spectra and Crystal Symmetry of ZrS_3
(b) Vol. **96**, No. 2, p. K117–K120 (1979).
See also *20.1*, p. *323*; 22, p. 407.

T. M. Haridasan and J. Govindarajan
*Harmonic and Anharmonic Properties of RbBr from Phonon Dispersion at Three Temperatures
(b) Vol. **96**, No. 2, p. K121–K124 (1979).
See also *8*, p. *84*; 22.5.2, p. 472.

A. Siems
Quantitative Inverstigation of Kikuchi Structures (II): Thermal Diffuse Scattering
(a) Vol. **56**, No. 2, p. 697–709 (1979).
See also *1.1*, p. *14*; 22.1.2, p. 426.

K. Zellama, J. F. Morhange, P. German, and J. C. Bourgoin
Growth Rate of Crystallization in Amorphous Germanium Produced by Ion Implantation. A Raman Spectroscopy Study
(a) Vol. **56**, No. 2, p. 717–723 (1979).
See also 1.2, p. 21; *2*, p. *50*; *11*, p. *126*; 20.1, p. 324; 22.1.1, p. 413.

7. Acoustic Properties

Original Papers and Short Notes

MAHIRADHWAJ SINGH
Scattering of Phonons and Ultrasonic Attenuation by Donors in Doped Diamond-Type Semiconductors
(b) Vol. 91, No. 2, p. 681 – 687 (1979).
See also *6*, p. *64*; 8, p. 80; 22.1.2, p. 416.

H. N. SPECTOR
*Electromagnetic Generation of Ultrasound in Semiconductors
(b) Vol. 91, No. 2, p. K161 – K163 (1979).
See also *6*, p. *65*; 22, p. 394.

S. GUHA and N. APTE
*Effect of Nonparabolic Band Structure of Semiconductors on Acoustoelectric Instability
(b) Vol. 92, No. 1, p. K47 – K50 (1979).
See also 14.3.4, p. 232; 22.2.3, p. 439.

S. P. SINGH and H. KUZMANY
*Amplification of Bleustein-Gulyaev Waves in a Piezoelectric Coated with a Semiconducting Film
(a) Vol. 52, No. 1, p. K93 – K96 (1979).
See also *14.3*, p. *211*.

S. GUHA, P. K. SEN, and S. GHOSH
*Parametric Instability of Acoustic Waves in Transversely Magnetised Piezoelectric Semiconductors
(a) Vol. 52, No. 2, p. 407 – 414 (1979).
See also *14.3.4*, p. *233*; 22.2.3, p. 439.

L. V. CHEBOTAREV
Surface Waves in Low-Dimensional Crystals
(b) Vol. 93, No. 1, p. 71 – 77 (1979).
See also *6*, p. *67*.

A. WITEK, J. WOSIK, and W. ZBIERANOWSKI
*The Elastic Constants and Debye Temperature of the Monocrystal $NaMgAl(C_2O_4)_3 \cdot 9\,H_2O$
(a) Vol. 53, No. 1, p. K27 – K29 (1979).
See also *12.2*, p. *134*; 22.8, p. 493.

A. F. BARABANOV and A. M. TSVELIK
On the Renormalization of Collective Excitations in Intermediate Valence Metals
(b) Vol. 93, No. 2, p. 817 – 822 (1979).
See also *13.5.2*, p. *192*; 21.4, p. 383.

P. HEGEDÜŠ and S. KOLNÍK
*Some Invariance Relations for Acoustic Waves in Crystals Deformed by Hydrostatic Pressure
(a) Vol. 54, No. 1, p. 349 – 353 (1979).
See also 12, p. 128.

M. S. Sodha and R. R. Sharma
Stimulated Brillouin Scattering of Helicon Waves in a Semiconductor
(b) Vol. **94**, No. 2, p. 667 – 671 (1979).
See also *13*, p. *140*; *20*, p. *301*; 22.2.3, p. 440.

A. I. Mitsek, D. I. Sirota, and N. P. Kolmakova
Phenomenological Theory of Electromagnetic and Sound Energy Attenuation in Many-Domain Magnetics
(b) Vol. **94**, No. 2, p. 739 – 747 (1979).
See also *18.2*, p. *277*; *18.4*, p. *287*; 20, p. 301; 21.1, p. 362; 22.5, p. 462; 22.6, p. 482.

R. Ranganathan and S. Dattagupta
*Theory of Ultrasonic Scattering by Grains in Polycrystals
(a) Vol. **54**, No. 2, p. 537 – 546 (1979).
See also 1.4, p. 29; 21, p. 347.

H. C. Kim and T. Kishi
The Grain Size and Flow Stress Dependence of Acoustic Emission Energy Release in Polycrystalline Aluminium
(a) Vol. **55**, No. 1, p. 189 – 195 (1979).
See also *10.1*, p. *98*; 21, p. 349.

A. V. Mitin and V. A. Tarasov
*Excitation of Magnetoacoustic Surface Waves by Bloch Walls
(b) Vol. **95**, No. 2, p. 461 – 472 (1979).
See also *18.3*, p. *285*.

E. L. Albuquerque
Surface Modes in Piezoelectric Crystals
(b) Vol. **96**, No. 1, p. 475 – 482 (1979).
See also *6*, p. *75*.

S. Guha and N. Apte
Stimulated Brillouin Scattering in a Magnetised Semiconductor Plasma
(b) Vol. **96**, No. 2, p. 569 – 574 (1979).
See also *13*, p. *142*.

8. Thermal Properties

Original Papers and Short Notes

V. N. KASHCHEEV
*Thermal Conductivity of Low-Dimensional Displacive-Type Ferroelectrics
(b) Vol. **92**, No. 1, p. K5 – K7 (1979).
See also *14.4.1*, p. *245*.

S. C. GOEL and V. S. NANDA
*Application of a Low Temperature Equation of State for Monovalent Metals
(b) Vol. **92**, No. 1, p. K65 – K67 (1979).
See also 21.1, p. 356; 21.6, p. 387.

R. P. SINGH
A Lattice Dynamical Study of Potassium Halide Solids
(b) Vol. **92**, No. 2, p. 557 – 563 (1979).
See also *4*, p. *56*; 6, p. 67; 22.5.2, p. 466.

S. SINGH and G. S. VERMA
*Resonance Scattering of Phonons in Inhomogeneously Strained n-Ge
(b) Vol. **92**, No. 2, p. 627 – 631 (1979).
See also *6*, p. *67*; 22.1.1, p. 411.

K. BARTKOWSKI, D. WŁOSEWICZ, and J. RAFAŁOWICZ
*Lorenz Function and Horizontal and Vertical Component of Thermal Resistivity of Tin Monocrystals
(a) Vol. **52**, No. 2, p. 397 – 400 (1979).
See also *14.1*, p. *200*; 21.7, p. 391.

M. E. GUREVICH, A. I. MITSEK, and A. F. ZHURAVLEV
Anomalies in Thermal Expansion and Heat Capacity of Ferromagnets
(b) Vol. **93**, No. 1, p. 127 – 134 (1979).
See also *1.2*, p. *17*; 12.1, p. 129; 18.2, p. 274; 21.1, p. 358.

C. MALINOWSKA-ADAMSKA and L. WOJTCZAK
*Influence of Lattice Constant Fluctuations on the Thermal Properties of Crystals
(b) Vol. **93**, No. 1, p. K5 – K8 (1979).
See also 1.2, p. 17.

M. DRULIS and Z. BIEGAŃSKI
*Low-Temperature Specific Heats of Cerium and Praseodymium Trihydrides. Second-Order Phase Transitions and Crystal Field Splitting Effects
(a) Vol. **53**, No. 1, p. 277 – 281 (1979).
See also 21.4, p. 383.

KH. I. AMIRKHANOV, YA. B. MAGOMEDOV, SH. M. ISMAILOV, and N. L. KRAMYNINA
*The Thermal Conductivity of Semiconductors of the System $As_2(Se_{1-x}Te_x)_3$ Depending on Temperature and Pressure
(a) Vol. **53**, No. 1, p. 361 – 366 (1979).
See also 2, p. 43; 6, p. 68; 22, p. 399.

N. I. OSTAPENKO, V. I. SUGAKOV, M. P. CHERNOMORETS, and M. T. SHPAK
*Spectra of Directionally Deformed Naphthalene Crystals and Exciton–Phonon Interaction Constants
(b) Vol. **93**, No. 2, p. 493 – 501 (1979).
See also 6, p. 68; *20.1*, p. *312*; 22.9, p. 511.

A. Freudenhammer and D. Deutges
*On the $1/T$-Behaviour of the Specific Heat of a RKKY Ising Spin Glass
(b) Vol. 95, No. 1, p. K55–K59 (1979).
See also 2, p. 47; 18, p. 263; 21, p. 348.

B. K. Samantaray and G. D. Nigam
A Simple Relation for the Additivity of the Debye Characteristic Temperature of Binary Alloys
(a) Vol. 55, No. 1, p. K23–K26 (1979).
See also *1.3*, p. *24*; *6*, p. *73*; 21, p. 349.

T. Suski and W. Müller-Lierheim
Anharmonic Effects in Ferroelectric $Pb_5Ge_3O_{11}$
(b) Vol. 95, No. 2, p. 349–357 (1979).
See also *6*, p. *73*; *14.4.1*, p. *246*; 22.8.1, p. 502.

A. P. Gupta and Jai Shanker
*Evaluation of the Anderson-Grüneisen Parameter for Alkali Halides
(b) Vol. 95, No. 2, p. K103–K106 (1979).
See also 22.5.2, p. 471.

J. Mucha and J. Rafałowicz
*Influence of the Lattice Component on the Thermal Conductivity Minimum of Aluminium
(a) Vol. 55, No. 2, p. 631–634 (1979).
See also 21, p. 350.

P. D. Pathak and N. P. Shah
*Debye Temperatures of Silver and Aluminium at High Temperatures. Some New Correlations
(a) Vol. 55, No. 2, p. K159–K162 (1979).
See also *6*, p. *75*; 21, p. 351; 21.6, p. 389.

E. Borchi, S. de Gennaro, G. Pelosi, and A. Rettori
*The Lattice Thermal Conductivity of Aluminium at Low Temperatures
(b) Vol. 96, No. 1, p. 425–428 (1979).
See also *6*, p. *75*; 21, p. 351.

E. Praveczki
*Pade Approximation Method as Renormalization Group Method
(b) Vol. 96, No. 1, p. 437–440 (1979).

Th. F. Nonnenmacher and P. F. Zweifel
A Boltzmann Equation for Phonons and Electrons
(b) Vol. 96, No. 2, p. 653–658 (1979).
See also 6, p. 76 ; *14*, p. *198*.

S. N. Maity, D. Roy, and S. Sengupta
Long-Range Many-Body Interaction and Monovacancy Heat of Formation of Rare-Gas Crystals
(b) Vol. 96, No. 2, p. 883–890 (1979).
See also 4, p. 58; *10.2*, p. *116*; 23, p. 516.

9. Diffusion

Original Papers and Short Notes

R. Kirchheim
*Zur Lösung des 2. Fickschen Gesetzes unter nicht-isothermen Bedingungen in Gegenwart von äußeren Kräften
(On the Solution of Fick's Second Law under Non-Isothermal Conditions in the Presence of External Forces)
(b) Vol. **91**, No. 1, p. 123–134 (1979).

M. P. Macht, G. Frohberg, and H. Wever
*Electro- and Thermotransport in Vanadium
(a) Vol. **51**, No. 1, p. 131–139 (1979).
See also 10.1, p. 92; 21.1, p. 354.

B. B. Yu and R. F. Davis
*Near-Surface Effect in Self-Diffusion in NbC_x
(a) Vol. **51**, No. 1, p. 261–267 (1979).
See also 21, p. 339.

S. K. Pabi
*Computer Simulation of the Two-Phase Diffusion-Controlled Dissolution in the Planar Finite Multilayer Couples
(a) Vol. **51**, No. 1, p. 281–289 (1979).
See also *1.2*, p. *15*.

W. Heywang, E. F. Krimmel, and H. Runge
Annealing Mechanism of Radiation Damage and Dopants in Pulsed Laser Light Irradiated Ion Implanted Layers
(a) Vol. **51**, No. 1, p. K79–K82 (1979).
See also *11*, p. *119*; 22.1.2, p. 415.

E. A. Lebedev, E. Nebauer, and P. Süptitz
*Broadening of Electrodiffusion Profiles at High Electric Fields in Glassy As_2Se_3:Cu
(a) Vol. **51**, No. 2, p. K207–K211 (1979).
See also *2*, p. *41*; 14.3.4, p. 232; 22, p. 395.

T. D. Dzhafarov, E. A. Skoryatina, E. S. Guds, and I. E. Moronchuk
*The Diffusion and Electric Activity of Co in Epitaxial GaAs
(a) Vol. **51**, No. 2, p. K221–K225 (1979).
See also 14.3, p. 211; 22.2.1, p. 433.

K.-H. Kraatz, G. Frohberg, and H. Wever
*Thermo- and Electromigration in High Purity Cobalt
(a) Vol. **52**, No. 1, p. 149–159 (1979).
See also 21.1, p. 356.

H. J. Schenk, H. J. Bauer, and B. Baranowski
Formation and Decomposition of the Hydride Phase in Ni–Mn Alloys in High-Pressure Gaseous Hydrogen Followed by Magnetic Measurements
(a) Vol. **52**, No. 1, p. 195–200 (1979).
See also *18.2*, p. *272*; 21.1, p. 356.

B. Rauschenbach and G. Blasek
*Determination of Very Low Diffusion Coefficients in Glass with Means of Secondary-Ion Mass Spectrometry
(a) Vol. **53**, No. 1, p. K11 – K13 (1979).
See also *2*, p. *44*; 17, p. 257; 22, p. 399.

F. Hashimoto, Y. Kamiura, and H. Takahashi
Effect of Quenching Temperature on the Annealing Rate of Quenched-In Acceptors in Germanium
(a) Vol. **53**, No. 1, p. K31 – K33 (1979).
See also *10.2*, p. *108*; 22.1.1, p. 411.

W. Schüle and R. Scholz
Properties of Vacancies and Divacancies in F.C.C. Metals
(b) Vol. **93**, No. 2, p. K119 – K123 (1979).
See also *10.1*, p. *96*; 14.1, p. 201; 21, p. 344; 21.1, p. 359; 21.6, p. 388.

R. Kutner and I. Sosnowska
*Nonlinear Equations of Correlated Jump Diffusion Derived in the Context of Hydrogen Migration in a Metal
(b) Vol. **93**, No. 2, p. K167 – K171 (1979).
See also 21, p. 344.

R. Kutner, I. Sosnowska, and W. Kley
*Erratum to: Next-Neighbour Jumps of Hydrogen in Quasi-Elastic Neutron Scattering from α-PdH_x
(b) Vol. **93**, No. 2, p. K187 (1979).
See also 21.6, p. 388.

K. Obrtlík and J. Kučera
*Diffusion of Vanadium in the Fe–V System
(a) Vol. **53**, No. 2, p. 589 – 597 (1979).
See also 21.1.1, p. 374.

M. G. Grimaldi, P. Baeri, S. U. Campisano, G. Foti, and E. Rimini
Laser Induced As Profile Broadening in Amorphous Silicon Layers
(a) Vol. **54**, No. 1, p. 55 – 59 (1979).
See also *2*, p. *45*; *11*, p. *122*; 22.1.2, p. 421.

N. D. Stojadinovič
*Conditions for the Creation of Dislocations by Diffusion of Phosphorus into Silicon
(a) Vol. **54**, No. 1, p. K5 – K10 (1979).
See also *10.2*, p. *110*; 22.1.2, p. 421.

D. R. Kaplan, C. Weigel, and J. W. Corbett
Calculations on the Properties of Helium in Silicon
(b) Vol. **94**, No. 2, p. 359 – 366 (1979).
See also *13.1*, p. *150*; *13.4*, p. *173*; 22.1.2, p. 422.

O. A. Troitskii, V. I. Spitsyn, M. M. Moiseenko, and A. D. Tsiganov
Fractionation of Stable Isotopes and Impurities during Zone Recrystallization of Metals and Preparation of Single Crystals Rich in Individual Isotopes
(a) Vol. 54, No. 2, p. 651 – 654 (1979).
See also *3*, p. *53*; 21, p. 347; 21.1, p. 363.

CH. KLEINT
*Stationary Stochastic Adparticle Flip-Flop as a Precursor to Surface Diffusion
(a) Vol. **55**, No. 2, p. 447—456 (1979).
See also *1.6*, p. *39*; 17.1, p. 260; 22.9, p. 513.

G. D. BAGRATISHVILI, R. B. DZHANELIDZE, D. A. JISHIASHVILI, L. V. PISKANOVSKII, and Z. N. SHIOLASHVILI
*Boron Diffusion from a Reactively Sputtered Glass Source in Si and SiO_2
(a) Vol. **56**, No. 1, p. 27—35 (1979).
See also 1.5, p. 36; 20.1, p. 322; 22.8.1, p. 503.

W. BUSSMANN, CH. HERZIG, W. REMPP, K. MAIER, and H. MEHRER
*Isotope Effect and Self-Diffusion in Face-Centred Cubic Cobalt
(a) Vol. **56**, No. 1, p. 87—97 (1979).
See also 10.1, p. 100; 21.1, p. 366.

M. B. DUTT, S. K. SEN, and A. K. BARUA
*The Diffusion of Nickel into Copper and Copper–Nickel Alloys
(a) Vol. **56**, No. 1, p. 149—155 (1979).
See also 21.1, p. 367.

E. YA. MIKHLIN
*Suppression of Diffusion Mobility of Small Gas Bubbles in Solids
(a) Vol. **56**, No. 2, p. 763—768 (1979).
See also 10.1, p. 102; 21.1, p. 369; 21.5, p. 386; 21.6, p. 390.

J. C. PFISTER and D. BENSAHEL
*Diffusion of Impurities Contained in Te Precipitates in ZnTe: Theoretical Model
(a) Vol. **56**, No. 2, p. K79—K82 (1979).
See also 22.4.3, p. 457.

10. Defect Properties (Irradiation Defects See 11)

Review Article

V. I. Dotsenko
*Stress Relaxation in Crystals
(b) Vol. **93**, No. 1, p. 11–43 (1979).
See also 12, p. 127.

Original Papers and Short Notes

V. A. Solovev and S. N. Zolotarev
*The Isotropic Limit in Anisotropic Dislocation Theory. Green's Functions in a Crystal Containing a Dislocation Pile-up
(b) Vol. **91**, No. 1, p. 319–323 (1979).
See also 12, p. 127.

E. Gerlach and K. Paier
*Dynamical Electron Scattering by a Layer Structure. Geometrical Resonances
(b) Vol. **92**, No. 2, p. 495–499 (1979).
See also *13*, p. *138*.

K. Jagannadham and M. J. Marcinkowski
*Discrete Dislocation Analysis of Cracks
(a) Vol. **52**, No. 2, p. 663–674 (1979).

H. Trinkaus
*Drift Diffusion in Grain Boundaries under the Influence of Stress Fields
(b) Vol. **93**, No. 1, p. 293–303 (1979).
See also *9*, p. *86*; 12, p. 127.

M. J. Marcinkowski
Unifying Concepts in the Theories of Eleasticity, Plasticity, and Dislocations
(b) Vol. **93**, No. 2, p. 703–711 (1979).
See also *1.4*, p. *28*.

S. D. Mamchich
*Effective Stress in the Eyring Creep Model
(a) Vol. **53**, No. 2, p. K181–K184 (1979).

G. Schoeck
*The Volume Change in Internal Stress Fields
(b) Vol. **94**, No. 1, p. 147–151 (1979).
See also *12*, p. *127*.

F. N. Chukhovskii and A. M. Arustamyan
*A Dynamical Analytical Theory of the Dislocation Image Contrast in Electron Transmission Microscopy
(a) Vol. **54**, No. 1, p. 45–54 (1979).
See also *1.4*, p. *29*.

V. A. Solovev
*The "Superdislocation" Model of a Plastic Zone near a Wedge-Shaped Crack
(a) Vol. **54**, No. 1, p. 297–304 (1979).

C. M. van Baal, R. Fastenau, and P. Penning
*The Probability of Visiting a New Site in a Random Walk on a B.C.C. Lattice
(a) Vol. **54**, No. 1, p. K45–K48 (1979).

H.-J. Burmeister and H. Hermann
*A Simple Model for an Inhomogeneous Dislocation Structure
(a) Vol. **54**, No. 1, p. K59–K61 (1979).
See also *12*, p. *128*.

L. S. Khazan, L. A. Matveeva, G. N. Semenova, and Yu. A. Tkhorik
*Account of Dislocation Structure at Stress Measurements in Heteroepitaxial Systems (I): Theory
(a) Vol. **54**, No. 2, p. 447–456 (1979).
See also *1.5*, p. *35*.

K. Jagannadham and M. J. Marcinkowski
*Dislocations, Disclinations, and Grain Boundaries in a Finite Solid
(a) Vol. **54**, No. 2, p. 715–727 (1979).
See also *1.4*, p. *30*.

S. N. Smirnov
*Interaction of a Dislocation with a Point Obstacle in the Non-Uniform Stress Field
(a) Vol. **55**, No. 1, p. 51–60 (1979).

D. Bonchev, O. Mekenyan, and H.-G. Fritsche
*A Topological Approach to Crystal Vacancy Studies (I): Model Crystallites with a Single Vacancy
(a) Vol. **55**, No. 1, p. 181–187 (1979).

P. Varotsos
*Comments on the Migration Volume of Vacancies in Solids
(a) Vol. **55**, No. 1, p. K63–K66 (1979).
See also 21, p. 349; 22, p. 404.

Y. P. Joshi
*Scattering of Phonons by Spherical Defect Aggregates
(b) Vol. **95**, No. 2, p. 627–634 (1979).
See also *6*, p. *74*.

P. Varotsos
*Entropy and Volume of Solution of Substitutional Impurities in Solids
(a) Vol. **55**, No. 2, p. K139–K142 (1979).
See also 21, p. 351; 22, p. 405.

L. S. Khazan, L. A. Matveeva, G. N. Semenova, and Yu. A. Tkhorik
*Account of Dislocation Structure at Stress Measurements in Heteroepitaxial Systems (II): Experiment
(a) Vol. **56**, No. 1, p. 47–54 (1979).
See also *1.5*, p. *37*.

B. Dietrich

*Sensitivity of X-Ray Diffraction Methods for the Measurement of Extended Deformation Fields

(a) Vol. 56, No. 1, p. 199—206 (1979).

See also *1*, p. *10*.

J. L. Martin and L. P. Kubin

*Optimum Conditions for Straining Experiments in the HVEM

(a) Vol. 56, No. 2, p. 487—494 (1979).

See also *1.4*, p. *32*.

O. Mekenyan, D. Bonchev, and H. Fritsche

*A Topological Approach to Crystal Vacancy Studies (II): Model Crystallites with Two and Three Vacancies

(a) Vol. 56, No. 2, p. 607—614 (1979).

10.1 Metals

Original Papers and Short Notes

A. ČÍŽEK, M. ŠOB, I. YA. DEKHTYAR, S. G. ŠACHAROVA, and R. G. FEDCHENKO
*Positron Annihilation in Aluminium, Subjected to the Thermocyclic Treatment
(a) Vol. **51**, No. 2, p. K113–K116 (1979).
See also 21, p. 340.

R. BALZER and H. SIGVALDASON
Equilibrium Vacancy Concentration Measurements on Tin Single Crystals
(b) Vol. **92**, No. 1, p. 143–147 (1979).
See also *8*, p. *80*; 21.7, p. 391.

W. F. W. M. VAN HEUGTEN, L. M. CASPERS, and J. TH. M. DE HOSSON
*Atomic Configuration of a $^1/_2$ ⟨111⟩ Screw Dislocation in Pure Mo and in Mo Containing He Interstitials
(b) Vol. **92**, No. 1, p. 199–209 (1979).
See also 21, p. 341.

C. N. TOMÉ, A. M. MONTI, and E. J. SAVINO
*Vacancy and Interstitial Configuration in Hexagonal Close Packed Metals
(b) Vol. **92**, No. 1, p. 323–334 (1979).
See also 21, p. 341.

A. M. MONTI and E. J. SAVINO
*Frenkel Pair Stability in a Hexagonal Close Packed Lattice
(b) Vol. **92**, No. 1, p. K39–K42 (1979).
See also 21.1, p. 356.

J. GASTALDI and C. JOURDAN
*Observation of the Annealing Twin Growth by Synchrotron Radiation X-Ray Topography
(a) Vol. **52**, No. 1, p. 139–148 (1979).
See also *1.4*, p. *27*; 21, p. 341.

G. DLUBEK, O. BRÜMMER, and ASHRAFUL ALAM
*A Study of Recovery and Recrystallization of Plastically Deformed $\alpha + \beta$ Brass by Positron Annihilation
(a) Vol. **52**, No. 1, p. 217–221 (1979).
See also 21, p. 341.

W. DECKER, J. DIEHL, A. DUNLOP, W. FRANK, H. KRONMÜLLER, W. MENSCH, H. E. SCHAEFER, B. SCHWENDEMANN, A. SEEGER, H. P. STARK, F. WALZ, and M. WELLER
*Interstitial Migration in Iron at 220 K
(a) Vol. **52**, No. 1, p. 239–248 (1979).
See also 11, p. 120; 14.1, p. 200; 21.1.1, p. 371.

F. VÁVRA
*Influence of Metallic Impurities on Defect Production and Recovery in Niobium Neutron-Irradiated at 4.6 K
(a) Vol. **52**, No. 1, p. 269–276 (1979).
See also 11, p. 120; 14.1, p. 200; 21, p. 341.

W. R. THORPE and I. O. SMITH
*A Model for High-Temperature Creep Incorporating Both Recovery and Thermally Activated Glide
(a) Vol. **52**, No. 2, p. 487–497 (1979).
See also *12.1*, p. *129*.

J. G. ERLINGS and F. W. SCHAPINK
Electron Diffraction from Periodic ⟨111⟩ Twist Grain Boundary Structures
(a) Vol. **52**, No. 2, p. 529–539 (1979).
See also *1.4*, p. *27*; 21.6, p. 387.

R. H. J. FASTENAU, C. M. VAN BAAL, P. PENNING, and A. VAN VEEN
*On the Sink Concentration Dependence of Reaction Constants for Point Defect Trapping
(a) Vol. **52**, No. 2, p. 577–586 (1979).
See also 21, p. 343.

O. V. PRESENYAKOVA, V. I. ZAITSEV, and N. A. DOROSHENKO
*Aluminium Polygonization under High Pressure
(a) Vol. **52**, No. 2, p. 627–633 (1979).
See also *1.4*, p. *28*; 21, p. 343.

J. DE GROOT, P. M. BRONSVELD, and J. TH. M. DE HOSSON
*Superlattice Dislocations in Cu_2NiZn
(a) Vol. **52**, No. 2, p. 635–645 (1979).
See also *1.3*, p. *23*; 21.1, p. 358.

G. KNUYT and L. M. STALS
*The Na Vacancy Formation Energy Value from Phonon Data
(b) Vol. **93**, No. 1, p. K9–K11 (1979).
See also 4, p. 56; 6, p. 68; 21.2, p. 379.

V. G. GAVRILYUK, V. G. PROKOPENKO, and O. N. RAZUMOV
Mössbauer Study of the Effect of Alloying Elements on the Carbide Phase in Cold-Worked Steel
(a) Vol. **53**, No. 1, p. 147–154 (1979).
See also *5*, p. *60*; 21.1.1, p. 373.

H. U. HABERMEIER, W. KLEIN, R. AOKI, and H. KRONMÜLLER
Magneto-Optical Investigation of Flux Penetration in Plastically Deformed Niobium Monocrystals
(a) Vol. **53**, No. 1, p. 225–235 (1979).
See also *18*, p. *262*; *20*, p. *300*; 21, p. 344.

M. I. DARBY, M. N. READ, and K. N. R. TAYLOR
*The Energy of Hydrogen in Zirconium–Rhodium
(a) Vol. **53**, No. 1, p. 391–395 (1979).
See also *13.4*, p. *170*; 21.1, p. 359.

H. O. K. KIRCHNER, H. MUGHRABI, F. PRINZ, and H. SCHMID
*Erratum to: Dislocation Loops in α-Iron
(a) Vol. **53**, No. 1, p. 411 (1979).
See also 12.1, p. 130; 21.1.1, p. 373.

D. H. Sastry and C. V. Shastry
*Cottrell-Stokes Law during High Temperature Deformation of α-Iron
(a) Vol. **53**, No. 1, p. K5–K7 (1979).
See also 21.1.1, p. 373.

R. H. J. Fastenau
*The Random Walk of a Cloud of Lattice Points as a Model for Trapping of Point Defects by Extended Sinks
(a) Vol. **53**, No. 1, p. K39–K42 (1979).
See also 21, p. 344.

J. L. Besson and J. Petit
*Accurate Determination of Activation Parameters of the Bordoni Peaks in Nickel
(a) Vol. **53**, No. 1, p. K61–K66 (1979).
See also 21.1, p. 359.

V. I. Zaitsev, E. I. Lyafer, and V. A. Fomchenko
*Influence of Compression Anisotropy Effects on Zn Creep
(a) Vol. **53**, No. 1, p. K99–K101 (1979).
See also *12.1*, p. *130*; 21.1, p. 359.

K. M. Kesharwani
Localized Phonons Due to Substitutional Impurities in Dilute Copper Alloys
(b) Vol. **93**, No. 2, p. 691–701 (1979).
See also *6*, p. *69*; 21.1, p. 359.

W. Schüle and R. Scholz
*Properties of Vacancies and Divacancies in F.C.C. Metals
(b) Vol. **93**, No. 2, p. K119–K123 (1979).
See also 9, p. 87; 14.1, p. 201; 21, p. 344; 21.1, p. 359; 21.6, p. 388.

R. Ciach, H. Löffler, J. Salawa, and G. Wendrock
Modulated Structures and Position of the Coherent Spinodal Curve in Al–Zn(22 at%)–Cu(x) Alloys
(a) Vol. **53**, No. 2, p. 441–446 (1979).
See also *1.2*, p. *18*; 21, p. 345; 21.1, p. 360.

A. A. Aly, G. N. Podus, and A. F. Sirenko
*Thermoactivated Creep and Stress Relaxation in Cu and Al
(a) Vol. **53**, No. 2, p. 461–467 (1979).
See also 21, p. 345; 21.1, p. 360.

M. Fumeron, J. P. Guillot, A. Dauger and J. Caisso
X-Ray Scattering Study of the Shape of Gunier-Preston Zones and Their Size Effect in an Al–6.8 at% Alloy
(a) Vol. **53**, No. 2, p. 605–610 (1979).
See also *1.3*, p. *23*; 12.1, p. 130; 21.1, p. 360.

E. Bonetti, E. Evangelista, and P. Gondi
*Internal Friction and Dynamic Modulus of Polycrystalline Al during Deformation in the Temperature Range of the Kê Peak
(a) Vol. **53**, No. 2, p. 653–658 (1979).
See also 12.1, p. 130; 21, p. 345.

P. Lukáč and Z. Trojanová
*Solid Solution Hardening of Cadmium Single Crystals
(a) Vol. **53**, No. 2, p. K143–K145 (1979).
See also 21, p. 345.

G. Dlubek, O. Brümmer, and N. Meyendorf
*Vacancy–Impurity Binding Enthalpy in NiSb by Means of Positron Annihilation
(a) Vol. **53**, No. 2, p. K157–K160 (1979).
See also 21, p. 345; 21.1, p. 360.

W. Kleinert and R. Franke
*Change of Electrical Resistivity and Stress Amplitude during Push-Pull Fatigue of Polycrystalline Nickel
(a) Vol. **53**, No. 2, p. K177–K179 (1979).
See also *14.1*, p. *201*; 21.1, p. 361.

F. F. Lavrentev, O. P. Salita, and S. Sokolski
*Jump-Like Plastic Deformation and Dislocation Structure Instability in Zinc Crystals with Forest Dislocations
(a) Vol. **54**, No. 1, p. 145–152 (1979).
See also 21.1, p. 361.

N. M. Ghoniem and D. D. Cho
The Simultaneous Clustering of Point Defects during Irradiation
(a) Vol. **54**, No. 1, p. 171–178 (1979).
See also *11*, p. *122*; 21, p. 346.

C. Holste, G. K. Schmidt, and R. Torber
*A Quantitative Estimation of Specific Internal Stress Fields in Cyclically Deformed Nickel
(a) Vol. **54**, No. 1, p. 305–314 (1979).
See also 21.1, p. 362.

G. Fuchs
Pinning Mechanism in Superconducting NbTi Alloys at High Magnetic Fields
(a) Vol. **54**, No. 1, p. 321–327 (1979).
See also *14.2*, p. *206*; 21, p. 346.

C. B. Carter
*The Extension of Jogs on Dissociated Dislocations in F.C.C. Metals
(a) Vol. **54**, No. 1, p. 395–406 (1979).
See also 21, p. 346.

G. Dlubek and O. Brümmer
*The Composition Dependence of the Effective Vacancy Formation Enthalpy in Copper-Rich Brass by Means of Positron Annihilation
(a) Vol. **54**, No. 1, p. K13–K16 (1979).
See also 21, p. 346; 21.1, p. 362.

P. Lukáč
*The Critical Resolved Shear Stress of Mg–Cd Single Crystals
(a) Vol. **54**, No. 1, p. K33–K35 (1979).
See also 21, p. 346; 21.3, p. 381.

C. RADOMSKY and H. LÖFFLER
On the Extra-Resistivity of Coherent Precipitates
(a) Vol. **54**, No. 2, p. 615–620 (1979).
See also *14.1*, p. *201*; 21.1, p. 363.

T. LEFFERS and B. N. SINGH
*Loop Growth and Point-Defect Profils during HVEM Irradiation
(a) Vol. **54**, No. 2, p. 621–631 (1979).
See also 11, p. 123; 21.1, p. 363.

V. NAVRÁTIL, P. LUKÁŠ, and V. P. SOLDATOV
*The Critical Resolved Shear Stress of Cd–Zn Single Crystals at Low Temperatures
(a) Vol. **54**, No. 2, p. K175–K176 (1979).
See also *12.1*, p. *131*; 21, p. 348; 21.1, p. 364.

V. M. CHERNOV and M. M. SAVIN
*Elastic Interaction of Edge and Screw Dislocations with Point Defects in Cubic Crystals
(a) Vol. **55**, No. 1, p. 113–122 (1979).
See also 21, p. 349.

H. C. KIM and T. KISHI
*The Grain Size and Flow Stress Dependence of Acoustic Emission Energy Release in Polycrystalline Aluminium
(a) Vol. **55**, No. 1, p. 189–195 (1979).
See also 7, p. 79; 21, p. 349.

L. M. SANDLER, V. P. POPOV, and YA. V. NAGLYUK
*Magnetic Anisotropy Induced by Rolling Single Crystals of Ferromagnetic Alloys (I): Face-Centered Cubic Single Crystals
(a) Vol. **55**, No. 1, p. 271–277 (1979).
See also *18.2*, p. *278*; 21.1, p. 364; 21.1.1, p. 375.

J. ECHIGOYA and S. HAYASHI
*Anomaly of Dislocation Motion in Fe–35%Ni Invar Alloys
(a) Vol. **55**, No. 1, p. 279–284 (1979).
See also 18.2, p. 278; 21.1, p. 364; 21.1.1, p. 375.

N. K. GILRA
*Dislocation Climb Rate and Core Diffusion
(a) Vol. **55**, No. 1, p. K31–K34 (1979).

N. IGATA, N. KURAMOCHI, N. MASAMURA, A. KOHYAMA, K. MIYAHARA, S. SATO, and H. YOSHIDA
*The α'-Peak of Deformed Niobium
(a) Vol. **55**, No. 1, p. K127–K129 (1979).
See also *12.1*, p. *131*; 21, p. 349.

A. K. TEWARI, S. K. ROY, and D. L. BHATTACHARYA
*Resonance Energy Shift for Mössbauer Impurity Atoms Trapped by Vacancies
(b) Vol. **95**, No. 2, p. 607–614 (1979).
See also *5*, p. *61*; 21, p. 350.

O. K. Alekseeva, V. A. Onishchuk, V. P. Shantarovich, I. Ya. Dekhtyar, and V. I. Shevchenko
*Influence of Absorbed Hydrogen on Positron Lifetime Spectra in Pd, Pd–Ag, Pd–Cu
(b) Vol. **95**, No. 2, p. K135–K139 (1979).
See also *4*, p. *57*; 21, p. 350.

A. Seeger
*A Theory of the Snoek-Köster Relaxation (Cold-Work Peak) in Metals
(a) Vol. **55**, No. 2, p. 457–468 (1979).
See also 21, p. 350; 21.1.1, p. 375.

G. Dlubek, O. Kabisch, O. Brümmer, and H. Löffler
*Precipitation and Dissolution Processes in Age-Hardenable Al Alloys. A Comparison of Positron Annihilation and X-Ray Small-Angle Scattering Investigations (I): Al–Zn(x) (x = 3, 4.5, 6, 10, and 18 at%)
(a) Vol. **55**, No. 2, p. 509–518 (1979).
See also 21, p. 350; 21.1, p. 365.

J. F. Dinhut, P. Lepinay et J. C. Desoyer
*Vieillissement dynamique dans l'alliage Fe–Co–2% V
Dynamic Aging of Fe–Co–2% V
(a) Vol. **55**, No. 2, p. 621–629 (1979).
See also 21.1, p. 365; 21.1.1, p. 376.

P. Bartuška, R. Kroggel, H. Löffler, M. Simerská, V. Syneček, and G. Wendrock
*On the Room Temperature Stability of Metastable Phases Formed at 200 °C in Al–Zn Alloys
(a) Vol. **55**, No. 2, p. 673–679 (1979).
See also *1.3*, p. *25*; 21, p. 350; 21.1, p. 365.

V. A. Goltsov, N. I. Timofeev, and I. Yu. Machikina
The Phenomenon of Controllable Hydrogen Phase "Naklep" of Metals and Alloys
(a) Vol. **55**, No. 2, p. 691–698 (1979).
See also *1.2*, p. *20*; *1.3*, p. *25*; 21, p. 351.

J. J. Jackson, E. A. Ryan, and Y. N. Lwin
*A Study of Thermal Conversion of Interstitials in Platinum by Means of Electrical Resistivity Measurements
(a) Vol. **55**, No. 2, p. 735–746 (1979).
See also 11, p. 124; 14.1, p. 202; 21.6, p. 389.

G. J. den Otter and A. van den Beukel
*Flow Stress and Activation Volume of Some Cold-Worked Copper-Based Solid Solutions
(a) Vol. **55**, No. 2, p. 785–792 (1979).
See also 21.1, p. 366.

H. Gollisch
*Density Functional Approach to Point Defect Properties of Copper–Germanium Alloys
(b) Vol. **96**, No. 1, p. 225–231 (1979).
See also 21.1, p. 366; 22.1.1, p. 412.

L. H. Dunleavy and H. J. Blythe
Magnetic Disaccommodation between 30 and 300 K in Low-Temperature Neutron-Irradiated Iron
(a) Vol. 56, No. 1, p. 237–246 (1979).
See also *11*, p. *125*; *18.2*, p. *279*; 21.1.1, p. 376.

M. Bernardin, A. Dupasquier, A. Gallone, and P. Pizzi
*Positron Annihilation Study of Ageing and Creep of Inconel X750
(a) Vol. 56, No. 1, p. 277–284 (1979).
See also 21.1, p. 367.

M. Pahutová and J. Čadek
*On Two Types of Creep Behaviour of F.C.C. Solid Solution Alloys
(a) Vol. 56, No. 1, p. 305–313 (1979).
See also 21, p. 352.

H. Rauh
Untersuchungen zum elektrischen Widerstand von Versetzungen in Metallen (I): Allgemeine Theorie (Investigation of the Electrical Resistance of Dislocations in Metals (I): General Theory)
(b) Vol. 96, No. 2, p. 627–640 (1979).
See also 13.2, p. 157; *14.1*, p. *203*; 21, p. 353.

C. S. B. Piani and J. Aspeling
*On the Existence of Two-Interstitial Species in Platinum after Thermal-Neutron Irradiation
(b) Vol. 96, No. 2, p. 817–822 (1979).
See also 11, p. 125; 21.6, p. 390.

W. Schüle and A. Panzarasa
*Properties of Vacancies and Divacancies in Copper-Gold Alloys
(b) Vol. 96, No. 2, p. K103–K105 (1979).
See also 21.1, p. 368; 21.6, p. 390.

L. Ya. Vinnikov and O. V. Zharikov
*Investigation of Pinning at Grain Boundaries in Pb–In Alloys
(a) Vol. 56, No. 2, p. 433–442 (1979).
See also *14.2*, p. *208*; 21, p. 353.

B. Tirbonod and B. Vittoz
*Introduction of a Dislocation–Defect Elastic Force in the Dragging Model of Internal Friction Peaks
(a) Vol. 56, No. 2, p. 501–504 (1979).
See also *12.1*, p. *132*; 21, p. 353.

S. M. Ohr
*Black–White Contrast of Dislocation Loops in Anisotropic Cubic Crystals
(a) Vol. 56, No. 2, p. 527–530 (1979).
See also *1.4*, p. *32*; 21, p. 353.

L. H. Dunleavy and H. J. Blythe
Magnetic Disaccommodation between 30 and 300 K in Low-Temperature Neutron-Irradiated Carbon-Doped Iron
(a) Vol. 56, No. 2, p. 669–677 (1979).
See also *11*, p. *126*; *18.2*, p. *280*; 21.1.1, p. 378.

10.2 Non-Metals

Original Papers and Short Notes

L. Bosi, A. Longoni, and M. Nimis
*Addendum to: Lifetime of Color Centers in NaBr
(b) Vol. **91**, No. 1, p. 377 (1979).
See also 13.4, p. 164; *20.3*, p. *327*; 22.5.2, p. 465.

A. Makosa and T. Figielski
*Spin-Dependent Recombination at Dislocations in Heat Treated Silicon
(b) Vol. **91**, No. 1, p. K65–K67 (1979).
See also 13.4, p. 164; *16*, p. *252*; 22.1.2, p. 415.

K. Kojima
*Slip Systems in Deformed Anthracene Crystals
(a) Vol. **51**, No. 1, p. 71–78 (1979).
See also 22.9, p. 509.

M. Suezawa, K. Sumino, and I. Yonenaga
*Dislocation Dynamics in the Plastic Deformation of Silicon Crystals (II): Theoretical Analysis of Experimental Results
(a) Vol. **51**, No. 1, p. 217–226 (1979).
See also 22.1.2, p. 415.

J. Daunay, J. Baillou, J. Daunay, and P. Bugnet
Application of Etch Figures to the Study of Growth Twins in Single-Crystalline ZnSe
(a) Vol. **51**, No. 1, p. K5–K8 (1979).
See also *3*, p. *52*; 22.4.2, p. 453.

T. G. M. Kleinpenning and C. A. J. Peters
*The Anion Mobility in β-PbF_2 from Noise Measurements
(a) Vol. **51**, No. 1, p. K105–K108 (1979).
See also *14.4*, p. *238*; 22.5, p. 460.

M. Ikeya and L. O. Schwan
*Extended Hückel Molecular Orbital Calculation of Defects in LiH Cluster Crystals
(b) Vol. **91**, No. 2, p. 433–441 (1979).
See also 13.4, p. 164; 22.5, p. 460.

J. M. de Siebenthal and H. Bill
*Investigation of Sulfur Molecular Paramagnetic Centers in $SrCl_2$
(b) Vol. **91**, No. 2, p. 479–486 (1979).
See also 19, p. 290; 20.1, p. 305; 22.5.3, p. 473.

Cc. Pawlaczyk, G. Luther, and H. E. Müser
Influence of Crystal Lattice Defects on the Ferroelectric Dispersion of Triglycine Sulphate
(b) Vol. **91**, No. 2, p. 627–631 (1979).
See also *14.4.1*, p. *244*; 22.9, p. 509.

M. Siu Li and M. de Souza
*Off-Center Cu^+ in Mixed Crystals
(b) Vol. 92, No. 1, p. 287–291 (1979).
See also *20.1*, p. *307*; 22.5.4, p. 475.

Zh. Egemberdiev, V. Seeman, S. Zazubovich, and N. Jaanson
*Polarized Luminescence and EPR Study of Sn^+ Centers in $KCl:SnCl_2$ Crystals
(b) Vol. 92, No. 1, p. 335–345 (1979).
See also 19, p. 291; 20.1, p. 308; *20.3*, p. *328*; 22.5.4, p. 475.

F. Agullo-López and M. Aguilar
*On the Space Distribution of F Centers: the Pair Correlation Function
(b) Vol. 92, No. 1, p. K43–K45 (1979).
See also 22.5.2, p. 466.

R. Birsoy and L. E. Murr
*Coloring of Natural Fluorites
(a) Vol. 52, No. 1, p. 77–84 (1979).
See also *20.1*, p. *308*; 22.5.3, p. 473.

R. J. Gaboriaud et J. Cadoz
*Etude par microscopie électronique à haute tension de la configuration des dislocations, dans l'oxyde d'yttrium (Y_2O_3) monocristallin déformé plastiquement
(High-Voltage Electron Microscopic Study of Dislocation Configurations in Plastically Deformed Yttrium Oxide (Y_2O_3) Single Crystals)
(a) Vol. 52, No. 1, p. 95–101 (1979).
See also 22.6, p. 478.

F. Fischer
*A Simple Model for Photo-Induced Polarization
(a) Vol. 52, No. 1, p. 189–194 (1979).
See also 13.4, p. 167; 14.4, p. 238; 16, p. 253; 22.5.2, p. 466.

Hj. Bernhardt
Defect Electron Centres in Lead Molybdate Crystals
(a) Vol. 52, No. 1, p. 223–229 (1979).
See also 13.5.3, p. 193; *20.1*, p. *308*; *20.3*, p. *328*; 22.8.1, p. 499.

M. E. van Hulle and W. Maenhout-van der Vorst
*Influence of Pb^{2+} Ions on the Ionic Conductivity and the Space Charge Characteristics of Cubic Silver Halide Microcrystals
(a) Vol. 52, No. 1, p. 277–282 (1979).
See also *14.4*, p. *239*; 22.5.1, p. 463.

J. Perez, C. Maï, J. Tatibouet et R. Vassoille
*Etude des joints de grains dans la glace I_h par mesure du frottement intérieur
(Study of Grain Boundaries of Ice I_h by Internal Friction Measurements)
(a) Vol. 52, No. 1, p. 321–330 (1979).
See also *12.2*, p. *133*; 22, p. 396.

A. Krell and D. Schulze
*$\langle 1\bar{2}10\rangle$-Zone Fracture Anisotropy of Sapphire
(a) Vol. 52, No. 1, p. K45–K48 (1979).
See also *12.2*, p. *133*; 22.6, p. 478.

J. ISOYA and J. A. WEIL
*Uncompensated Titanium (3+) Center in α-Quartz
(a) Vol. 52, No. 2, p. K193–K196 (1979).
See also 11, p. 121; *19*, p. *291*; 22.6, p. 479.

F. LOHSE and J. M. SPAETH
*Fluorescence of Interstitial Atomic Hydrogen in Alkali Halides Doped with I^-
(b) Vol. 93, No. 1, p. 153–160 (1979).
See also 13.4, p. 169; *20.3*, p. *330*; 22.5.2, p. 467.

L. BOSI and M. NIMIS
*On the Interaction between F- and α-Centres in KBr
(b) Vol. 93, No. 1, p. 183–187 (1979).
See also *20.3*, p. *330*; 22.5.2, p. 467.

CHAIK CHANG, J. NEUBERGER, and R. FISCHER
*Activation Energies for Migration of K^+ and N^0 in KN_3
(b) Vol. 93, No. 1, p. 255–259 (1979).
See also 22, p. 398.

E. P. ROTH and A. C. ANDERSON
*Phonon Scattering at a Low-Angle Grain Boundary in Silicon
(b) Vol. 93, No. 1, p. 261–268 (1979).
See also *6*, p. *68*; 22.1.2, p. 419.

A. A. KISELEV and K. LÜDERS
*On the Symmetry Groups of Non-Rigid Impurity Centres in Crystals
(b) Vol. 93, No. 1, p. 285–291 (1979).
See also 6, p. 68; 22.5.2, p. 467.

F. DOCHY
*Theoretical Calculation of the Absorption Energy of the F-Centre in Alkali Halides
(b) Vol. 93, No. 1, p. 325–335 (1979).
See also 13.4, p. 169; *20.1*, p. *311*; 22.5.2, p. 467.

E. GERLACH and H. KLUTTIG
*Electron Scattering by a Layer of Dislocation Lines in a Semiconductor
(b) Vol. 93, No. 1, p. 443–446 (1979).
See also 13.5.2, p. 191; *14.3*, p. *212*; 22, p. 398.

K. K. KIM, J. TOULOUSE, and A. S. NOWICK
*Anelastic Relaxation of a Li^+–Li^+ Split Interstitial Defect in LiF-Doped MnF_2 Crystals
(b) Vol. 93, No. 1, p. K1–K4 (1979).
See also 22.5, p. 461.

A. P. BATRA and L. M. SLIFKIN
Contributions from Vacancy Pairs and Single Vacancies to Chloride Impurity Diffusion in AgBr
(b) Vol. 93, No. 1, p. K77–K80 (1979).
See also *9*, p. *86*; 22.5.1, p. 463.

E. M. Trukhanov, S. I. Stenin, and A. G. Noskov
*Dislocations and Stresses in a Crystal with an Island Film
(a) Vol. **53**, No. 2, p. 433–440 (1979).
See also *1.5*, p. *35*; 22.1.1, p. 411.

P. Köhler, W. Möhling, and M. Pasemann
Microdefects in Silicon Different from Swirls
(a) Vol. **53**, No. 2, p. 509–517 (1979).
Erratum
(a) Vol. **56**, No. 2, p. 769 (1979).
See also *1.4*, p. *28*; 22.1.2, p. 420.

F. Walz, H. Deusch, and H. Kronmüller
Investigation of Magnetic After-Effects in Magnetite near the Verwey Temperature
(a) Vol. **53**, No. 2, p. 519–528 (1979).
See also *18.3*, p. *284*; 22.6, p. 480.

A. George and G. Champier
*Velocities of Screw and 60° Dislocations in n- and p-Type Silicon
(a) Vol. **53**, No. 2, p. 529–540 (1979).
See also 22.1.2, p. 420.

A. R. Lakshmanan, R. C. Bhatt, and K. G. Vohra
γ-Radiation-Induced Sensitization, Supralinearity, and Photostimulated Thermoluminescence in LiF TLD-100. Correlation Studies
(a) Vol. **53**, No. 2, p. 617–625 (1979).
See also 13.4, p. 171; *20.3*, p. *332*; 22.5.2, p. 468.

A. E. Kiv and V. N. Soloviev
*The Grasshopper Effect in the Diamond Lattice
(b) Vol. **94**, No. 1, p. K91–K95 (1979).
See also 22.1, p. 408.

E. V. Suvorov, O. S. Gorelik, V. M. Kaganer, and V. L. Indenbom
*Shape of Extinction Fringes and Determination of the Burgers Vector in Single-Crystal Interferometric Techniques
(a) Vol. **54**, No. 1, p. 29–35 (1979).
See also 22.1.2, p. 421.

G. M. Renfro and H. J. Fischbeck
Optically Stimulated Exoelectron Emission from Potassium Chloride Single Crystals
(a) Vol. **54**, No. 1, p. 85–92 (1979).
See also *17*, p. *257*; 22.5.2, p. 468.

M. Pasemann and P. Werner
*Identification of Small Defects in Silicon
(a) Vol. **54**, No. 1, p. 179–188 (1979).
See also *1.4*, p. *29*; 22.1.2, p. 421.

I. V. Gridneva, Yu. V. Milman, V. I. Trefilov, and S. I. Chugunova
*Analysis of Dislocation Mobility under Concentrated Loads at Indentations of Single Crystals
(a) Vol. **54**, No. 1, p. 195–206 (1979).
See also 12.2, p. 135; 22.1.1, p. 411; 22.1.2, p. 421; 22.2.1, p. 434; 22.2.3, p. 440.

D. M. G. Pinatti, S. C. Zílio, and M. de Souza
Point Defects in NaCl Crystals Doped with Divalent Ions. The NaCl–NaF:Pb^{++} Crystal
(b) Vol. 94, No. 2, p. 749–755 (1979).
See also *20.1*, p. *317*; 22.5.2, p. 469.

S. B. S. Sastry and S. Sapru
*Thermoluminescence Glow and Emission Studies on Suprapure Rubidium Bromide Irradiated at Room Temperature
(b) Vol. 94, No. 2, p. K149–K154 (1979).
See also *20.3*, p. *333*; 22.5.2, p. 469.

A. V. Dvurechenskii, L. P. Potapova, and V. V. Kalinin
Efficiency of Defect Build-up in Ion-Implanted Polycrystalline Silicon Films
(a) Vol. 54, No. 2, p. 457–462 (1979).
See also *1.5*, p. *35*; *11*, p. *123*; 19, p. 294; 22.1.2, p. 422.

V. D. Verner, D. K. Nichugovskii, and V. Ya. Kovyasin
*Investigation of Internal Friction in GaP
(a) Vol. 54, No. 2, p. 593–596 (1979).
See also *12.2*, p. *135*; 22.2.2, p. 438.

Hj. Bernhardt
*On the Polaron Character of the Coloration Centres in $LiNbO_3$ Crystals
(a) Vol. 54, No. 2, p. 597–603 (1979).
See also *13.5.3*, p. *193*; 20.1, p. 317; 20.3, p. 334; 22.8.2, p. 506.

D. Lapraz, A. Baumer, and P. Iacconi
*On the Thermoluminescent Properties of Hydroxyapatite $Ca_5(PO_4)_3OH$
(a) Vol. 54, No. 2, p. 605–613 (1979).
See also *20.3*, p. *334*; 22.8.1, p. 501.

E. Lilley and J. E. Strutt
*Bulk and Grain Boundary Ionic Conductivity in Polycrystalline β''-Alumina
(a) Vol. 54, No. 2, p. 639–650 (1979).
See also *14.4*, p. *241*; 22.6, p. 482.

J. Prakash
*Photo-Induced Polarization and Photo-Induced Depolarization of S^{2-}–Anion Vacancy Dipoles in Potassium Iodide
(a) Vol. 54, No. 2, p. 681–688 (1979).
See also 14.4, p. 242; 20.1, p. 317; 22.5.2, p. 469.

Y. Kamiura and F. Hashimoto
*Degree of Copper Contamination in Germanium Heat-Treated at High Temperatures
(a) Vol. 54, No. 2, p. 697–700 (1979).
See also *13.4*, p. *175*; 22.1.1, p. 412.

A. Iida
*Applications of X-Ray Triple Crystal Diffractometry to Studies on the Diffusion-Induced Defects in Silicon Crystals
(a) Vol. 54, No. 2. p. 701–706 (1979).
See also *1.1*, p. *13*; 22.1.2, p. 423.

D. BENSAHEL and M. DUPUY
*SEM and TEM: Diffusion of Lithium in ZnTe
(a) Vol. **55**, No. 1, p. 203–210 (1979).
See also *1.4*, p. *30*; 22.4.3, p. 456.

D. BENSAHEL, M. DUPUY, and J. C. PFISTER
*Electrial, SEM, and TEM Studies in ZnTe (I): Impurity Segregation during Short Annealings and Quenching
(a) Vol. **55**, No. 1, p. 211–221 (1979).
See also *1.4*, p. *30*; 14.3, p. 215; 22.4.3, p. 456.

R. ERDMANN and H. ALEXANDER
*Photo-EPR of Dislocations in Silicon
(a) Vol. **55**, No. 1, p. 251–259 (1979).
See also 13.4, p. 176; 19, p. 295; 22.1.2, p. 423.

M. SATO and K. SUMINO
*Motion of Extended Dislocations in Silicon Crystals Observed by HVEM
(a) Vol. **55**, No. 1, p. 297–306 (1979).
See also 22.1.2, p. 423.

N. D. STOJADINOVIĆ and R. S. POPOVIĆ
*A New Method for Elimination of Emitter Edge Dislocations of Silicon Planar NPN Transistors
(a) Vol. **55**, No. 1, p. 307–313 (1979).
See also *14.3.3*, p. *229*; 22.1.2, p. 424.

L. NIKOLOVA, T. TODOROV, D. POPOV, and K. TERSHSKI
*Laser Induced Fast Reversible Processes in Additively Coloured KCl Containing F_A Centres
(a) Vol. **55**, No. 1, p. 333–337 (1979).
See also *20.1*, p. *319*; 22.5.2, p. 470.

R. K. SINHA, R. K. GARTIA, and B. S. ACHARYA
*Thermoluminescence of Z_1 Centers in Ca-Doped $KCl_{0.5}Br_{0.5}$ Mixed Crystals
(a) Vol. **55**, No. 1, p. K109–K113 (1979).
See also *20.3*, p. *335*; 22.5.4, p. 476.

V. Z. POLINGER and G. I. BERSUKER
Multimode Jahn-Teller Effect for an E Term with Strong Vibronic Coupling (I): Local and Resonant States
(b) Vol. **95**, No. 2, p. 403–411 (1979).
See also *6*, p. *74*; *13.4*, p. *176*.

J. P. ESTIENNE and J. L. FARVACQUE
*Experimental Evidence for Piezoelectrical Free-Carrier Scattering by Screw Dislocations in Tellurium
(b) Vol. **95**, No. 2, p. 593–600 (1979).
See also *14.3*, p. *215*; 22.1.3, p. 429.

R. WĘCŁAWIK, J. WAWRYSZCZUK, C. RYBKA, and T. GOWOREK
*Positronium and Thermal Defects in Molecular Crystals
(b) Vol. **95**, No. 2, p. K113–K116 (1979).
See also 22.9, p. 512.

D. Kashchiev
On the Saturation Cluster Density in Deposition of Thin Films from Vapours
(a) Vol. 55, No. 2, p. 369–374 (1979).
See also *1.5*, p. *36*; *3.* p. *53*; 21.6, p. 389; 22.5.2, p. 471.

H.-J. Möller, P. Haasen, and H. Ewaldt
*Cross Slip of Single Dissociated Screw Dislocations in Silicon and Germanium
(a) Vol. 55, No. 2, p. 469–478 (1979).
See also 22.1.1, p. 412; 22.1.2, p. 424.

P. Grau and F. Fröhlich
*Aggregation and Precipitation Hardening in NaCl:Ca Crystals
(a) Vol. 55, No. 2, p. 479–487 (1979).
See also 22.5.2, p. 471.

D. Nenow and S. Gueleva
Formation of Radiation-Induced Voids on Dislocations in LiF
(a) Vol. 55, No. 2, p. 489–496 (1979).
See also *11*, p. *124*; 22.5.2, p. 471.

S. I. Stenin
*Problems of the Formation of Dislocation Structures in Heteroepitaxial Layers
(a) Vol. 55, No. 2, p. 519–527 (1979).
See also *1.5*, p. *36*; 22.1, p. 408; 22.2, p. 431.

F. Appel, U. Messerschmidt, and M. Kuna
*Crack Propagation in MgO during In-Situ Deformation in the High-Voltage Electron Microscope
(a) Vol. 55, No. 2, p. 529–536 (1979).
See also 1.4, p. 31; 22.6, p. 484.

A. Krell and D. Schulze
*Microstructure Dependence of the Brittle Fracture of Alumina
(a) Vol. 55, No. 2, p. 537–544 (1979).
See also 12.2, p. 135; 22.6, p. 484.

V. D. Negryi and Yu. A. Osipyan
*Dislocation Emission in CdS
(a) Vol. 55, No. 2, p. 583–588 (1979).
See also *20.3*, p. *335*; 22.4.1, p. 451.

J. Auleytner, Z. Lilienthal, E. Mizera, and T. Warmiński
*On X-Ray and Electron Microscopic Characterization of Lattice Defects in $A^{II}B^{VI}$ Semiconductor Compounds
(a) Vol. 55, No. 2, p. 603–609 (1979).
See also 22.4.3, p. 456.

H. Blumtritt, R. Gleichmann, J. Heydenreich, and H. Johansen
*Combined Scanning (EBIC) and Transmission Electron Microscopic Investigations of Dislocations in Semiconductors
(a) Vol. 55, No. 2, p. 611–620 (1979).
See also *1.4*, p. *31*; 14.3.3, p. 230; 22.1.2, p. 424.

Z. Lilienthal
*The Effect of Ag Impurity on the Real Structure of ZnTe
(a) Vol. **55**, No. 2, p. 639–645 (1979).
See also **22.4.3**, p. **456**.

C. Manolikas, J. van Landuyt, R. de Ridder, and S. Amelinckx
Electron Microscopic Study of the Domain Structure and of the Transition State in $Cu_{0.5}.In_{2.5}Se_4$
(a) Vol. **55**, No. 2, p. 709–722 (1979).
See also *1.4*, p. *31*; **22.8**, p. **496**.

E. Pink, H. Bäck, and B. Ortner
*The Stress Relaxation of Polyvinyl Chloride in the Macrodeformation Range
(a) Vol. **55**, No. 2, p. 751–761 (1979).
See also **22.9**, p. **513**.

A. Ourmazd and G. R. Booker
*The Electrical Recombination Efficiency of Individual Edge Dislocations and Stacking Fault Defects in n-Type Silicon
(a) Vol. **55**, No. 2, p. 771–784 (1979).
See also **14.3.2**, p. **224**; **22.1.2**, p. **425**.

S. V. Moharil, V. S. Kamavisdar, and B. T. Deshmukh
Thermoluminescence of Z_1 Centres in NaCl:Ca
(a) Vol. **55**, No. 2, p. K167–K172 (1979).
See also *20.3*, p. *336*; **22.5.2**, p. **471**.

Z. Lilienthal
The Effect of Real Structure on Optical Properties of ZnTe
(a) Vol. **55**, No. 2, p. K173–K178 (1979).
See also *20.1*, p. *320*; *20.3*, p. *336*; **22.4.3**, p. **456**.

A. Audouard
*Activation thermique de la déformation plastique de l'oxyde cuivreux
(Thermal Activation of Plastic Deformation of Copperoxide)
(a) Vol. **55**, No. 2, p. K191–K196 (1979).
See also **22.6**, p. **484**.

A. Shluger, E. Kotomin, and Yu. Dzelme
*Quantum-Chemical Calculations of Electronic and Hole Centres and Surface of NaCl Crystals (I)
(b) Vol. **96**, No. 1, p. 91–99 (1979).
See also **13.3**, p. **161**; *13.4*, p. *177*; **22.5.2**, p. **471**.

G. Reuter, F. Lohse, and J. M. Spaeth
*Generation of Dichroic H Centres in KCl by Polarized Optical Bleaching of Interstitial Atomic Hydrogen
(b) Vol. **96**, No. 1, p. 243–250 (1979).
See also **13.4**, p. **178**; **19**, p. **296**; *20.1*, p. *321*; **22.5.2**, p. **471**.

B. N. Mukashev, K. H. Nussupov, and M. F. Tamendarov
*Divacancy in Silicon Irradiated by Protons
(b) Vol. **96**, No. 1, p. K17–K19 (1979).
See also *11*, p. *125*; 20.1, p. 322; 22.1.2, p. 425.

E. M. Trukhanov, S. I. Stenin, and A. G. Noskov
*Dislocation Interactions in Germanium Covered with Dielectric Films
(a) Vol. **56**, No. 2, p. **443**—450 (1979).
See also *1.5*, p. *37*; 2, p. 50; 22, p. 407; 22.1.1, p. 413; 22.6, p. 485.

V. Z. Bengus and T. P. Kovalenko
*The Spectrum of Glide Band Sources and Work-Hardening in LiF Single Crystals
(a) Vol. **56**, No. 2, p. 473—480 (1979).
See also 22.5.2, p. 472.

W. Raupach
*Influence of the Interionic Potentials on the Core Configuration of Screw Dislocations in NaCl Single Crystals
(a) Vol. **56**, No. 2, p. 535—540 (1979).
See also 22.5.2, p. 472.

H.-J. Kos and R. Nink
*Correlation of Thermoluminescence and Optical Absorption of Z_3 Centers in LiF:Mg, Ti Crystals
(a) Vol. **56**, No. 2, p. 593—596 (1979).
See also *20.1*, p. *324*; 20.3, p. 337; 22.5.2, p. 472.

V. A. Krylov, W. Ulrici, and L. S. Sochava
*Optical Spectra of Off-Centre Ni^{2+} Ions in SrO
(a) Vol. **56**, No. 2, p. 615—621 (1979).
See also *20.1*, p. *324*; 22.6, p. 485.

G. Dhalenne, A. Revcolevschi, and C. Monty
*Grain Boundaries in NiO (II): Determination of Mass Transport Mechanisms by Thermal Grooving
(a) Vol. **56**, No. 2, p. 623—636 (1979).
See also *1.4*, p. *32*; 22.6, p. 485.

P. V. Petrashen and E. N. Kislovskii
*Experimental Study of the X-Ray Diffraction in a Bent Crystal with a Large Strain Gradient
(a) Vol. **56**, No. 2, p. 663—667 (1979).
See also *1.1*, p. *14*; 22.1.2, p. 426.

Z. Liliental, M. Pasemann, and P. Werner
*Electron Microscopic Investigations of Precipitates in Cu-Doped ZnTe
(a) Vol. **56**, No. 2, p. 679—685 (1979).
See also 22.4.3, p. 457.

K. Somaiah, P. Veeresham, K. L. N. Prasad, and V. Hari Babu
*Thermoluminescence and Optical Absorption of BaFCl Crystals
(a) Vol. **56**, No. 2, p. 737—742 (1979).
See also 20.1, p. 324; *20.3*, p. *337*; 22.5.4, p. 476.

V. F. Antyushin, B. I. Sysoev, and V. F. Synorov
*Identification of Alkali Metal Ions in Silicon Dioxide Films
(a) Vol. **56**, No. 2, p. K91—K95 (1979).
See also *14.4*, p. *243*; 22.6, p. 486.

11. Irradiation Effects
(X-Ray Diffraction Investigations See 1 and 10)

Original Papers and Short Notes

G. Müller, M. Trapp, R. Schimko, and C. E. Richter
*Measurement of Range Distributions of Zinc and Nitrogen Ions in Multiple-Layer Substrates with the Secondary Ion Microprobe
(a) Vol. **51**, No. 1, p. 87–92 (1979).
See also 22.6, p. 477.

W. Heywang, E. F. Krimmel, and H. Runge
*Annealing Mechanism of Radiation Damage and Dopants in Pulsed Laser Light Irradiated Ion Implanted Layers
(a) Vol. **51**, No. 1, p. K79–K82 (1979).
See also 9, p. 85; 22.1.2, p. 415.

V. I. Shulga
*On a Property of the Continuum Planar Potential
(b) Vol. **91**, No. 2, p. K165–K167 (1979).

S. Wasim and R. Nava
Thermal Conductivity of γ-Irradiated α-Quartz below 10 K
(a) Vol. **51**, No. 2, p. 359–366 (1979).
See also *8*, p. *80*; 22.6, p. 478.

O. Alstrup
*Investigation of Radiation Damage in Ion Implanted Silicon Crystals by Pendellösung Topography
(a) Vol. **51**, No. 2, p. 407–418 (1979).
See also 1.1, p. 11; 10.2, p. 104; 22.1.2, p. 416.

S. Hinckley, H. Hora, and J. C. Kelly
*Subthreshold Defect Generation and Annealing in Silicon by Intense Electron Beam Bombardment
(a) Vol. **51**, No. 2, p. 419–427 (1979).
See also 8, p. 80; 10.2, p. 104; 22.1.2, p. 416.

A. Dunlop, N. Lorenzelli et J. C. Jousset
*Irradiation du fer à 20 K par les neutrons rapides et par les produits de fission de l'uranium 235
(Investigation of Iron at 20 K by Fast Neutrons and Fission Fragments of Uranium 235)
(a) Vol. **51**, No. 2, p. 479–485 (1979).
See also 21.1.1, p. 371.

G. Debras, J. M. Gilles, and A. A. Lucas
*Observation of Overpressurized Bubbles in Helium-Irradiated Aluminium Films
(a) Vol. 51, No. 2, p. 559–565 (1979).
See also 1.5, p. 34; 21, p. 340.

M. Fähnle
*Irradiation Effects in Nb_3Sn — New Theoretical Results
(b) Vol. **92**, No. 2, p. K127—K129 (1979).
See also *14.2*, p. *205*; 21, p. 342.

M. F. Vereshchak, A. K. Zhetbaev, and Sh. Sh. Ibragimov
*A Mössbauer Study of the Effect of γ-Rays and Neutron Irradiation on Iron Oxalates
(a) Vol. **52**, No. 2, p. 415—419 (1979).
See also *5*, p. *60*; 22.8, p. 493.

V. L. Vinetskii, G. N. Eritsyan, and R. A. Melkonyan
*On the Nature of the "Limiting" Position of the Fermi Level in Irradiated Silicon
(a) Vol. **52**, No. 2, p. 441—446 (1979).
See also *13.4*, p. *168*; 14.3, p. 212; 20.1, p. 310; 22.1.2, p. 418.

H. F. Kappert, G. Sixt, and G. H. Schwuttke
*Minority Carrier Lifetime in Silicon after Ar^+ and Si^+ Implantation
(a) Vol. **52**, No. 2, p. 463—474 (1979).
See also 14.3.3, p. 227; 22.1.2, p. 418.

C. S. Pande
*Transmission Electron Microscopy of Radiation Damage in Nb_3Sn
(a) Vol. **52**, No. 2, p. 687—696 (1979).
See also *1.4*, p. *28*; 14.2, p. 205.

J. Isoya and J. A. Weil
Uncompensated Titanium (3+) Center in α-Quartz
(a) Vol. **52**, No. 2, p. K193—K196 (1979).
See also *10.2*, p. *107*; *19*, p. *291*; 22.6, p. 479.

E. Tamura and Y. H. Ohtsuki
Diffusion of the Laue Spots in Electron Diffraction
(b) Vol. **93**, No. 1, p. 167—173 (1979).
See also *1*, p. *9*; 9, p. 86.

V. D. Korolev, S. A. Peskovatskii, and V. M. Shulga
Spin-Lattice Relaxation in Fast Neutron Irradiated Quartz at Helium Temperatures and below 1 K
(b) Vol. **93**, No. 1, p. K45—K48 (1979).
See also *19*, p. *292*; 22.6, p. 480.

Y. Ya. Chumanov, Sh. Z. Izmailov, G. P. Pokhil, E. I. Sirotinin, and A. F. Tulinov
*On the Determination of Energy Losses by Charged Particles from the Backscattered Energy Spectra
(a) Vol. **53**, No. 1, p. 51—62 (1979).
See also 21, p. 343.

V. A. Ryabov
*Shift of Axial Blocking of Scattered Nuclear Products
(a) Vol. **53**, No. 1, p. 155—160 (1979).

A. E. Karkin, V. E. Arkhipov, V. A. Marchenko, and B. N. Goshchitskii
Electrical Resistivity of V_3Si and Nb_3Sn under Neutron Radiation
(a) Vol. 54, No. 1, p. K53—K58 (1979).
See also *14.2*, p. *207*; 21, p. 346.

A. V. Dvurechenskii, L. P. Potapova, and V. V. Kalinin
*Efficiency of Defect Build-up in Ion-Implanted Polycrystalline Silicon Films
(a) Vol. 54, No. 2, p. 457—462 (1979).
See also *1.5*, p. *35*; 10.2, p. 111; 19, p. 294; 22.1.2, p. 422.

P. Revesz and J. W. Mayer
*Regrowth Behaviour of Ge Implanted ⟨100⟩ Si
(a) Vol. 54, No. 2, p. 513—516 (1979).
See also *3*, p. *53*; 22.1.2, p. 422.

T. Leffers and B. N. Singh
Loop Growth and Point-Defect Profiles during HVEM Irradiation
(a) Vol. 54, No. 2, p. 621—631 (1979).
See also *10.1*, p. *98*; 21.1, p. 363.

J. Z. Kristapson, V. I. Ozerskii, and V. A. Chadaev
*Annealing of NaCl Crystals Irradiated by High Electron Doses at High Temperatures
(a) Vol. 54, No. 2, p. K119—K121 (1979).
See also 22.5.2, p. 469.

O. Greschner and R. Wedell
*Comparison between Spontaneous Radiation in Channelling, Coherent Bremsstrahlung, and Synchrotron Radiation
(b) Vol. 95, No. 1, p. 137—140 (1979).

M. D. Varentsov
Quantum Current Oscillations in γ-Irradiated n-Si with Layer Inhomogeneity
(b) Vol. 95, No. 1, p. K47—K49 (1979).
See also *14.3*, p. *215*; 22.1.2, p. 423.

V. G. Chudinov, N. V. Moseev, B. N. Goshchitskii, and V. I. Protasov
*On the Influence of Irradiation Type on the Formation of Radiation Defects in Nb-Based Superconductors with A-15 Structure
(a) Vol. 55, No. 1, p. 109—112 (1979).
See also *14.2*, p. *207*; 21, p. 348.

N. M. Bezdetny, V. G. Silvestrov, S. V. Safonova, A. Kh. Zeinally, and A. L. Timofeev
*The Influence of γ-Irradiation on Photorefraction in Iron Doped Lithium Niobate
(a) Vol. 55, No. 1, p. K45—K47 (1979).
See also *20.1*, p. *319*; 22.8.1, p. 502.

M. A. Chubisov and A. F. Akkerman
*Computer Simulation of Low Energy Hydrogen and Helium Ion Reflection from Monocrystalline Surfaces
(a) Vol. 55, No. 1, p. K53—K58 (1979).
See also 21, p. 349; 22.1.2, p. 424.

D. E. Popov, V. V. Kaplin, and S. A. Vorobiev
*Flux Redistribution of Channeled Fast Electrons
(b) Vol. 96, No. 1, p. 263–269 (1979).
See also 13, p. 141; 22.1.2, p. 425.

V. V. Rozhkov
*Theory of Dechanneling
(b) Vol. 96, No. 1, p. 463–468 (1979).

B. N. Mukashev, K. H. Nussupov, and M. F. Tamendarov
*Divacancy in Silicon Irradiated by Protons
(b) Vol. 96, No. 1, p. K17–K19 (1979).
See also *10.2*, p. *115*; 20.1, p. 322; 22.1.2, p. 425.

K. Urban
Radiation-Induced Processes in Experiments Carried out In-situ in the High-Voltage Electron Microscope
(a) Vol. 56, No. 1, p. 157–168 (1979).
See also *10.1*, p. *100*; 21, p. 352; 21.1.1, p. 376; 21.6, p. 389.

Y. Ohmura, T. Shigematsu, and R. Nakayama
*Effects of Irradiation Time and Subsequent Heat Treatment on Nd:YAG Laser Annealing of High-Dose As-Implanted Silicon
(a) Vol. 56, No. 1, p. 225–230 (1979).
See also 14.3, p. 217; 22.1.2, p. 425.

L. H. Dunleavy and H. J. Blythe
*Magnetic Disaccommodation between 30 and 300 K in Low-Temperature Neutron-Irradiated Iron
(a) Vol. 56, No. 1, p. 237–246 (1979).
See also 10.1, p. 101; *18.2*, p. *279*; 21.1.1, p. 376.

T. I. Kolchenko, V. M. Lomako, and I. G. Tarutin
*Introduction of Defect Clusters in GaAs by High-Energy γ-Rays
(a) Vol. 56, No. 1, p. 387–393 (1979).
See also 14.3, p. 217; 22.2.1, p. 435.

F. F. Komarov
*Energy Loss of Relativistic Particles in Solids
(b) Vol. 96, No. 2, p. 555–568 (1979).
See also 13, p. 142; 21.6, p. 389; 22.1, p. 408.

C. S. B. Piani and J. Aspeling
On the Existence of Two-Interstitial Species in Platinum after Thermal-Neutron Irradiation
(b) Vol. 96, No. 2, p. 817–822 (1979).
See also *10.1*, p. *101*; 21.6, p. 390.

S. M. Darbinian and K. A. Ispirian
*On High Energy Particle Channeling
(b) Vol. 96, No. 2, p. 835–842 (1979).
See also 13, p. 142; 22.1.1, p. 413.

12. Mechanical Properties (Plastic Deformations See 10)

Review Article

V. I. Dotsenko
Stress Relaxation in Crystals
(b) Vol. **93**, No. 1, p. 11–**43** (1979).
See also *10*, p. *89*.

Original Papers and Short Notes

V. A. Solovev and S. N. Zolotarev
The Isotropic Limit in Anisotropic Dislocation Theory. Green's Functions in a Crystal Containing a Dislocation Pile-up
(b) Vol. **91**, No. 1, p. 319–323 (1979).
See also *10*, p. *89*.

M. B. Geilikman
*Elastic Properties of Polytypes
(b) Vol. **91**, No. 2, p. 725–733 (1979).

B. Perrin
*Cauchy Relations Revisited
(b) Vol. **91**, No. 2, p. K115–K120 (1979).
See also *4*, p. *55*.

E. V. Zarochentsev, S. M. Orel, and V. N. Varyukhin
*Elastic Constants of a Stressed Crystal (I): General Theory
(a) Vol. **52**, No. 2, p. 455–462 (1979).

H. Trinkaus
Drift Diffusion in Garin Boundaries under the Influence of Stress Fields
(b) Vol. **93**, No. 1, p. 293–303 (1979).
See also *9*, p. *86*; *10*, p. *89*.

E. V. Zarochentsev, S. M. Orel, and V. N. Varyukhin
*Elastic Constants of a Stressed Crystal (II): Cubic Crystal
(a) Vol. **53**, No. 1, p. 75–85 (1979).

G. Schoeck
*The Volume Change in Internal Stress Fields
(b) Vol. **94**, No. 1, p. 147–151 (1979).
See also *10*, p. *89*.

O. Eberhardt
*Ebene Elastostatik anisotroper Medien mit antisymmetrischen Spannungen und ihre Anwendung auf die Theorie des Flußliniengitters
(Plane Elasticity of Anisotropic Media with Antisymmetric Stress and Its Application to the Theory of the Flux Line Lattice)
(b) Vol. **94**, No. 1, p. 171–179 (1979).
See also 14.2, p. 206.

12.1 Metals

Original Papers and Short Notes

K. N. Khanna and P. K. Sharma
*Binding Energy, Compressibility, Interionic Potential, and Transition Temperature of Simple Metals Obtained by the Pseudopotential Method
(b) Vol. **91**, No. 1, p. 251–254 (1979).
See also *4*, p. *55*; 14.2, p. 204; 21.2, p. 379.

S. A. Mahmoud, K. H. Georgy, F. M. Mansy, and R. Kamel
*Effect of Cold-Work on the Mechanism Controlling the High-Temperature Creep in Al–2 wt% Mg
(a) Vol. **51**, No. 1, p. 257–260 (1979).
See also *10.1*, p. *92*; 21, p. 339.

A. Korner, H. Schmid, and F. Prinz
*Asymmetrical Equilibrium Positions and Line Tensions of Composite Dislocations
(a) Vol. **51**, No. 2, p. 613–621 (1979).
See also *10.1*, p. *92*; 21.6, p. 387.

T. Soma
Equation of State under High Pressure for Metallic Si and Ge
(b) Vol. **92**, No. 1, p. K51–K53 (1979).
See also *1.2*, p. *16*; 4, p. 55; 22.1.1, p. 410; 22.1.2, p. 417.

I. G. Ritchie, J. F. Dufresne, and P. Moser
*Internal Friction in Deformed Hydrogen-Doped Iron
(a) Vol. **52**, No. 1, p. 331–340 (1979).
See also *10.1*, p. *94*; 21.1.1, p. 371.

F. Fouquet, P. Merle, and J. Merlin
Characterization of Stages during the Precipitation of the Θ′-Phase in an Al–4 wt% Cu Alloy
(a) Vol. **52**, No. 1, p. K1–K5 (1979).
See also *1.3*, p. *22*; 10.1, p. 94; 21, p. 341; 21.1, p. 357.

D. K. Bowen and V. Vitek
*The Construction of Central-Force Interatomic Potentials from Whisker Deformation Data
(b) Vol. **92**, No. 2, p. 645–652 (1979).
See also 4, p. 56; *10.1*, p. *94*; 21.1.1, p. 372.

W. R. Thorpe and I. O. Smith
*A Model for High-Temperature Creep Incorporating Both Recovery and Thermally Activated Glide
(a) Vol. **52**, No. 2, p. 487–497 (1979).
See also *10.1*, p. *95*.

M. E. Gurevich, A. I. Mitsek, and A. F. Zhuravlev
Anomalies in Thermal Expansion and Heat Capacity of Ferromagnets
(b) Vol. **93**, No. 1, p. 127–134 (1979).
See also *1.2*, p. *17*; 8, p. 81; 18.2, p. 274; 21.1, p. 358.

J. D. STEPHENSON, T. TUOMI, V. KELHÄ, and M. TILLI
*Quasi-Lifetime Topography of Magnetic Domain Movements in Grain Oriented (100) [001] Fe–3 wt% Si Polycrystals Using White Synchrotron Radiation (II): The Influence of Small Tensile Stress
(a) Vol. **53**, No. 1, p. 271–276 (1979).
Erratum
(a) Vol. **55**, No. 1, p. 339 (1979).
See also *1.4*, p. *28*; *18.2*, p. *274*; 21.1.1, p. 373.

N. RUSOVIĆ and H. WARLIMONT
*Young's Modulus of β_2-NiAl Alloys
(a) Vol. **53**, No. 1, p. 283–288 (1979).
See also 21.1, p. 359.

H. O. K. KIRCHNER, H. MUGHRABI, F. PRINZ, and H. SCHMID
Erratum to: Dislocation Loops in α-Iron
(a) Vol. **53**, No. 1, p. 411 (1979).
See also *10.1*, p. *95*; 21.1.1, p. 373.

V. I. ZAITSEV, E. I. LYAFER, and V. A. FOMCHENKO
*Influence of Compression Anisotropy Effects on Zn Creep
(a) Vol. **53**, No. 1, p. K99–K101 (1979).
See also *10.1*, p. *96*; 21.1, p. 359.

T. TUOMI, J. D. STEPHENSON, M. TILLI, and V. KELHÄ
Application of Synchrotron Topography to the Study of Magnetic Domain Movements in (100) Fe–3 wt% Si Produced by a Mechanical Stress
(a) Vol. **53**, No. 2, p. 571–576 (1979).
See also *1.4*, p. *29*; *18.2*, p. *275*; 21.1.1, p. 374.

M. FUMERON, J. P. GUILLOT, A. DAUGER, and J. CAISSO
X-Ray Scattering Study of the Shape of Gunier-Preston Zones and Their Size Effect in an Al–6.8 at% Zn Alloy
(a) Vol. **53**, No. 2, p. 605–610 (1979).
See also *1.3*, p. *23*; 10.1, p. 96; 21.1, p. 360.

E. BONETTI, E. EVANGELISTA, and P. GONDI
Internal Friction and Dynamic Modulus of Polycrystalline Al during Deformation in the Temperature Range of the Kê Peak
(a) Vol. **53**, No. 2, p. 653–658 (1979).
See also *10.1*, p. *96*; 21, p. 345.

E. TÖRÖK and G. HAUSCH
*Bulk Modulus Anomaly of Fe–Ni and Fe–Pt Invar Alloys
(a) Vol. **53**, No. 2, p. K147–K151 (1979).
See also 18.1, p. 267; *18.2*, p. *276*; 21.1, p. 360; 21.1.1, p. 374; 21.6, p. 388.

H. H. WAWRA
*On the Role of Grüneisen's Elastic Parameters in the Theory of Polycrystal Elasticity. A Critical Assessment
(a) Vol. **54**, No. 1, p. 291–296 (1979).
See also 21.1, p. 362.

G. A. Hassan
*Effect of the Type of Pre-Torsional Deformation on the Creep Rates of an Aluminium–Silicon–Zinc Alloy
(a) Vol. **54**, No. 1, p. K31–K32 (1979).
See also 21, p. 346.

R. Ramji Rao and A. Rajput
*Third-Order Elastic Constants, Thermal Expansion, and Bulk Modulus of Body-Centered Cubic Crystals
(b) Vol. **94**, No. 2, p. 691–699 (1979).
See also 21, p. 347; 21.2, p. 379.

L. D. Livshits and A. G. Fomichev
*Micro-Indentation of Aluminium under Hydrostatic Pressures up to 8500 kp/cm²
(a) Vol. **54**, No. 2, p. K123–K127 (1979).
See also 21, p. 347.

V. Navrátil, P. Lukáš, and V. P. Soldatov
*The Critical Resolved Shear Stress of Cd–Zn Single Crystals at Low Temperatures
(a) Vol. **54**, No. 2, p. K175–K176 (1979).
See also *10.1*, p. *98*; 21, p. 348; 21.1, p. 364.

O. Buck, L. A. Ahlberg, L. J. Graham, G. A. Alers, C. A. Wert, and K. C. Hsieh
*Harmonic and Anharmonic Elastic Properties of the α-Nb–H System
(a) Vol. **55**, No. 1, p. 223–230 (1979).
See also 21, p. 349.

N. Igata, N. Kuramochi, N. Masamura, A. Kohyama, K. Miyahara, S. Sato, and H. Yoshida
*The α'-Peak of Deformed Niobium
(a) Vol. **55**, No. 1, p. K127–K129 (1979).
See also *10.1*, p. *98*; 21, p. 349.

H. R. Hilzinger, H. Hillmann, and A. Mager
Magnetostriction Measurements on Co-Base Amorphous Alloys
(a) Vol. **55**, No. 2, p. 763–769 (1979).
See also *2*, p. *48*; *18.2*, p. *278*; 21.1, p. 366; 21.1.1, p. 376.

D. M. Esterling and A. Swaroop
Interatomic Potentials from Experimental Phonon Spectra (I): Prototypes
(b) Vol. **96**, No. 2, p. 401–411 (1979).
See also 2, p. 48; *4*, p. *57*; 6, p. 75; 21.1, p. 366; 21.1.1, p. 376; 21.2, p. 379.

A. van den Beukel and G. J. den Otter
*Modulus Effect and Internal Friciton of Copper and Some Copper Alloys Cold-Worked at 78 K
(a) Vol. **56**, No. 1, p. 117–127 (1979).
See also *10.1*, p. *100*; 21.1, p. 367.

A. Swaroop and D. M. Esterling
Interatomic Potentials from Experimental Phonon Spectra (II): General Results
(b) Vol. **96**, No. 2, p. 691–696 (1979).
See also 2, p. 49; *4*, p. *58*; 6, p. 76; 21, p. 353; 21.1, p. 368; 21.2, p. 380; 21.6, p. 390.

12.2 Non-Metals

Review Article

J. Fousek and J. Petzelt
Changes of Refractive Indices of Crystals Induced by Structural Phase Transitions
(a) Vol. **55**, No. 1, p. 11–40 (1979).
See also *1.2*, p. *15*; *14.4.1*, p. *244*; 20, p. 298; 22.8.1, p. 498.

Original Papers and Short Notes

J. Shanker, A. K. G. Lashkari, and V. P. Gupta
*Effect of Three-Body Interactions on the Electronic Polarizabilities and Photoelastic Behaviour of Alkali Halides
(b) Vol. **91**, No. 1, p. 263–268 (1979).
See also *4*, p. *55*; 14.4, p. 238; 22.5.2, p. 465.

A. Ya. Altshul, V. K. Bashenov, and A. G. Petukhov
*Relation between Elastic Constants from Keating's Valence-Force Field
(b) Vol. **91**, No. 2, p. K121–K123 (1979).
See also *4*, p. *55*; 22, p. 394.

V. K. Bashenov, I. Baumann, and D. I. Marvakov
*Elastic Properties of Layer Gallium Selenide
(b) Vol. **91**, No. 2, p. K125–K128 (1979).
See also 6, p. 65; 22.3, p. 444.

S. E. Horan and L. M. Slifkin
*An Internal Friction Study of Stress-Induced Migration and Aggregation of Solute Ions in Silver Bromide Single Crystals
(a) Vol. **51**, No. 2, p. 467–478 (1979).
See also *10.2*, p. *104*; 22.5.1, p. 463.

B. N. Oshcherin
On Surface Energy and Compressibility of $A^2B^4C_2^5$ Semiconducting Compounds
(a) Vol. **51**, No. 2, p. K175–K179 (1979).
See also *1.6*, p. *38*; *4*, p. *55*; 22.8, p. 492.

J. Perez, C. Maî, J. Tatibouet et R. Vassoille
*Etude des joints de grains dans la glace I_h par mesure du frottement intérieur
(Study of Grain Boundaries of Ice I_h by Internal Friction Measurements)
(a) Vol. **52**, No. 1, p. 321–330 (1979).
See also *10.2*, p. *105*; 22, p. 396.

A. Krell and D. Schulze
*$\langle 1\bar{2}10 \rangle$-Zone Fracture Anisotropy of Sapphire
(a) Vol. **52**, No. 1, p. K45–K48 (1979).
See also *10.2*, p. *105*; 22.6, p. 478.

S. Lehri and M. P. Verma
Phonon Dispersion in Rare Gas Solids
(b) Vol. **92**, No. 2, p. 363–370 (1979).
See also 4, p. 56; *6*, p. *66*; 23, p. 515.

G. Weber and W. Rühle
Effect of Uniaxial Stress on Free and Bismuth-Bound Excitons in InP
(b) Vol. 92, No. 2, p. 425—431 (1979).
See also *13.4*, p. *167*; *13.5.1*, p. *184*; 20.3, p. 329; 22.2.2, p. 437.

S. C. Goyal and V. Kumar
*Strain Derivatives of the Static Dielectric Constant of Silver Halides
(b) Vol. 92, No. 2, p. K77—K80 (1979).
See also *14.4*, p. *239*; 22.5.1, p. 463.

S. E. Horan and L. M. Slifkin
*Stress-Induced Anomalies in the Internal Friction of Strontium-Doped Silver Bromide
(a) Vol. 52, No. 2, p 657—662 (1979).
See also *10.2*, p. *106*; 22.5.1, p. 463.

J. Shanker and V. C. Jain
Analysis of the Crystal Binding of Alkali Hydrides
(a) Vol. 52, No. 2, p. 675—680 (1979).
See also *4*, p. *56*; 22, p. 397.

M. S. Kushwaha and S. S. Kushwaha
The Lattice Dynamics of Quantum Crystals of F.C.C. ^4He
(b) Vol. 93, No. 1, p. 237—241 (1979).
See also *6*, p. *68*; 23, p. 515.

R. P. Singh and J. Shanker
*Evaluation of Third-Order Elastic Constants and Pressure Dependence of Second-Order Elastic Constants of Ionic Crystals
(b) Vol. 93, No. 1, p. 373—380 (1979).
See also 22.5.2, p. 467; 22.6, p. 480.

B. Chevalier, G. Demazeau, J. Etourneau, and P. Hagenmuller
Pressure Dependence of the Resistivity of the Rare Earth Oxide Nitrides $Eu^{2+}_{1-x}Ln^{3+}_xN_xO_{1-x}$ (Ln = Nd, Eu)
(b) Vol. 93, No. 1, p. K63—K66 (1979).
See also 14.1, p. 200; *14.3*, p. *212*; 22.8, p. 493.

A. Witek, J. Wosik, and W. Zbieranowski
*The Elastic Constants and Debye Temperature of the Monocrystal $NaMgAl(C_2O_4)_3 \cdot 9\ H_2O$
(a) Vol. 53, No. 1, p. K27—K29 (1979).
See also *7*, p. *78*; 22.8, p. 493.

F. Dochy
Theoretical Calculation of the Electron–Phonon Coupling Parameters of the F-Centre in Alkali Halides
(b) Vol. 93, No. 2, p. 743—753 (1979).
See also *6*, p. *69*; *13.4*, p. *170*; 22.5.2, p. 468.

U. Steigenberger and M. von Ortenberg
Cyclotron Resonance in Uniaxially Stressed Tellurium
(b) Vol. 93, No. 2, p. K139—K141 (1979).
See also *13.1*, p. *149*; *19*, p. *293*; 22.1.3, p. 429.

S. S. BEDI and M. P. VERMA
*On the Long-Wavelength Behaviour of Solids with Fluorite Structure
(b) Vol. **94**, No. 1, p. 297–300 (1979).
See also 6, p. 70; 22.5.3, p. 473.

I. V. GRIDNEVA, YU. V. MILMAN, V. I. TREFILOV, and S. I. CHUGUNOVA
Analysis of Dislocation Mobility under Concentrated Loads at Indentations of Single Crystals
(a) Vol. **54**, No. 1, p. 195–206 (1979).
See also *10.2*, p. *109*; 22.1.1, p. 411; 22.1.2, p. 421; 22.2.1, p. 434; 22.2.3, p. 440.

V. M. FRIDKIN, R. NITSCHE, N. KORCHAGINA, N. A. KOSONOGOV, R. MAGOMADOV, A. I. RODIN, and K. A. VERKHOVSKAYA
Photoferroelastic Phenomena in Sb_5O_7I Crystals
(a) Vol. **54**, No. 1, p. 231–237 (1979).
See also *16*, p. *253*; 22.8, p. 495.

Y. MOTOHASHI, M. G. BLANCHIN, E. VICARIO, G. FONTAINE, and S. OTAKE
*Elastic Parameters, Elastic Energy, and Stress Fields of Dislocations in TiO_2 Rutile Crystals
(a) Vol. **54**, No. 1, p. 355–364 (1979).
See also *10.2*, p. *110*; 22.6, p. 482.

S. GHOSH, A. N. BASU, and S. SENGUPTA
*An Extended Deformable Shell Model for Ionic Crystals Including Quadrupolar Deformation of the Charge Cloud
(b) Vol. **94**, No. 2, p. 673–683 (1979).
See also *6*, p. *72*; 22.5.2, p. 469.

V. D. VERNER, D. K. NICHUGOVSKII, and V. YA. KOVYASIN
*Investigation of Internal Friction in GaP
(a) Vol. **54**, No. 2, p. 593–596 (1979).
See also *10.2*, p. *111*; 22.2.2, p. 438.

S. TSUNEKAWA and M. TANAKA
*Temperature Dependence of *w'* Walls in Ferroelastic $LaNbO_4$ Crystals
(a) Vol. **55**, No. 1, p. K1–K5 (1979).
See also *1.4*, p. *30*; 22.8.1, p. 502.

T. SOMA
*Volume Effect on the Bulk Modulus of Si–Ge Solid Solutions
(b) Vol. **95**, No. 2, p. K117–K119 (1979).
See also 13, p. 141; 22.1.1, p. 412; 22.1.2, p. 424.

A. KRELL and D. SCHULZE
Microstructure Dependence of the Brittle Fracture of Alumina
(a) Vol. **55**, No. 2, p. 537–544 (1979).
See also *10.2*, p. *114*; 22.6, p. 484.

13. Electron States

Review Articles

Original Papers and Short Notes

M. V. MEDVEDEV
*Short-Range Magnetic Order in a Spin Glass with Nearest-Neighbour Randon Exchange Bonds
(b) Vol. **91**, No. 2, p. 713–724 (1979).
See also *18.1*, p. *265*.

D. I. KHOMSKII
*Local Effects and Valence Transitions in Doped Sm Monochalcogenides
(b) Vol. **91**, No. 2, p. K99–K103 (1979).
See also 22.7, p. 488.

S. GUHA, P. K. SEN, and S. GHOSH
*Modulational Instability of a Finite Amplitude Alfvén Wave in Semimetals
(b) Vol. **91**, No. 2, p. K135–K138 (1979).
See also *20*, p. *299*; 21.7, p. 391.

T. O. GEGECHKORI and Z. S. KACHLISHVILI
*Influence of the Capture of Charge Carriers upon the I–U Curves of Hot Electron Semiconductors with Account of Phonon Heating
(a) Vol. **51**, No. 2, p. 333–336 (1979).
See also 6, p. 65; *14.3.4*, p. *232*; 22, p. 395.

M. A. KUMAKHOV and R. WEDELL
Theory of Channelling of Relativistic Particles
(b) Vol. **92**, No. 1, p. 65–71 (1979).
See also *11*, p. *120*; 22.1.2, p. 417.

K. N. SHRIVASTAVA
*Numerical Solution of the Summation Process in Spin–Lattice Relaxation
(b) Vol. **92**, No. 1, p. K19–K21 (1979).
See also *6*, p. *66*.

T. A. ABRAMOVSKAYA and M. D. ZVIADADZE
*On Effects of the Dipole Reservoir in Radiofrequency Discrete Saturation
(a) Vol. **52**, No. 1, p. 111–120 (1979).
See also *18*, p. *261*; 19, p. 291.

I. V. ABARENKOV and I. M. ANTONOVA
*The Electronic Structure of MgO (I): Calculation Method
(b) Vol. **92**, No. 2, p. 389–396 (1979).
See also 22.6, p. 478.

V. R. NAGIBAROV and O. KH. KHASANOV
*The Effect of Retarded Interaction on the Self-Induced Transparency Phenomenon
(b) Vol. **92**, No. 2, p. 467–472 (1979).
See also 13.5, p. 181.

H. KÖHLER and M. ROOS
*The Influence of the Valley Splitting and Lineshape in the Density of States on the Quantum Oscillations of n-Type (100) Inverted Silicon MOSFET Surfaces
(b) Vol. **92**, No. 2, p. 489–494 (1979).
See also *14.3.2*, p. *223*; 22.1.2, p. 418.

A. M. Berezhkovskii and A. A. Ovchinnikov
*Certain Features of Semiconductor Behaviour in Strong Parallel Magnetic and Electric Fields (III): Interband Absorption
(b) Vol. **93**, No. 2, p. 549–551 (1979).
See also 20.1, p. 313; 22, p. 399.

K. H. J. Buschow, U. Goebel, and E. Dormann
*Valence Instabilities in $CeSn_3$ and $YbAl_3$
(b) Vol. **93**, No. 2, p. 607–615 (1979).
See also 18, p. 262; 19, p. 292; 22.7, p. 489.

J. Katriel
*Explicit Expressions for the Coefficients in Bosons Series Expansions of Arbitrary Spin Operators
(b) Vol. **93**, No. 2, p. K177–K179 (1979).
See also *18.2*, p. *275*.

A. P. Pathak
"Simplicity Considerations" in Interatomic Potential for Channeling Calculations
(b) Vol. **93**, No. 2, p. K181–K185 (1979).
See also 4, p. 57; *11*, p. *122*.

V. A. Shenderovskii and P. N. Gorley
*Warm Electron Conductivity in Lead Chalcogenides
(a) Vol. **53**, No. 2, p. K189–K194 (1979).
See also *14.3.4*, p. *234*; 22, p. 400.

A. M. Berezhkovskii
*Certain Features of Semiconductor Behaviour in Strong Parallel Magnetic and Electric Fields (IV): Zener Tunnelling
(b) Vol. **94**, No. 1, p. 87–94 (1979).
See also 14, p. 197; 22, p. 400; 22.2.3, p. 440.

S. Szuszkiewicz, B. Rozenfeld, and M. Szuszkiewcz
*Positron Annihilation Studies of the Electronic Structure of B.C.C. Fe
(b) Vol. **94**, No. 1, p. 273–278 (1979).
See also 13.1, p. 150; 13.2, p. 156; 21.1.1, p. 374.

G. I. Georgiev and M. P. Sandalski
*One-Dimensional Ising Model with an Infinite Radius of Interaction in a Transverse Magnetic Field. Some Thermodynamic Properties of the Rigorously Solvable Part
(b) Vol. **94**, No. 2, p. 477–486 (1979).
See also 8, p. 82; *18*, p. *263*.

P. M. van den Broek
*The Clebsch-Gordan Coefficients of the Crystallographic Space Groups
(b) Vol. **94**, No. 2, p. 487–496 (1979).
See also *1*, p. *10*.

K. Berndt, U. Marx, and O. Brümmer
*Photoelectron Spectra of Core Electrons from Palladium-Silver Alloys
(b) Vol. **94**, No. 2, p. 541–548 (1979).
See also 17, p. 257; 21, p. 347; 21.6, p. 388.

P. Vorderwisch, S. Hautecler, and J. B. Suck
*Inelastic Neutron Scattering by Cerium Hydrides (IV): Crystal Field
(b) Vol. **94**, No. 2, p. 569–576 (1979).
See also 18, p. 263; 21.4, p. 384.

W. John and W. Keller
*Density of States of Expanded Liquid Mercury. An Effective Medium Approach
(b) Vol. **94**, No. 2, p. 603–609 (1979).
See also *2*, p. *46*; 21, p. 347.

B. K. Chakrabarti, B. P. Barua, and S. K. Sinha
Critical Temperature of a Compressible Ising Magnet
(b) Vol. **94**, No. 2, p. 621–625 (1979).
See also 6, p. 72; *18.2*, p. *277*.

G. Duggan, G. J. Morgan, and A. Lettington
*The Effective Screening Function in Si, Ge, and GaAs and the Effects of Exchange
(b) Vol. **94**, No. 2, p. 659–665 (1979).
See also *14.4*, p. *241*; 22.1.1, p. 412; 22.1.2, p. 422; 22.2.1, p. 434.

M. S. Sodha and R. R. Sharma
*Stimulated Brillouin Scattering of Helicon Waves in a Semiconductor
(b) Vol. **94**, No. 2, p. 667–671 (1979).
See also 7, p. 79; *20*, p. *301*; 22.2.3, p. 440.

I. M. Antonova
*The Electronic Structure of CaO
(b) Vol. **94**, No. 2, p. K165–K168 (1979).
See also 22.6, p. 482.

V. S. Sundaram, C. L. Barreto, J. D. Rogers, and G. G. Kleiman
*ESCA Studies of Pt–Cu Alloys
(b) Vol. **94**, No. 2, p. K191–K193 (1979).
See also 21.1, p. 363; 21.6, p. 388.

J. Kötzler and W. Scheithe
*Determination of Crystalline Field Interaction in $GdCl_3$
(b) Vol. **95**, No. 1, p. K1–K4 (1979).
See also *18*, p. *263*; 22.5, p. 462.

N. Shamir, M. Melamud, and H. Shaked
*Crystalline Electric Field Studies of Uranium Compounds Using Neutron Inelastic Scattering at the NRCN
(a) Vol. **55**, No. 1, p. K49–K52 (1979).
See also 18, p. 263; 21.5, p. 386; 22.1.2, p. 424; 22.5, p. 462; 22.6, p. 483.

K. I. Wysokiński and M. Piłat
*Effect of Temperature on the Off-Diagonal Alloy Properties
(b) Vol. **95**, No. 2, p. 369–375 (1979).
See also *14.1*, p. *202*; 21.6, p. 389.

T. Soma
The Electronic Theory of Si–Ge Solid Solutions
(b) Vol. **95**, No. 2, p. 427–431 (1979).
See also *4*, p. *57*; 22.3, p. 445.

T. Soma
Volume Effect on the Bulk Modulus of Si–Ge Solid Solutions
(b) Vol. **95**, No. 2, p. K117–K119 (1979).
See also *12.2*, p. *135*; 22.1.1, p. 412; 22.1.2, p. 424.

I. M. Dykman and L. A. Prokhnitskii
*Electromagnetic Waves in an Inhomogeneous Semiconductor under Electric and Magnetic Fields
(b) Vol. **96**, No. 1, p. 55–60 (1979).
See also *14.3*, p. *216*; 20, p. 302; 22, p. 405.

H.-W. Neumann and A. Rauh
*Diamagnetic Band Structure up to Second Order in the Crystal Potential
(b) Vol. **96**, No. 1, p. 233–242 (1979).
See also 18, p. 264.

K. Elk
On the Susceptibility of Strongly Correlated Electrons
(b) Vol. **96**, No. 1, p. 251–257 (1979).
See also *18.1*, p. *269*.

D. E. Popov, V. V. Kaplin, and S. A. Vorobiev
Flux Redistribution of Channeled Fast Electrons
(b) Vol. **96**, No. 1, p. 263–269 (1979).
See also *11*, p. *125*; 22.1.2, p. 425.

V. L. Bonch-Bruevich, V. D. Karaivanov, and Ya. G. Proykova
*Random Field Parameters and Pseudopotential
(b) Vol. **96**, No. 1, p. 271–278 (1979).
See also *2*, p. *48*; 14, p. 198; 20, p. 302.

H. Eschrig
*Readily Computable Expressions for LCAO Two-Centre Integrals over Slater-Type Orbitals with Arbitrarily High Quantum Numbers
(b) Vol. **96**, No. 1, p. 329–342 (1979).

H.-J. Wünsche
*Mean Component Approximation for the Correlation Energy of Multicompenent Electron–Hole Plasmas
(b) Vol. **96**, No. 1, p. 379–387 (1979).
See also 22, p. 405.

M. Załużny
*On the Optical Orientation of Free Carriers in Size Quantized $Pb_{1-x}Sn_xTe$ Films
(b) Vol. **96**, No. 1, p. K51–K55 (1979).
See also *20*, p. *302*; 22, p. 406.

F. F. Komarov
Energy Loss of Relativistic Particles in Solids
(b) Vol. **96**, No. 2, p. 555–568 (1979).
See also *11*, p. *125*; 21.6, p. 389; 22.1, p. 408.

S. Guha and N. Apte
*Stimulated Brillouin Scattering in a Magnetised Semiconductor Plasma
(b) Vol. **96**, No. 2, p. 569–574 (1979).
See also 7, p. 49.

K. Seeger and H. Vana
*Hot-Carrier Infrared Absorption in n-Type Germanium
(b) Vol. **96**, No. 2, p. 605–610 (1979).
See also 14.3.4, p. 237; *20.1*, p. *322*; 22.1.1, p. 413.

B. Lorenz and D. Ihle
Electron–Phonon versus Coulomb Interaction Effects at the Verwey Transition of Fe_3O_4
(b) Vol. **96**, No. 2, p. 659–669 (1979).
See also *1.2*, p. *21*; 4, p. 58; 6, p. 76; 22.6, p. 485.

S. Guha and P. Namjoshi
*Parametric Decay of a Helicon in n-InSb
(b) Vol. **96**, No. 2, p. 713–716 (1979).
See also *20*, p. *302*; 22.2.3, p. 441.

S. M. Darbinian and K. A. Ispirian
*On High Energy Particle Channeling
(b) Vol. **96**, No. 2, p. 835–842 (1979).
See also *11*, p. *125*; 22.1.1, p. 413.

S. Guha and P. K. Sen
*Effect of Transverse Magnetic Field on the Electromagnetic Wave Propagation in Semiconductor Plasmas at High Electric Field
(a) Vol. **56**, No. 2, p. K117–K121 (1979).
See also 14.3, p. 218; *20*, p. *303*; 22.2.3, p. 441.

13.1 Band Structure

Review Articles

A. A. Katsnelson, V. M. Silonov, and Farid A. Khwaja
Electronic Theory of Short-Range Order in Alloys Using the Pseudopotential Approximation and Its Comparison with Experiments
(b) Vol. **91**, No. 1, p. 11–33 (1979).
See also 1.3, p. 22; *4*, p. *55*; *13*, p. *136*; 21, p. 338.

R. S. Brazis, J. K. Furdyna, and J. K. Požela
Microwave Effects in Narrow-Gap Semiconductors (I)
(a) Vol. **53**, No. 1, p. 11–41 (1979).
See also *13*, p. *136*; 14.3, p. 209; 18.1, p. 265; *20*, p. *298*; 22.4.3, p. 455; 22.4.4, p. 458; 22.7, p. 488.

M. Averous
*Symmetry-Induced Zero-Gap Semiconductors
(b) Vol. **95**, No. 1, p. 9–28 (1979).
See also 13.4, p. 163; 14.3, p. 209; 22.1.3, p. 428; 22.4, p. 448.

Original Papers and Short Notes

V. A. Vilkotskii, D. S. Domanevskii, R. D. Kakanakov, V. V. Krasovskii, and V. D. Tkachev
*Burstein-Moss Effect and Near-Band-Edge Luminescence Spectrum of Highly Doped Indium Arsenide
(b) Vol. **91**, No. 1, p. 71–81 (1979).
See also *20.3*, p. *327*; 22.2.1, p. 432.

D. Ihle and Th. Eifrig
Thermoelectric Power near the Atomic Limit of the Hubbard Model
(b) Vol. **91**, No. 1, p. 135–140 (1979).
See also *15*, p. *249*.

K.-R. Schulze
*On the Dielectric Response of Solids (I): New Approximations for the Treatment of Local Field Effects
(b) Vol. **91**, No. 1, p. 211–221 (1979).
See also 14.4, p. 238; 21, p. 338; 22, p. 393.

E. Tegeler, N. Kosuch, G. Wiech, and A. Faessler
*On the Electronic Structure of Hexagonal Boron Nitride
(b) Vol. **91**, No. 1, p. 223–231 (1979).
See also 20, p. 298; 22.2, p. 430.

O. Breitenstein and K. Unger
*Distribution of the Perturbation Potential and the Density of States in High-Doped Semiconductors
(b) Vol. **91**, No. 1, p. 243–250 (1979).
See also *14.3.4*, p. *231*; 22, p. 393.

N. I. KULIKOV
*Band Structure and Electronic Properties of Transition Metal Hydrides
(b) Vol. 91, No. 2, p. 753–762 (1979).
See also 14.2, p. 204; 18.2, p. 271; 20.1, p. 306; 21.1, p. 355.

V. K. MALYUTENKO and A. I. LIPTUGA
The Determination of the Intrinsic Concentration and Effective Mass of Carriers in Semiconductors by an Optical Method
(a) Vol. 51, No. 2, p. K137–K139 (1979).
See also 14.3, p. 210; *20.1*, p. *306*; 22.2.3, p. 439.

M. DEB ROY
Effect of Random Variation in Impurity Concentration on Fermi Level in n-Type Silicon
(a) Vol. 51, No. 2, p. K147–K149 (1979).
See also *14.3*, p. *210*; 22.1.2, p. 417.

P. R. WALLACE
*Electronic g-Factor in Cd_3As_2
(b) Vol. 92, No. 1, p. 49–55 (1979).
See also *18.1*, p. *266*; 22.3, p. 444.

A. VAIDYANATHAN, Y. F. TSAY, and S. S. MITRA
*Electronic Structure and Optical Properties of Rhombohedral Arsenic and Their Variations with the Internal Displacement Parameter
(b) Vol. 92, No. 1, p. 73–81 (1979).
See also 14.4, p. 238; 20.1, p. 307; 22.1, p. 408.

R. STEGMANN, G. OELGART, and R. HERRMANN
*Propagation of Electromagnetic Waves in a Direction Perpendicular to the Magnetic Field in Semiconductive Bi–Sb Alloys (I): $\boldsymbol{B} \perp \boldsymbol{C}_3 \parallel \boldsymbol{k}$
(b) Vol. 92, No. 1, p. 133–141 (1979).
See also 13.2, p. 156; 19, p. 290; 22.7, p. 489.

T. JUNG and W. BRAUNE
*On the Deformation Model of the Carrier Spectrum for Bismuth-Type Crystals (II): Holes
(b) Vol. 92, No. 1, p. 171–176 (1979).
See also 13.2, p. 156; 21.7, p. 391.

K. TAKITA, K. ONABE, and S. TANAKA
*Anomalous Magnetoresistance and Band Crossing in Uniaxially Compressed HgTe
(b) Vol. 92, No. 1, p. 297–306 (1979).
See also *14.3*, p. *211*; 22.4.3, p. 455.

D. M. EAGLES
*Erratum to: Widths of 5d Energy Bands in the Heavy Rare-Earth Metals
(b) Vol. 92, No. 1, p. 353 (1979).
See also 21.4, p. 382.

W. HÖRIG, W. MÖLLER, H. NEUMANN, E. RECCIUS, and G. KÜHN
*Optical Properties of $CuIn_{1-x}Ga_xSe_2$ Mixed Crystals
(b) Vol. 92, No. 1, p. K1–K4 (1979).
See also *20.1*, p. *308*; 22.8, p. 492.

J. M. Pawlikowski, J. Misiewicz, B. Sujak-Cyrul, and J. Wróbel
*Absorption Edge of Zn_3As_2 at 5, 80, and 300 K
(b) Vol. **92**, No. 2, p. K123–K125 (1979).
See also *20.1*, p. *310*; 22.3, p. 445.

I. Gorczyca
*Spin Dependent Wave Functions at Some Points of the Brillouin Zone of InSb
(b) Vol. **92**, No. 2, p. K143–K146 (1979).
See also 22.2.3, p. 439.

K. Hübner
*Chemical Bond and Related Properties of SiO_2 (VI): Electronic Structure of SiO_x
(a) Vol. **52**, No. 2, p. 541–547 (1979).
See also *4*, p. *56*; 14.4, p. 239; 22.6, p. 479.

N. M. Ravindra, Sushil Auluck, and V. K. Srivastava
*Temperature Dependence of the Energy Gap in PbS, PbSe, and PbTe
(a) Vol. **52**, No. 2, p. K151–K155 (1979).
See also 22, p. 397.

A. Toneva
*Effective Mass of Holes in $Pb_{1-x}Sn_xTe$
(a) Vol. **52**, No. 2, p. K197–K200 (1979).
See also 14.3, p. 212; 15, p. 250; 22, p. 398.

J. Stankiewicz
*Transport Properties of $Cd_vHg_{1-v}Se$ Crystals
(b) Vol. **93**, No. 1, p. 113–118 (1979).
See also *14.3*, p. *212*; 22.4.4, p. 458.

C. Boiziau, V. Dose, and H. Scheidt
*The Density of Unoccupied Electronic States in Tantal
(b) Vol. **93**, No. 1, p. 197–201 (1979).
See also 21, p. 343.

N. C. Halder
*Self-Consistency in the Calculation of Electron Effective Mass
(b) Vol. **93**, No. 1, p. 363–372 (1979).
See also 13.2, p. 156; 22, p. 398.

Ya. M. Ksendzov
*The Electron Structure of 3d Metals
(b) Vol. **93**, No. 1, p. 415–425 (1979).
See also 13.5.2, p. 191; 21.1, p. 358; 21.1.1, p. 372.

J. V. McCanny
*The Electronic Structure of Lithiated SnS_2
(b) Vol. **93**, No. 1, p. K29–K34 (1979).
See also 22, p. 398.

W. BLAU, J. WEISBACH, G. MERZ, and K. KLEINSTÜCK
*X-Ray Spectra and Electronic Structure of Some CsCl-Type Transition Metal–Aluminium Alloys
(b) Vol. **93**, No. 2, p. 713—720 (1979).
See also *20*, p. *300*; 21, p. 344.

A. BIANCONI, S. STIZZA, R. BERNARDINI, and S. NANNARONE
*Electronic Structure of Ti_2O_3 from Single Crystal Thermoreflectance Spectroscopy
(b) Vol. **93**, No. 2, p. 767—774 (1979).
See also *20.1*, p. *313*; 22.6, p. 480.

A. BERNDTSSON, R. NYHOLM, N. MÅRTENSSON, R. NILSSON, and J. HEDMAN
*ESCA Studies of Rhenium and Osmium
(b) Vol. **93**, No. 2, p. K103—K105 (1979).
See also 17, p. 257; 21, p. 344.

V. K. BASHENOV, I. K. DOICHO, and A. G. PETUKHOV
*Selection between Various Local Pseudopotentials for Silicon
(b) Vol. **93**, No. 2, p. K131—K134 (1979).
See also *4*, p. *57*; 22.1.2, p. 420.

U. STEIGENBERGER and M. VON ORTENBERG
*Cyclotron Resonance in Uniaxially Stressed Tellurium
(b) Vol. **93**, No. 2, p. K139—K141 (1979).
See also 12.2, p. 134; *19*, p. *293*; 22.1.3, p. 429.

W. JOHN and D. HAMANN
*Relativistic Electron–Phonon Interaction in Transition Metals
(b) Vol. **93**, No. 2, p. K143—K148 (1979).
See also *6*, p. *70*; 21.1, p. 359.

N. M. RAVINDRA, SUSHIL AULUCK, and V. K. SRIVASTAVA
*On the Penn Gap in Semiconductors
(b) Vol. **93**, No. 2, p. K155—K160 (1979).
See also *14.4*, p. *240*; 22, p. 400.

O. V. FARBEROVICH, S. I. KURGANSKII, and E. P. DOMASHEVSKAYA
*Problems of the OPW Method (I): Transition Metals
(b) Vol. **94**, No. 1, p. 51—62 (1979).
See also 21.1, p. 361.

H. SCHMIDT and E. MANN
*Anisotropy of the Electron–Phonon Interaction in Copper (II): Scattering and Transport Properties
(b) Vol. **94**, No. 1, p. 95—105 (1979).
See also *6*, p. *70*; 14.1, p. 201; 21.1, p. 361.

M. BÜNGER and R. J. JELITTO
On the Phase Diagram of Itinerant Magnets in the Presence of Unmagnetic Electron Bands
(b) Vol. **94**, No. 1, p. 191—198 (1979).
See also *18.1*, p. *267*; *18.2*, p. *276*.

A. H. EDWARDS
*F-Center Photoionization in the Semicontinuum Model
(b) Vol. **94**, No. 2, p. 367–370 (1979).
See also 20.1, p. 316; 22.5.2, p. 469.

R. M. LUCHYTSKII, V. S. MANZHARA, and P. M. STARYK
Investigation of n-$Pb_{1-x}Sn_xSe$ Absorption Spectra
(b) Vol. **94**, No. 2, p. 381–386 (1979).
See also 6, p. 71; *20.1*, p. *316*; 22, p. 402.

A. RICHTER, G. RÖPKE, and F. GOEDSCHE
*Functional Integral Approach for a Doubly Degenerate Hubbard Model
(b) Vol. **94**, No. 2, p. 437–450 (1979).
See also 18.1, p. 268; 18.2, p. 277; 18.4, p. 287.

C. ANCE
*Energie dispersive dans les semiconducteurs
(Dispersion Energy in Semiconductors)
(b) Vol. **94**, No. 2, p. 557–568 (1979).
See also *14.4*, p. *241*; 22, p. 402; 22.2.4, p. 442.

R. L. SARKAR and S. CHATTERJEE
*Electronic Structure and Optical Properties of AlP, AlAs, and AlSb
(b) Vol. **94**, No. 2, p. 641–647 (1979).
See also *20.1*, p. *317*; 22.2.1, p. 434; 22.2.2, p. 438; 22.2.3, p. 440.

J. F. MCGILP and A. W. PARKE
*Angle-Resolved UPS of GaS, GaSe, and $GaS_{0.6}Se_{0.4}$
(b) Vol. **94**, No. 2, p. 685–689 (1979).
See also *17*, p. *257*; 22.3, p. 445; 22.3.1, p. 447.

G. DRÄGER, O. BRÜMMER, and J. BONITZ
*Polarized K-Emission Valence Bands, Structure, and Bonding of Some Hexagonal Close-Packed Metals
(b) Vol. **94**, No. 2, p. K111–K114 (1979).
See also *4*, p. *57*; 20, p. 301; 21.1, p. 362; 21.3, p. 381.

E. SOBCZAK
X-Ray Bremsstrahlung Isochromat and X-Ray Photoelectron Investigations of Ion Implanted Germanium
(a) Vol. **54**, No. 2, p. K151–K153 (1979).
See also *17*, p. *258*; *20*, p. *301*; 22.1.1, p. 412.

P. PLENKIEWICZ and B. DOWGIAŁŁO-PLENKIEWICZ
*Energy Band Structure of Cd_3P_2 for Real D_{4h}^{15} Symmetry (II): Symmetry Properties
(b) Vol. **95**, No. 1, p. 29–37 (1979).
See also 22.3, p. 445.

F. POPEZ-AGUILAR
*A Simple Linear Method in Band Theory. Application to B.C.C. Tantalum
(b) Vol. **95**, No. 1, p. 171–178 (1979).
See also 21, p. 348.

F. Bechstedt and R. Enderlein
*Electronic Polarization (Relaxation) Effects in the Core Level Spectra of Semiconductors (II): Application to Ga3d and Si2p Levels
(b) Vol. 95, No. 1, p. 185–194 (1979).
See also 14.4, p. 242; 22.1.2, p. 423; 22.2, p. 430.

Ch. Müller, H. Wonn, W. Blau, P. Ziesche, and V. P. Krivitskii
*Band Structure and X-Ray Investigation of Transition Metal Aluminides
(b) Vol. 95, No. 1, p. 215–225 (1979).
See also *20.1*, p. *318*; 21.1, p. 364.

S. Szuszkiewicz, B. Rozenfeld, and M. Szuszkiewicz
*Investigation of the Electronic Structure of F.C.C. $Co_{92}Fe_8$ by Measuring the Angular Correlation of Positron Annihilation γ-Quanta
(b) Vol. 95, No. 1, p. K51–K53 (1979).
See also *13.2*, p. *157*; 21.1, p. 364.

K. B. Tolpygo and S. M. Zubkova
On the Indirect Exciton Energy Levels in Crystals of Diamond and Sphalerite Type with Spin–Orbit Interaction
(b) Vol. 95, No. 2, p. 341–348 (1979).
See also *13.5.1*, p. *188*; 22.1.1, p. 412; 22.1.2, p. 424; 22.2, p. 431.

G. Borstel, W. Braun, M. Neumann, and G. Seitz
*Band Structure of Rhodium and Photoemission from Its Low Index Surfaces
(b) Vol. 95, No. 2, p. 453–460 (1979).
See also 17, p. 258; 21, p. 349.

M. Ostrow and A. Goldmann
*Density of Valence States of the Superionic Conductor AgI
(b) Vol. 95, No. 2, p. 509–516 (1979).
See also 17, p. 258; 22.5.1, p. 464.

K. Heidrich and G. Trendel
Optical Studies of Alkali Trihalogeno-Plumbate Mixed Crystals
(b) Vol. 95, No. 2, p. 527–532 (1979).
See also *20.1*, p. *320*; 22.5.4, p. 476.

R. Podloucky and A. Neckel
*Compton Profiles of FeAl
(b) Vol. 95, No. 2, p. 541–548 (1979).
See also 20, p. 301; 21.1.1, p. 375.

H. Köhler and R. Weidner
*Electronic Properties of n-Type $(Bi_{1-x}Sb_x)_2Te_3$ from Magneto-Quantum Oscillations
(b) Vol. 95, No. 2, p. 635–641 (1979).
See also 13.2, p. 157; 14.3, p. 216; 22.7.1, p. 490.

G. Ambrazevičius, G. Babonas, and A. Šileika
*Reflectance Spectra of $CdSiP_2$ and $ZnSiP_2$
(b) Vol. 95, No. 2, p. 643–647 (1979).
See also *20.1*, p. *320*; 22.8, p. 496.

D. GENZOW, A. G. MIRONOV, and O. ZIEP
Erratum to: On the Interband Absorption in Lead Chalcogenides
(b) Vol. **95**, No. 2, p. 649 (1979).
See also *20.1*, p. *320*; 22, p. 404.

W. CORDTS, A. HEINRICH, and J. MONECKE
*Electronic Band Structure of Pseudodirect Chalcopyrite Semiconductors (I): $CdSiP_2$ and $ZnSiP_2$
(b) Vol. **96**, No. 1, p. 201–209 (1979).
See also 22.8, p. 496.

E. DONATO, E. S. GIULIANO, R. RUGGERI, and A. STANCANELLI
*Appearance Potential Spectroscopy in Metals and Alloys. Multiple Scattering Approach
(b) Vol. **96**, No. 1, p. 313–320 (1979).
See also 20, p. 302; 21, p. 351.

K. HÜBNER and F. BECHSTEDT
*Temperature Dependence of Atomic Core Levels in Solids (I): Separation of the Temperature Dependence of Conduction-Band and Core-Exciton Energies
(b) Vol. **96**, No. 1, p. 351–357 (1979).
See also *13.5.1*, p. *189*; 22.1.2, p. 425; 22.2.2, p. 438; 22.6, p. 484.

O. ZIEP and D. GENZOW
*Calculation of the Interband Absorption in Lead Chalcogenides Using a Multiband Model
(b) Vol. **96**, No. 1, p. 359–368 (1979).
See also *20.1*, p. *321*; 22, p. 405.

S. TAKEYAMA and R. R. GAŁĄZKA
*Band Structure of $Hg_{1-x}Mn_xSe$ from Anomalous Shubnikov-de Haas Effect
(b) Vol. **96**, No. 1, p. 413–423 (1979).
See also 14.3, p. 216; 18, p. 264; 22.4.4, p. 459.

W. C. LEIPOLD and T. E. FEUCHTWANG
*Calculation of the Electronic Energy Spectrum of a Thin Disordered Solid Film
(b) Vol. **96**, No. 1, p. 453–461 (1979).
See also 1.5, p. 36.

D. R. MAŠOVIĆ and S. ZEKOVIĆ
*Model Pseudopotential for Elementary Semiconductors
(b) Vol. **96**, No. 1, p. 469–474 (1979).
See also 21.7, p. 392; 22.1.1, p. 413; 22.1.2, p. 425.

O. ZIEP and D. GENZOW
*On the Concentration Dependence of the Susceptibility Effective Mass in $Pb_{0.8}Sn_{0.2}Te$
(b) Vol. **96**, No. 1, p. K25–K27 (1979).
See also 22, p. 405.

L. TICHÝ, M. FRUMAR, and J. KLIKORKA
Some Electrical Properties of $GeBi_2Te_4$ Single Crystals
(a) Vol. **56**, No. 1, p. 323–326 (1979).
See also *14.3*, p. *217*; *15*, p. *251*; 22.8, p. 496.

A. K. Saxena
*Non-Shallow Levels and the Conduction Band Structure of $Ga_{1-x}Al_xAs$
(b) Vol. **96**, No. 2, p. K77–K82 (1979).
See also 13.4, p. 180; 22.2.4, p. 443.

V. K. Bashenov, Y. K. Markolenko, and A. G. Petukhov
*Electronic Structure of Silicon Using Non-Spherical Local Pseudopotential
(b) Vol. **96**, No. 2, p. K91–K95 (1979).
See also 22.1.2, p. 426.

K. Elk and M. Raab
*Absence of Ferromagnetism in the Hybridized Hubbard Model
(b) Vol. **96**, No. 2, p. K113–K116 (1979).
See also *18.2*, p. *280*.

S. G. Stoyanov and I. S. Gorban
*Diffuse Reflectance and Thermoreflectance Spectra of Pure and Ga-Diluted YIG near the Curie Point
(a) Vol. **56**, No. 2, p. 655–661 (1979).
See also 18.3, p. 285; 20.1, p. 324; 22.8.2, p. 508.

13.2 Fermi Surfaces

Original Papers and Short Notes

H. Köhler and M. Roos
*Quantitative Determination of the Valley Splitting in n-Type Inverted Silicon (100) MOSFET Surfaces
(b) Vol. **91**, No. 1, p. 233–241 (1979).
See also *14.3.3*, p. *225*; 18, p. 261; 22.1.2, p. 415.

T. Jung and W. Braune
On the Deformation Model of the Carrier Spectrum for Bismuth-Type Crystals (I): Electrons
(b) Vol. **91**, No. 2, p. 563–569 (1979).
See also *13.1*, p. *144*; 21.7, p. 391.

R. Stegmann, G. Oelgart, and R. Herrmann
Propagation of Electromagnetic Waves in a Direction Perpendicular to the Magnetic Field in Semiconductive Bi–Sb Alloys (I): $\boldsymbol{B} \perp C_3 \parallel \boldsymbol{k}$
(b) Vol. **92**, No. 1, p. 133–141 (1979).
See also *13.1*, p. *145*; 19, p. 290; 22.7, p. 489.

T. Jung and W. Braune
On the Deformation Model of the Carrier Spectrum for Bismuth-Type Crystals (II): Holes
(b) Vol. **92**, No. 1, p. 171–176 (1979).
See also *13.1*, p. *145*; 21.7, p. 391.

N. C. Halder
*Self-Consistency in the Calculation of Electron Effective Mass
(b) Vol. **93**, No. 1, p. 363–372 (1979).
See also *13.1*, p. *147*; 22, p. 398.

V. A. Gasparov and M. H. Harutunian
*Precision Investigations of the Fermi Surface of Molybdenum
(b) Vol. **93**, No. 1, p. 403–414 (1979).
See also 19, p. 292; 21, p. 343.

R. Stegmann, G. Oelgart, and R. Herrmann
Propagation of Electromagnetic Waves in a Direction Perpendicular to the Magnetic Field in Semiconductive Bi–Sb Alloys (II): $\boldsymbol{B} \perp C_2 \parallel \boldsymbol{k}$
(b) Vol. **93**, No. 2, p. 553–560 (1979).
See also *13.1*, p. *148*; 19, p. 292; 22.7, p. 488.

S. Szuszkiewicz, B. Rozenfeld, and M. Szuszkiewicz
Positron Annihilation Studies of the Electronic Structure of B.C.C. Fe
(b) Vol. **94**, No. 1, p. 273–278 (1979).
See also *13*, p. *139*; 13.1, p. 150; 21.1.1, p. 374.

H. Sato
Sondheimer Oscillation in Aluminium Single Crystals
(b) Vol. **94**, No. 1, p. 309–315 (1979).
See also *14.1*, p. *201*; 21, p. 345.

I. Mertig and E. Mrosan
*Calculation of Cyclotron Masses of Copper Using a Tetrahedron Method
(b) Vol. **94**, No. 1, p. K23–K25 (1979).
See also **21.1**, p. **361**.

I. Mertig and E. Mrosan
*Calculation of Dingle Temperatures for Ni Impurities in Copper Using a Tetrahedron Method
(b) Vol. **94**, No. 2, p. K119–K122 (1979).
See also *13.4*, p. *174*; 21.1, p. 362.

K. N. Khanna and P. L. Srivastava
*Fermi Surface Distortion in Na and K
(b) Vol. **94**, No. 2, p. K169–K172 (1979).
See also 21.2, p. 379.

H. Köhler and M. Roos
*Evidence for a Landé-Factor $g = 2$ in n-Type Inverted Silicon MOSFET Surfaces
(b) Vol. **95**, No. 1, p. 107–115 (1979).
See also *14.3.3*, p. *229*; 18, p. 263; 22.1.2, p. 423.

S. Szuszkiewicz, B. Rozenfeld, and M. Szuszkiewicz
*Investigation of the Electronic Structure of F.C.C. $Co_{92}Fe_8$ by Measuring the Angular Correlation of Positron Annihilation γ-Quanta
(b) Vol. **95**, No. 1, p. K51–K53 (1979).
See also *13.1*, p. *152*; 21.1, p. 364.

H. Köhler and R. Weidner
Electronic Properties of n-Type $(Bi_{1-x}Sb_x)_2Te_3$ from Magneto-Quantum Oscillations
(b) Vol. **95**, No. 2, p. 635–641 (1979).
See also *13.1*, p. *152*; 14.3, p. 216; 22.7.1, p. 490.

H. Rauh
Investigation of the Electrical Resistance of Dislocations in Metals (I): General Theory
(b) Vol. **96**, No. 2, p. 627–640 (1979).
See also *10.1*, p. *101*; *14.1*, p. *203*; 21, p. 353.

13.3 Surface and Interface States

Original Papers and Short Notes

L. C. R. Alfred
*Interaction between a Metal Surface and an External Molecular Complex
(b) Vol. **91**, No. 1, p. 113–121 (1979).
See also 13.4, p. 163; 21, p. 338.

M. Nishida and T. Schimizu
*Local Densities of States for the Overlayer of Al Chemisorbed on the Si (111) Surface
(b) Vol. **91**, No. 1, p. K5–K8 (1979).
See also 22.1.2, p. 415.

K. Breuer
*On Size Effects in Small Gold Particles by Electron Energy Loss Analysis
(b) Vol. **91**, No. 1, p. K69–K72 (1979).
See also 14.4, p. 238; 21.6, p. 387.

H. Kirihata and J. Akutsu
Formation of Exoelectron Emission Centers by Slow Electron Impact
(a) Vol. **51**, No. 2, p. K109–K112 (1979).
See also *17*, p. *256*; 21.1.1, p. 371.

V. G. Litovchenko, I. P. Lisovskii, R. O. Litvinov, H. Flietner, and W. Füssel
*Specific Memory Effect in MOS Structures
(a) Vol. **51**, No. 2, p. K161–K164 (1979).
See also 11, p. 120; 14.3.2, p. 223; 22.1.2, p. 417.

N. Lieske and R. Hezel
*Surface States on Si(111) 7 × 7 Studied by Low-Energy Electron Loss Spectroscopy
(b) Vol. **92**, No. 1, p. 159–170 (1979).
See also *14.3.2*, p. *223*; 22.1.2, p. 417.

J. Heinrichs
*The Virial Theorem for the Jellium Model of Metal Surfaces
(b) Vol. **92**, No. 1, p. 185–192 (1979).
See also 14.1, p. 199.

H. Ueba and S. Ichimura
*Theory of Local Electronic States in a Semi-Infinite Mixed Crystal
(b) Vol. **92**, No. 1, p. 307–316 (1979).
See also 13.4, p. 166; 22, p. 396.

G. Kaden and H. Reimer
*Impulsstromverhalten von MOS-Strukturen mit zum Substrat kurzgeschlossenem externem pn-Übergang bei linearer Gate-Spannungsansteuerung
(Impulse-Current Behaviour of MOS Structures with External p–n Junction Connected to the Substrate for Linear Gate Voltage Selection)
(a) Vol. **52**, No. 1, p. 35–46 (1979).
See also *14.3.3*, p. *226*; 22, p. 396.

H. Kirihata
Effects of Gas Pressure and Sample Temperature during Slow Electron Excitation on Exoelectron Emission
(a) Vol. **52**, No. 1, p. K97 – K99 (1979).
See also *17*, p. *256*; 21.1.1, p. 372.

Nguyen van Trong
*Theory of Inelastic Reflection of Particles by Rayleigh Waves
(b) Vol. **93**, No. 1, p. 45 – 52 (1979).
See also *20*, p. *299*.

J. S. Nkoma and S. K. Rwaboona
*Frequency Dependence of Some Functions Associated with Surface Polaritons
(b) Vol. **93**, No. 1, p. 397 – 402 (1979).
See also *13.5*, p. *182*; 22.2.2, p. 437.

G. Kaden and H. Reimer
*MOS-Strukturen mit sperrseitig vorgespanntem externem pn-Übergang bei linearer Gatespannungsansteuerung (I): Impulsstromverhalten
(MOS Structures with Reverse-Biased External p–n Junction for Linear Gate Voltage Selection (I): Impulse-Current Behaviour)
(a) Vol. **53**, No. 1, p. 195 – 206 (1979).
See also *14.3.3*, p. *227*; 14.3.4, p. 234; 22, p. 399.

I. P. Lisovskii, V. G. Litovchenko, and R. O. Litvinov
*Effect of UV Illumination on the Electrial Properties of MOS Layer Structures
(a) Vol. **53**, No. 1, p. 253 – 262 (1979).
See also *14.3.1*, p. *219*; 22.1.2, p. 419.

J. Szuber and B. Salamon
Electron Spin Resonance Study on Low-Pressure-Baked Silicon Surfaces
(a) Vol. **53**, No. 1, p. 289 – 295 (1979).
See also *19*, p. *292*; 22.1.2, p. 419.

A. Wolkenberg, S. Mączyński, and B. Wasilewska
*Electrochemical Method to Determine Physical Surface Properties of Rare-Earth and Manganese Fluoride Doped CdF_2 Crystals
(a) Vol. **53**, No. 2, p. 447 – 453 (1979).
See also 14.4, p. 240; 20.3, p. 332; 22.5.3, p. 473.

E. W. Kreutz and P. Schroll
*Field Effect on Real GaAs Surfaces
(a) Vol. **53**, No. 2, p. 499 – 508 (1979).
See also *14.3.2*, p. *224*; 22.2.1, p. 433.

V. T. Cherepin, M. A. Vasilev, and O. A. Yakubtsov
*Appearance Potential Spectroscopy Study of the Oxidation of CoNi Alloy Surfaces
(a) Vol. **53**, No. 2, p. K207 – K210 (1979).
See also 17, p. 257; 21.1, p. 361.

K. Hess, J. F. Detry, and C. T. Sah
*Trapping Noise in Charge Coupled Devices
(a) Vol. **55**, No. 1, p. 243–249 (1979).
See also *14.3.3*, p. *229*.

J. Szuber and B. Długosz
*Dominant Interactions and Average Distances between Surface Paramagnetic Centers on Low Pressure Baked (111) Si Surface
(a) Vol. **55**, No. 1, p. K35–K38 (1979).
See also *19*, p. *295*; 22.1.2, p. 424.

V. T. Cherepin, M. A. Vasilev, and O. A. Yakubtsov
*Auger Electron and Appearance Potential Spectroscopic Study of Fe–Cr Alloy Surfaces
(a) Vol. **55**, No. 1, p. K83–K86 (1979).
See also 17, p. 258; 21.1, p. 365; 21.1.1, p. 375.

Kin-ichi Masuda
*Chemisorption of Ordered Atomic Layers on a Model Transition Metal
(b) Vol. **95**, No. 2, p. 413–425 (1979).
See also 21.1, p. 365.

Th. Klupsch
*On Pair Distribution Functions and Free Energy Functionals of the Crystal–Liquid Phase Boundary
(b) Vol. **95**, No. 2, p. K147–K150 (1979).
See also *1.6*, p. *39*.

A. Shluger, E. Kotomin, and Yu. Dzelme
Quantum-Chemical Calculations of Electronic and Hole Centres and Surface of NaCl Crystals (I)
(b) Vol. **96**, No. 1, p. 91–99 (1979).
See also *10.2*, p. *115*; *13.4*, p. *177*; 22.5.2, p. 471.

A. V. Kobelev, R. M. Kobeleva, and V. F. Ukhov
*On the Adhesion Theory for Two Metallic Surfaces
(b) Vol. **96**, No. 1, p. 169–176 (1979).
See also 21, p. 351; 21.1, p. 366; 21.2, p. 379.

M. Kolář
*On Surface Density of States of More Complex Tight-Binding Systems.
(b) Vol. **96**, No. 1, p. 447–452 (1979).

A. A. Guzev, V. A. Gurtov, A. V. Rzhanov, and A. A. Frantsuzov
*Value and Gradient Distribution of the Quasi-Fermi Level in MOS Transistor Channels
(a) Vol. **56**, No. 1, p. 37–46 (1979).
See also *14.3.3*, p. *230*; 14.3.4, p. 236.

J. Szuber and B. Salamon
Electron Spin Resonance Study on Low-Pressure-Baked (111) Si Surfaces Interacting with Adsorbed Gases
(a) Vol. **56**, No. 1, p. 111–115 (1979).
See also *19*, p. *296*; 22.1.2, p. 425.

13.4 Impurity and Defect States

Review Articles

O. A. Golikova
Boron and Boron-Based Semiconductors
(a) Vol. **51**, No. 1, p. 11–40 (1979).
See also *1*, p. *9*; *14.3*, p. *209*; 14.3.4, p. 231; 15, p. 249; 20.1, p. 304; 22, p. 393; 22.1.3, p. 428; 22.2, p. 430.

M. Averous
Symmetry-Induced Zero-Gap Semiconductors
(b) Vol. **95**, No. 1, p. 9–28 (1979).
See also *13.1*, p. *143*; 14.3, p. 209; 22.1.3, p. 428; 22.4, p. 448.

Original Papers and Short Notes

G. Lautenschläger and E. Mrosan
*Localization of Additional Screening Charge around 3d-Transition Metal Impurities in Al
(b) Vol. **91**, No. 1, p. 109–112 (1979).
See also 21, p. 338.

L. C. R. Alfred
Interaction between a Metal Surface and an External Molecular Complex
(b) Vol. **91**, No. 1, p. 113–121 (1979).
See also *13.3*, p. *158*; 21, p. 338.

K. Szlenk, Z. Dziuba, and R. R. Gałązka
Contribution of Resonant States to the Conductivity of HgTe
(b) Vol. **91**, No. 1, p. 255–261 (1979).
See also *14.3*, p. *209*; 22.4.3, p. 455.

S. A. Marshall, T. Marshall, S. A. Zager, and D. N. Zimmerman
*Electron Spin Resonance Absorption Spectrum of Divalent Manganese in Single Crystal Calcite
(b) Vol. **91**, No. 1, p. 275–282 (1979).
See also *19*, p. *289*; 22.8.1, p. 498.

H.-J. Wünsche and K. Henneberger
*Bound Excitons in Local Density Approximation
(b) Vol. **91**, No. 1, p. 331–337 (1979).
See also *13.5.1*, p. *183*; 22, p. 393.

S. Prakash and P. Lucasson
*Size Effect, Valency Factor, and Point Defect Interactions in Dilute Alloys
(b) Vol. **91**, No. 1, p. 339–348 (1979).
See also *10.1*, p. *92*; 21, p. 338; 21.1, p. 354.

R. Sebastian, G. A. Savari Raj, and T. M. Haridasan
*U Centres in Alkali Iodides
(b) Vol. **91**, No. 1, p. 371–376 (1979).
See also *20.1*, p. *304*; 22.5.2, p. 465.

S. Muramatsu and N. Sakamoto
*Calculation of the A Band Shape in KCl: Au^- in the D-Mode Model
(b) Vol. **91**, No. 2, p. K87 – K90 (1979).
See also *20.1*, p. *306*; 22.5.2, p. 465.

P. U. Arifov, N. Yu. Arutyunov, and A. S. Baltenkov
*The Annihilation of Positrons with Electrons Bound in Edge Dislocations
(b) Vol. **91**, No. 2, p. K91 – K94 (1979).
See also *10.2*, p. *104*; 22.1.1, p. 410.

D. P. Madacsi, L. A. Kappers, and J. F. Houlihan
*Crystal Field Splitting in Rutile-Structure Oxides from Correlation of ESR and Optical Results: V^{4+} in Tetragonal GeO_2
(b) Vol. **91**, No. 2, p. K105 – K109 (1979).
See also 19, p. 290; 20.1, p. 306; 22.6, p. 477.

X. Aymerich-Humet, and F. Serra-Mestres
Resonant Tunneling Current for General Junction Potential Barrier
(a) Vol. **51**, No. 2, p. 583 – 592 (1979).
See also *14.3.3*, p. *226*; 22, p. 395.

Y. Kamiura
*New Shallow Acceptors Produced in Germanium by Quenching
(a) Vol. **51**, No. 2, p. 601 – 607 (1979).
See also *14.3*, p. *210*; 22.1.1, p. 410.

K. D. Glinchuk, A. V. Prokhorovich, and V. I. Vovnenko
*Study of Non-Linear Extrinsic Luminescence in GaAs (II): The Role of Auger Recombination
(a) Vol. **51**, No. 2, p. 645 – 655 (1979).
See also *20.3*, p. *327*; 22.2.1, p. 432.

Y. Raman, S. Sivaraman, and J. D. Ranade
*Emission of CaO Phosphors Activated by Eu and Tb
(a) Vol. **51**, No. 2, p. 683 – 686 (1979).
See also *20.3*, p. *328*; 22.6, p. 478.

M. Godlewski
Microwave Conductivity Measurements in CdTe
(a) Vol. **51**, No. 2, p. K141 – K145 (1979).
See also *14.3*, p. *210*; 22.4.3, p. 455.

H.-J. Hoffmann and H. Reisser
*Investigation of Trap Levels in GaAs Schottky Diodes by Admittance Spectroscopy
(a) Vol. **51**, No. 2, p. K171 – K174 (1979).
See also 14.3.3, p. 226; 22.2.1, p. 432.

M. Samoć and A. Samoć
*Thermally Stimulated Current with the Participation of Two Trapping Levels
(a) Vol. **51**, No. 2, p. K195 – K199 (1979).
See also 14.3, p. 211; 22.9, p. 510.

F. Fischer
A Simple Model for Photo-Induced Polarization
(a) Vol. 52, No. 1, p. 189–194 (1979).
See also *10.2*, p. *105*; 14.4, p. 238; 16, p. 253; 22.5.2, p. 466.

J. C. Bourgoin, J. Krynicki, and B. Blanchard
*Boron Concentration and Impurity-to-Band Activation Energy in Diamond
(a) Vol. 52, No. 1, p. 293–298 (1979).
See also *14.3*, p. *211*; 22.1, p. 408.

L. A. Demberel, A. S. Popov, D. B. Kushev, and N. N. Zheleva
*Deep Levels in Fe-Doped InP
(a) Vol. 52, No. 1, p. 341–345 (1979).
See also *14.3.4*, p. *232*; 20.1, p. 308; 20.3, p. 328; 22.2.2, p. 437.

Ch. Schnittler
The Influence of a Spatial Trap Distribution on the SCLC Characteristic of Insulating Films and Related Problems of Trap Analysis
(a) Vol. 52, No. 1, p. K53–K56 (1979).
See also *14.3.4*, p. *232*; 22, p. 396.

W. Rühle, K. Lösch, and J. U. Fischbach
*A New Optical Method to Investigate Deep Traps in GaAs and InP
(a) Vol. 52, No. 1, p. K65–K67 (1979).
See also 20.1, p. 309; 22.2.1, p. 433; 22.2.2, p. 437.

J. Świątek
Thermally Stimulated Currents in Thin Polycrystalline p-Polyphenyl Layers
(a) Vol. 52, No. 1, p. K69–K71 (1979).
See also *14.3.1*, p. *219*; 22.9, p. 510.

Yu. E. Perlin, A. A. Kaminskii, V. N. Enakii, and D. N. Vylegzhanin
*Nonradiative Multiphonon Relaxation in $Y_3Al_5O_{12}:Nd^{3+}$
(b) Vol. 92, No. 1, p. 403–410 (1979).
See also 6, p. 66; *20.3*, p. *329*; 22.8.2, p. 505.

G. Weber and W. Rühle
*Effect of Uniaxial Stress on Free and Bismuth-Bound Excitons in InP
(b) Vol. 92, No. 2, p. 425–431 (1979).
See also 12.2, p. 134; *13.5.1*, p. *184*; 20.3, p. 329; 22.2.2, p. 437.

W. H. Lamfried
Temperature-Dependent Density of States of n-GaAs
(b) Vol. 92, No. 2, p. 433–439 (1979).
See also *13.1*, p. *146*; 22.2.1, p. 433.

J. Morkowski and A. M. Oleś
*On the Origin of Localized Magnetic States in Transition Metal Alloys (I): Wolff Model
(b) Vol. 92, No. 2, p. 525–536 (1979).
See also *18*, p. *261*; 21.1, p. 357.

G. Rudolf, J. Becherer, and H. Glaefeke
*A Remark on Determining Activation Energies by Fractional Glow Technique (FGT) in Case of Complex Glow Spectra – A Procedure of Improving This Method
(a) Vol. **52**, No. 2, p. K137–K141 (1979).
See also *17*, p. *256*; 22.6, p. 479.

R. Baubinas, F. Senulis, and J. Vaitkus
*Temperature Dependence of the Optical Ionization Energy of Deep Local Levels in CdSe Single Crystals
(a) Vol. **52**, No. 2, p. K143–K146 (1979).
See also *16*, p. *253*; 22.4.2, p. 453.

F. Adduci, L. Baldassarre, G. Maggipinto, A. Minafra, and F. Levy
*Defect Levels in PbI_2 by TSC Measurements
(a) Vol. **52**, No. 2, p. K173–K178 (1979).
See also 14.3, p. 212; 22.5, p. 461.

R. Bindemann, E. Hempel, and K. Kreher
*Determination of the Dielectric Constant of GaP from S–C Donor–Acceptor Pair Spectra
(a) Vol. **52**, No. 2, p. K201–K203 (1979).
See also 14.4, p. 239; *20.3*, p. *329*; 22.2.2, p. 437.

F. Lohse and J. M. Spaeth
Fluorescence of Interstitial Atomic Hydrogen in Alkali Halides Doped with I^-
(b) Vol. **93**, No. 1, p. 153–160 (1979).
See also *10.2*, p. *107*; *20.3*, p. *330*; 22.5.2, p. 467.

A. M. Oles and J. Morkowski
*On the Origin of Localized Magnetic States in Transition Metal Alloys (II): Doubly Degenerate Orbitals
(b) Vol. **93**, No. 1, p. 243–253 (1979).
See also *18*, p. *261*; 21.1, p. 358.

F. Dochy
Theoretical Calculation of the Absorption Energy of the F-Centre in Alkali Halides
(b) Vol. **93**, No. 1, p. 325–335 (1979).
See also *10.2*, p. *107*; *20.1*, p. *311*; 22.5.2, p. 467.

A. A. Serikov and Yu. M. Khomenko
*The Reabsorption Mechanism of the Donor–Acceptor Energy Transfer
(b) Vol. **93**, No. 1, p. 427–436 (1979).
See also *20.3*, p. *330*; 22, p. 398.

D. N. Popov, K. I. Terzijsky, and N. S. Backalova
*Spectral and Temperature Dependence of the Initial Yield of the Z_3-Center Formation in Additively Coloured KCl:Ca
(b) Vol. **93**, No. 1, p. K35–K37 (1979).
See also *20.1*, p. *311*; 22.5.2, p. 467.

H. Haberlandt, H.-J. Weckner, G. Gobsch, and J. Reinhold
*The *g*-Tensor of the Interstitial Oxygen and the Broken Si–O Bond in Silicon Dioxide – A MO Model Calculation
(b) Vol. **93**, No. 1, p. K71–K75 (1979).
See also *19*, p. *292*; 22.6, p. 480.

A. SZADKOWSKI
*Field Effect Studies of the Electron Transport in β-Rhombohedral Boron
(a) Vol. **53**, No. 1, p. 95 – 104 (1979).
See also *14.3.4*, p. *233*; 22.1, p. 408.

J. L. JANSON and Z. A. RACHKO
*Nature of Impurity-Induced UV Luminescence of MgO Crystals
(a) Vol. **53**, No. 1, p. 121 – 126 (1979).
See also *20.3*, p. *330*; 22.6, p. 480.

C. M. SUNTA
*Mechanism of Phototransfer of Thermoluminescence Peaks in Natural CaF_2
(a) Vol. **53**, No. 1, p. 127 – 135 (1979).
See also *20.3*, p. *330*; 22.5.3, p. 473.

A. G. MAKHANEK and G. A. SKRIPKO
Application of Two-Photon Spectroscopy in the Study of Trivalent Rare-Earth Ions in Crystals
(a) Vol. **53**, No. 1, p. 243 – 252 (1979).
See also *20.1*, p. *312*; *20.3*, p. *331*; 22.5.3, p. 473.

M. I. DARBY, M. N. READ, and K. N. R. TAYLOR
*The Energy of Hydrogen in Zirconium–Rhodium
(a) Vol. **53**, No. 1, p. 391 – 395 (1979).
See also *10.1*, p. *95*; 21.1, p. 359.

H. FEICHTINGER, A. GSCHWANDTNER, and J. WALTL
*Interstitial Iron and Thermal Defects in Silicon
(a) Vol. **53**, No. 1, p. K71 – K74 (1979).
See also *10.2*, p. *108*; 22.1.2, p. 419.

A. I. RYSKIN, A. V. VASILEV, and A. L. NATADZE
Stress-Induced Alignment of Jahn-Teller Centres Cr^{2+} in ZnSe Crystals. Optical Study
(b) Vol. **93**, No. 2, p. 459 – 467 (1979).
See also *20.1*, p. *312*; 22.4.2, p. 453.

W. BUSSE, H.-E. GUMLICH, A. GEOFFROY, and R. PARROT
*Spectral Distribution and Decay Times of the Luminescence of Mn^{2+} on Different Lattice Sites in ZnS
(b) Vol. **93**, No. 2, p. 591 – 596 (1979).
See also *20.3*, p. *331*; 22.4.1, p. 450.

F. DOCHY
*Theoretical Calculation of the Electron–Phonon Coupling Parameters of the F-Centre in Alkali Halides
(b) Vol. **93**, No. 2, p. 743 – 753 (1979).
See also *6*, p. *69*; 12.2, p. 134; 22.5.2, p. 468.

M. CASALBONI, F. CRISANTI, U. M. GRASSANO, C. MANFRONCELLI, A. SCACCO, and A. TANGA
*Luminescence of KBr:In^+
(b) Vol. **93**, No. 2, p. 755 – 759 (1979).
See also *20.3*, p. *331*; 22.5.2, p. 468.

G. P. Srivastava
*Electronic Structure of a Neutral Phosphorus Vacancy in GaP and InP
(b) Vol. **93**, No. 2, p. 761 – 765 (1979).
See also 22.2.2, p. 438.

S. J. Pačesová and L. Jastrabík
*The Energy Spectrum of a Deep Impurity – Oxygen in AlN
(b) Vol. **93**, No. 2, p. K111 – K114 (1979).
See also 20.1, p. 314; 20.3, p. 332; 22.2, p. 430.

V. B. Sharan, D. N. S. Srivastwa, and S. C. Sen
*Absorption Spectra of RbCl:Tl
(a) Vol. **53**, No. 2, p. 469 – 472 (1979).
See also *20.1*, p. *314*; 22.5.4, p. 475.

J. E. Ralph
Photoluminescence of CdS:Cu Doped by Diffusion
(a) Vol. **53**, No. 2, p. 611 – 615 (1979).
See also *20.3*, p. *332*; 22.4.1, p. 451.

A. R. Lakshmanan, R. C. Bhatt, and K. G. Vohra
γ-Radiation-Induced Sensitization, Supralinearity, and Photostimulated Thermoluminescence in LiF TLD-100. Correlation Studies
(a) Vol. **53**, No. 2, p. 617 – 625 (1979).
See also 10.2, p. 109; *20.3*, p. *332*; 22.5.2, p. 468.

A. le Bloa, P. N. Favennec, and Y. Colin
*Erratum to: Determination of Complementary Transition Energy Threshold by Photocapacitance Measurements
(a) Vol. **53**, No. 2, p. 681 (1979).
See also 14.3.3, p. 228; 22.2.1, p. 434.

G. Weimann
Electrical Characterization of n- and p-Type GaAs Films Grown by Molecular Beam Epitaxy
(a) Vol. **53**, No. 2, p. K173 – K176 (1979).
See also *14.3.1*, p. *220*; 22.2.1, p. 434.

M. D. Glinchuk, M. F. Deigen, and V. S. Rozhko
*Local Fields and Electric Dipole Moment of Localized Impurities in Crystals
(b) Vol. **94**, No. 1, p. 63 – 68 (1979).
See also 19, p. 293; 22, p. 400.

F. Mokross and H. Büttner
*The Exciton–Neutral Donor and Biexciton Complexes in Polar Semiconductors
(b) Vol. **94**, No. 1, p. 107 – 114 (1979).
See also *13.5.1*, p. *186*; 22, p. 401.

Doan Nhat Quang, B. Esser, and R. Keiper
*On the Theory of the Complex Dielectric Function of a Heavily Doped Degenerate Semiconductor
(b) Vol. **94**, No. 1, p. 115 – 123 (1979).
See also *20.1*, p. *314*; 22, p. 401.

A. van Calster
Barrier-Limited Conductivity in Thin Semiconducting Films
(a) Vol. **54**, No. 1, p. 207 – 216 (1979).
See also *14.3.1*, p. *220*; 22.4.2, p. 454.

G. Kühnel, W. Siegel, and E. Ziegler
*Deep Levels in $ZnSiP_2$ Determined by Schottky TSC Measurements
(a) Vol. **54**, No. 1, p. 315 – 320 (1979).
See also *14.3*, p. *214*; 22.8, p. 495.

J. A. Groenink, C. Hakfoort, and G. Blasse
*The Luminescence of Calcium Molybdate
(a) Vol. **54**, No. 1, p. 329 – 336 (1979).
See also *20.3*, p. *333*; 22.8.1, p. 500.

B. V. R. Chowdari and Y. Ravi Sekhar
*Electrical Conductivity and Thermal Depolarisation Studies of CrO_4^{2-} and $Cr_2O_7^{2-}$ Doped KDP and ADP Crystals
(a) Vol. **54**, No. 1, p. 413 – 419 (1979).
See also *14.3*, p. *214*; 22.8.1, p. 501.

M. A. S. Sweet and D. Urquhart
*Shallow Traps in ZnS Single Crystals
(a) Vol. **54**, No. 1, p. K81 – K84 (1979).
See also 20.3, p. 333; 22.4.1, p. 451.

D. R. Kaplan, C. Weigel, and J. W. Corbett
*Calculations on the Properties of Helium in Silicon
(b) Vol. **94**, No. 2, p. 359 – 366 (1979).
See also 9, p. 87; *13.1*, p. *150*; 22.1.2, p. 422.

P. Petelenz
*Mixing of Frenkel Excitons and Ionic Excited States in a Linear Crystal of Donor–Acceptor Complex
(b) Vol. **94**, No. 2, p. 371 – 379 (1979).
See also *13.5.1*, p. *186*; 22.9, p. 511.

A. P. Loghinov
Local and Resonance Vibrations in Defect Polymer Solids
(b) Vol. **94**, No. 2, p. 397 – 403 (1979).
See also *6*, p. *71*; 10.2, p. 110; 22.9, p. 512.

M. Friedrich and W. Karthe
*Structure and Kinetics of Fe^{3+}-Doped $PbMoO_4$
(b) Vol. **94**, No. 2, p. 451 – 456 (1979).
See also *19*, p. *294*; 22.8.1, p. 501.

Ch. Hoentzsch and J. M. Spaeth
*Dynamical Superhyperfine Interaction of Atomic Hydrogen in KCl and RbCl
(b) Vol. **94**, No. 2, p. 497 – 503 (1979).
See also *19*, p. *294*; 22.5.2, p. 469.

L. M. Lambert
*Theory of ISC Measurements for p-Type Schottky Diodes with Deep Acceptor Impurities
(a) Vol. **54**, No. 2, p. 665—673 (1979).
See also *14.3.3*, p. *229*; 22, p. 403.

Y. Kamiura and F. Hashimoto
*Degree of Copper Contamination in Germanium Heat-Treated at High Temperatures
(a) Vol. **54**, No. 2, p. 697—700 (1979).
See also *10.2*, p. *111*; 22.1.1, p. 412.

Z. N. Chigogidze and N. P. Khuchua
*Study of Electrical Breakdown Peculiarities in High-Resistivity Compensated p-InSb
(a) Vol. **54**, No. 2, p. 773—781 (1979).
See also *14.3.4*, p. *235*; 22.2.3, p. 441.

M. Balarin
*Temperature Dependence of the Pre-Exponential Factor in the Glow Curve Theory
(a) Vol. **54**, No. 2, p. K137—K140 (1979).
Erratum
(a) Vol. **56**, No. 2, p. K149 (1979).
See also 22, p. 403.

D. P. Bhattacharya and Z. S. Kachlishvili
*Nonradiative Multiphonon Carrier Capture by Deep Neutral Traps
(b) Vol. **95**, No. 1, p. 117—130 (1979).
See also *14.3.4*, p. *235*; 22, p. 404.

I. P. Zvyagin
On the Hopping Mechanism of Dispersive Transport
(b) Vol. **95**, No. 1, p. 227—235 (1979).
See also *2*, p. *47*; *14.3*, p. *214*.

L. Kubler, A. Jaéglé, and J. J. Koulmann
*The Effect of Oxygen on the Properties of Evaporated Amorphous Silicon. ESR Studies
(b) Vol. **95**, No. 1, p. 307—315 (1979).
See also 2, p. 47; *19*, p. *295*; 22.1.2, p. 423.

B. A. Orlowski
*Photoemission Evidence of Mn(3d^5) Band in $Cd_{0.4}Mn_{0.6}Te$ Crystals
(b) Vol. **95**, No. 1, p. K31—K35 (1979).
See also *17*, p. *258*; 22.4.4, p. 459.

D. N. Popov and K. I. Terzijsky
Electron Trapping of α_A- and F_A-Centers in KCl:Na
(b) Vol. **95**, No. 1, p. K43—K46 (1979).
See also *10.2*, p. *112*; *16*, p. *253*; 22.5.2, p. 470.

I. V. Mochalov
Spectroscopic Characteristics of the Erbium Ions and Laser Action at the Transition $^4S_{3/2} \rightarrow {}^4I_{9/2}$ ($\lambda_e \approx 1.66$ μm) in the Crystals of Gadolinium, Yttrium, and Lutecium Orthoaluminates
(a) Vol. **55**, No. 1, p. 79—87 (1979).
See also 20.1, p. 319; *20.2*, p. *326*; 20.3, p. 335; 22.8.1, p. 502.

G. Pauli, K. F. Renk, G. Klimer, and H. J. Kreuzer
*Resonance Interaction of Electronic Two-Level States and Short-Wave Phonon Radiation in Ruby
(b) Vol. **95**, No. 2, p. 503–508 (1979).
See also *6*, p. *74*; 20.3, p. 335; 22.6, p. 484.

L. Bosi and M. Nimis
*To What Extent Are the Observed Lifetimes of F-Centers Intrinsic?
(b) Vol. **95**, No. 2, p. 615–619 (1979).
See also *20.3*, p. *335*; 22.5.2, p. 470.

K. L. Brower and J. Vitko, Jr.
Comment on "A New Concept of the Model for the Paramagnetic E′-Centre in the SiO_{2-x} System"
(b) Vol. **95**, No. 2, p. K141–K144 (1979).
See also *2*, p. *48*; *19*, p. *296*; 22.8.1, p. 502.

A. V. Shendrik and D. M. Yudin
Reply to Comment on "A New Concept of the Model for the Paramagnetic E′-Centre in the SiO_{2-x} System"
(b) Vol. **95**, No. 2, p. K145–K146 (1979).
See also *2*, p. *48*; *19*, p. *296*; 22.8.1, p. 502.

A. Yu. Leiderman
*On the Theory of Double Injection in Semiconductors with Continuous Distribution of Trapping Levels within the Gap
(a) Vol. **55**, No. 2, p. 661–671 (1979).
See also *14.3.4*, p. *236*; 22, p. 405.

P. Gaworzewski and K. Schmalz
*On the Electrical Activity of Oxygen in Silicon
(a) Vol. **55**, No. 2, p. 699–707 (1979).
See also *14.3*, p. *216*; 22.1.2, p. 425.

Sh. N. Gifeisman and Yu. G. Malkoch
The Theory of Magneto-Optical Effects in Shallow F-Centres
(b) Vol. **96**, No. 1, p. 73–90 (1979).
See also 13.5.3, p. 193; *18*, p. *263*; *20*, p. *302*; 22, p. 405.

A. Shluger, E. Kotomin, and Yu. Dzelme
*Quantum-Chemical Calculations of Electronic and Hole Centres and Surface of NaCl Crystals (I)
(b) Vol. **96**, No. 1, p. 91–99 (1979).
See also *10.2*, p. *115*; 13.3, p. 161; 22.5.2, p. 471.

V. K. Dugaev and P. P. Petrov
*Impurity Band in $Hg_{1-x}Cd_xTe$ Alloys
(b) Vol. **96**, No. 1, p. 129–136 (1979).
See also 14.3, p. 216; 22.4.4, p. 459.

V. Z. Polinger and G. I. Bersuker
*Multimode Jahn-Teller Effect for an E-Term with Strong Vibronic Coupling (II): Band Shapes of the Infrared and Raman Spectra
(b) Vol. **96**, No. 1, p. 153–161 (1979).
See also *6*, p. *75*; 20.1, p. 321.

G. Lautenschläger and E. Mrosan
*Pre-Asymptotic Charge Oscillations around 3d Transition Metal Impurities in Aluminium
(b) Vol. **96**, No. 1, p. 183–187 (1979).
See also 21, p. 351.

S. Nakashima and S. Yasuda
*Excitation and Recombination of Donor–Acceptor Pairs in ZnTe
(b) Vol. **96**, No. 1, p. 211–218 (1979).
See also *20.3*, p. *336*; 22.4.3, p. 456.

G. Reuter, F. Lohse, and J. M. Spaeth
Generation of Dichroic H Centres in KCl by Polarized Optical Bleaching of Interstitial Atomic Hydrogen
(b) Vol. **96**, No. 1, p. 243–250 (1979).
See also *10.2*, p. *115*; 19, p. 296; *20.1*, p. *321*; 22.5.2, p. 471.

V. I. Sugakov and V. I. Tovstenko
Exciton States in Semi-Infinite Molecular Crystals with a Point Defect near the Surface
(b) Vol. **96**, No. 1, p. 307–312 (1979).
See also *13.5.1*, p. *188*.

T. Tsuboi
Identification of the Absorption Bands in KI:Pb^{2+} Crystals
(b) Vol. **96**, No. 1, p. 321–328 (1979).
See also *20.1*, p. *321*; 22.5.2, p. 472.

V. M. Fomin
Non-Linear Optical Properties of Band Charge Carriers Due to Scattering by Lattice Defects
(b) Vol. **96**, No. 1, p. 389–399 (1979).
See also 6, p. 75; *20.2*, p. *326*; 22, p. 405.

A. P. Marchetti and M. Scozzafava
The Optically Detected Magnetic Resonance of Nickel-Doped Silver Chloride
(b) Vol. **96**, No. 1, p. 441–445 (1979).
See also *19*, p. *296*; 20, p. 302; 22.5.1, p. 464.

T. Taguchi
*Ionization Energy of the Complex Acceptor in ZnTe
(b) Vol. **96**, No. 1, p. K33–K36 (1979).
See also 20.3, p. 336; 22.4.3, p. 456.

J. K. Srivastava and B. W. Dale
Effect of Annealing Temperature on Cross Relaxation in the Cr^{3+}-Doped Magnetically Dilute Paramagnetic Al_2O_3–Fe_2O_3 System
(b) Vol. **96**, No. 1, p. K47–K50 (1979).
See also *5*, p. *61*; 22.6.1, p. 487.

C. T. Hsing, D. P. Kennedy, A. D. Sutherland, and K. M. van Vliet
*Quantum Mechanical Determination of the Potential and Carrier Distributions in the Inversion Layer of Metal–Oxide–Semiconductor Devices
(a) Vol. **56**, No. 1, p. 129–141 (1979).
See also *14.3.3*, p. *230*; 22, p. 406.

D. Valentovič, J. Červenák, S. Luby, M. L. Aldea, and T. Boţilă
Some Non-Equilibrium Phenomena in Sputtered CdTe Thin Films
(a) Vol. **56**, No. 1, p. 341–347 (1979).
See also *1.5*, p. *37*; 16, p. 254; 22.4.3, p. 457.

H. van Cong
*Impurity Band Structure in Degenerate Semiconductors for Both Dense Donors and Acceptors
(a) Vol. **56**, No. 1, p. 395–405 (1979).
See also *13.1*, p. *154*; 22.2.1, p. 436.

P. Krispin
*Electron Traps around Dislocations in GaP
(a) Vol. **56**, No. 1, p. K15–K19 (1979).
See also *10.2*, p. *116*; 22.2.2, p. 438.

V. B. Sulimov
*Balance Equation for Non-Equilibrium Electrons in the Mobility Gap of a Disordered Semiconductor in the Presence of Light
(b) Vol. **96**, No. 2, p. 507–515 (1979).
See also *2*, p. *49*; *14.3*, p. *217*; 22, p. 406.

S. P. Reifman and B. V. Shulichenko
*The Calculation of the Electronic Structure of a Vacancy by the Green's Function Theory
(b) Vol. **96**, No. 2, p. 537–544 (1979).
See also 23, p. 516.

W. Pickin
*The Kinetics of Electron Transitions for Donors with Excited States
(b) Vol. **96**, No. 2, p. 617–626 (1979).
See also *16*, p. *254*; 22.1.1, p. 413.

M. Moreno
The Relation between Bonding and *g*-Shift for Tl^{2+} and Pb^{3+} in Alkali Halides
(b) Vol. **96**, No. 2, p. 647–651 (1979).
See also *19*, p. *297*; 22.5.2, p. 472.

C. A. Moyer, J. R. Kelly, and S. Arajs
Model for Local Moments in a Spin Density Wave Host
(b) Vol. **96**, No. 2, p. 775–783 (1979).
See also *18.2*, p. *279*; *18.4*, p. *288*; 21.1, p. 368.

S. K. Misra and B. Mikolajczak
EPR of Mn^{2+} Doped Single Crystals of $Ni(NH_4)_2(SO_4)_2 \cdot 6\,H_2O$ and $NiSO_4 \cdot 7\,H_2O$
(b) Vol. **96**, No. 2, p. 807–815 (1979).
See also *19*, p. *297*; 22.8.1, p. 503.

A. K. Saxena
Non-Shallow Levels and the Conduction Band Structure of $Ga_{1-x}Al_xAs$
(b) Vol. 96, No. 2, p. K77 – K82 (1979).
See also *13.1*, p. *155*; 22.2.4, p. 443.

J. Pebler, K. Schmidt and G. Weiser
Magnetic Susceptibility and Electrical Conductivity Measurements of $V_{1-x}Fe_xO_{2-x}F_x$ in the Concentration Range $0 < x < 0.20$
(a) Vol. 56, No. 2, p. 457 – 465 (1979).
See also *1.2*, p. *21*; 14.3, p. 218; 18.1, p. 269; 22.6.1, p. 487.

D. Wruck and P. Gaworzewski
*Electrical and Infrared Spectroscopic Investigations of Oxygen-Related Donors in Silicon
(a) Vol. 56, No. 2, p. 557 – 564 (1979).
See also 14.3, p. 218; 20.1, p. 323; 22.1.2, p. 426.

D. Haack and H. G. Wagemann
*Electronic Conductivity in Highly Doped SiO_2 by Means of Ion Implantation
(a) Vol. 56, No. 2, p. 581 – 591 (1979).
See also *14.3.4*, p. *237*; 22.6, p. 485.

P. G. Di Marco and G. Giro
*Hole Photogeneration and Mobility in α-Form Crystals of 9,10-Dichloroanthracene
(a) Vol. 56, No. 2, p. K133 – K136 (1979).
See also *14.3*, p. *218*; 22.9, p. 514.

H. Neumann, R. D. Tomlinson, E. Nowak, and N. Avgerinos
*Electrical Properties of p-Type $CuInSe_2$ Single Crystals
(a) Vol. 56, No. 2, p. K137 – K140 (1979).
See also *14.3*, p. *218*; 22.8, p. 497.

13.5 Elementary Excitations (Phonons See 6)

Original Papers and Short Notes

E. HEINER
*An Information Theoretical Method (ITM) Applied to the Metal–Insulator Transition of the Neutral Hubbard Model
(b) Vol. **91**, No. 1, p. 87–98 (1979).
See also 14.1, p. 199; 14.3, p. 209.

S. P. IONOV, L. A. MANAKOVA, and B. G. ZEMSKOV
The Electronic Spectra and Parametric Instabilities in Charge-Ordered Crystals with Intraband Electron Attraction
(b) Vol. **91**, No. 2, p. 389–400 (1979).
See also *13.1*, p. *144*; 13.4, p. 164.

G. O. MÜLLER, H. H. WEBER, and I. HÖRICKE
LinearOptical Response of Highly Excited Semiconductors. Electron–Hole Liquid in CdS
(b) Vol. **91**, No. 2, p. 531–539 (1979).
See also *20.1*, p. *305*; 22.4.1, p. 450.

I. V. STASYUK, R. R. LEVITSKII, and N. A. KORINEVSKII
*Collective Vibrations of Protons in Compounds of the KH_2PO_4-Type. The Cluster Approximation
(b) Vol. **91**, No. 2, p. 541–550 (1979).
See also *14.4.1*, p. *244*; 22.8.1, p. 498.

L. N. OVANDER and N. S. TYU
*Theory of Raman Scattering on Polar Phonons
(b) Vol. **91**, No. 2, p. 763–772 (1979).
See also 6, p. 65; *20*, p. *299*.

V. R. NAGIBAROV and O. KH. KHASANOV
The Effect of Retarded Interaction on the Self-Induced Transparency Phenomenon
(b) Vol. **92**, No. 2, p. 467–472 (1979).
See also *13*, p. *137*.

G. RÖPKE and R. DER
*The Influence of Two-Particle States (Excitons) on the Dielectric Function of the Electron–Hole Plasma
(b) Vol. **92**, No. 2, p. 501–510 (1979).
See also *13.5.1*, p. *184*; 22, p. 396.

K. SATO, M. KOBAYASHI, S. IWABUCHI, and I. YOKOTA
*TO Phonon–Helicon Coupling in Polar Semiconductors
(b) Vol. **93**, No. 1, p. 79–87 (1979).
See also *6*, p. *67*; 14.3, p. 212; 22.2.3, p. 440.

E. SZCZEPANIAK
*On Singlet–Triplet Ferromagnet Dynamics
(b) Vol. **93**, No. 1, p. 89–95 (1979).
See also *18.2*, p. *273*.

J. Wenk and L. Merten
Stimulated Raman Effect in Piezoelectric Cubic Crystals
(b) Vol. **93**, No. 1, p. 175–181 (1979).
See also *6*, p. *68*; *20*, p. *299*.

J. S. Nkoma and S. K. Rwaboona
*Frequency Dependence of Some Functions Associated with Surface Polaritons
(b) Vol. **93**, No. 1, p. 397–402 (1979).
See also *13.3*, p. *159*; 22.2.2, p. 437.

A. Stahl
*Polariton Structure of Interband Transitions in Semiconductors
(b) Vol. **94**, No. 1, p. 221–230 (1979).
See also 20, p. 300; 22, p. 401.

V. G. Neudachin, N. P. Yudin, and F. A. Zhivopistsev
*A Study of the Collective Properties of the Degenerate Electronic Gas in Metals Using the (e, 2e), (e, 3e), and (γ, 2e) Processes at High Energies 10^3 to 10^4 eV
(b) Vol. **95**, No. 1, p. 39–46 (1979).
See also 21, p. 348.

U. Lindner and V. K. Fedyanin
*Solitary Solutions in a Modified Hubbard Chain
(b) Vol. **95**, No. 1, p. K83–K87 (1979).
See also 18.4, p. 288.

A. A. Kazakov and N. A. Litvinenko
*Energy Spectrum, Magnetization, and Critical Field of the Uniaxial Two-Sublattice Ferrimagnet by the Green Function Method
(b) Vol. **95**, No. 2, p. 333–340 (1979).
See also *18.3*, p. *284*.

R. Claus
*On the Lattice Dynamical Background of Optical Activity in Crystals
(b) Vol. **96**, No. 2, p. 545–554 (1979).
See also 6, p. 76; *20*, p. *302*.

13.5.1 Excitons

Original Papers and Short Notes

J. Voigt, F. Spiegelberg, and M. Senoner
*Band Parameters of CdS and CdSe Single Crystals Determined from Optical Exciton Spectra
(b) Vol. **91**, No. 1, p. 189—199 (1979).
See also *20.1*, p. *304*; 22.4.1, p. 450; 22.4.2, p. 453.

H.-J. Wünsche and K. Henneberger
*Bound Excitons in Local Density Approximation
(b) Vol. **91**, No. 1, p. 331—337 (1979).
See also *13.4*, p. *163*; 22, p. 393.

B. A. Orłowski and J. M. Langer
Determination of the CdF_2 Valence Band Structure by Photoemission Measurement
(b) Vol. **91**, No. 1, p. K53—K56 (1979).
See also *13.1*, p. *144*; 17, p. 256; 22.5, p. 460.

J. Flohrer, E. Jahne, and M. Porsch
*Energy Levels of A and B Excitons in Wurtzite-Type Semiconductors with Account of Electron–Hole Exchange Interaction Effects
(b) Vol. **91**, No. 2, p. 467—478 (1979).
See also 22.4.1, p. 450.

G. Munschy and L. Stauffer
*Separation of Angles in the Problem of an Exciton Bound to an Ionized Donor in the Case of Anisotropic Effective Masses and Dielectric Constant
(b) Vol. **91**, No. 2, p. 633—639 (1979).
See also 22.4, p. 448.

A. D. Brothers and J. B. Brungardt
*Excitons in WS_2 Films. Pressure and Temperature Effects
(b) Vol. **91**, No. 2, p. 675—679 (1979).
See also *20.1*, p. *305*; 22, p. 394.

D. D. Kolendritskii, M. V. Kurik, and Yu. P. Piryatinskii
*Exciton Reflection Spectra of Tetracene Single Crystals
(b) Vol. **91**, No. 2, p. 741—751 (1979).
See also *20.1*, p. *306*; 22.9, p. 509.

A. S. Gaievskii, V. A. Lisovenko, A. N. Faidysh, and L. B. Yankovskaya
*Luminescence from Anthracene Single Crystals with Photoinduced Local Centres
(b) Vol. **92**, No. 1, p. 31—37 (1979).
See also *20.3*, p. *328*; 22.9, p. 510.

A. Stahl
Boundary Conditions for Polarization Waves in Spatially Dispersive Dielectrics
(b) Vol. **92**, No. 1, p. 113—122 (1979).
See also *6*, p. *65*; *20.1*, p. *307*.

Yu. B. Gaididei and A. S. Trofimov
*Theory of Exciton Migration in a Critical Region
(b) Vol. **93**, No. 1, p. 97 – 103 (1979).
See also 18.1, p. 267; 18.2, p. 273.

Z. A. Rachko and J. A. Valbis
*Luminescence of Free and Relaxed Excitons in MgO
(b) Vol. **93**, No. 1, p. 161 – 166 (1979).
See also 20.3, p. 330; 22.6, p. 479.

W. Ekardt and H.-J. Brocksch
*Phonon-Assisted Two-Photon Absorption of the Excitonic Molecule in CuCl (I): Bare-Exciton Approach
(b) Vol. **93**, No. 1, p. 203 – 211 (1979).
See also 20.1, p. 311; 22.5, p. 461.

J. Röseler and K. Henneberger
*Magnetopolaritons in Wurtzite-Type Crystals
(b) Vol. **93**, No. 1, p. 213 – 222 (1979).
See also 20, p. 300; 22.4.1, p. 450; 22.4.2, p. 453.

J. Gallardo and D. Mattis
*Green's Theorem Calculation of Energy of Exciton near Hard Walls
(b) Vol. **93**, No. 1, p. 229 – 235 (1979).
See also 14.3.2, p. 223; 20.1, p. 311.

Y. Itoh, S. Adachi, and C. Hamaguchi
Analysis of Resonant Brillouin Scattering in GaAs
(b) Vol. **93**, No. 1, p. 381 – 389 (1979).
See also *20.1*, p. *311*; 22.2.1, p. 433.

G. Behnke and H. Büttner
*Erratum to: The Exciton Dispersion in Polar Semiconductors
(b) Vol. **93**, No. 1, p. 447 (1979).
See also 13.5.3, p. 193; 22.5, p. 461.

A. V. Lelyakov
*Absorption Band Shape of Free Excitons in Indirect Band Gap Semiconductors at High Excitation Density
(b) Vol. **93**, No. 2, p. 523 – 530 (1979).
See also 20.1, p. 313; 22.5.1, p. 463.

R. Enderlein, P. Renner, and M. Scheele
Electro-Optic Effects Due to the Stark Splitting of Energy Bands
(b) Vol. **93**, No. 2, p. 539 – 547 (1979).
See also 13.1, p. 148; *14.3.4*, p. *234*; *20.1*, p. *313*; 22.6, p. 480.

J. M. Hvam
*Excitonic Molecule Transitions in ZnO
(b) Vol. **93**, No. 2, p. 581 – 590 (1979).
See also 20.1, p. 313; 20.3, p. 331; 22.4, p. 448.

B. DIETRICH and J. VOIGT
*Spatial Dispersion Effects in the Transmission Spectra of Semiconductors in the Region of Exciton Resonances
(b) Vol. **93**, No. 2, p. 669–677 (1979).
See also 20.1, p. 313; 22.4.1, p. 451.

H. SCHREY and C. KLINGSHIRN
*Ground and Excited States of the Biexcition and the M-Band Luminescence in Highly Excited ZnO
(b) Vol. **93**, No. 2, p. 679–689 (1979).
See also *20.3*, p. *331*; 22.4, p. 448.

A. RAINO, A. M. MANCINI, R. MURRI, A. RIZZO, and L. VASANELLI
*Cathodoluminescence of GaSe Single Crystals
(a) Vol. **53**, No. 2, p. 487–492 (1979).
See also *20.3*, p. *332*; 22.3, p. 445.

J. WAGNER and R. SAUER
*Electron–Hole Drops in Uniaxially Stressed Silicon
(b) Vol. **94**, No. 1, p. 69–78 (1979).
See also *20.3*, p. *332*; 22.1.2, p. 420.

F. MOKROSS and H. BÜTTNER
*The Exciton–Neutral Donor and Biexciton Complexes in Polar Semiconductors
(b) Vol. **94**, No. 1, p. 107–114 (1979).
See also 13.4, p. 171; 22, p. 401.

H. STOLZ and R. ZIMMERMANN
*Correlated Pairs and a Mass Action Law in Two-Component Fermi Systems. Excitons in an Electron–Hole Plasma
(b) Vol. **94**, No. 1, p. 135–146 (1979).
See also 13.5.2, p. 192; 22, p. 401.

CH. UIHLEIN and S. FEIERABEND
*Magnetoreflectance of Exciton-Polaritons in Cubic II–VI Compounds (I): Comparison of the Magnetic Field Behaviour of Excitons and Polaritons
(b) Vol. **94**, No. 1, p. 153–160 (1979).
See also 20.1, p. 314; 22.4, p. 448.

J. PULS and J. VOIGT
*Excitation Spectroscopy of the I_2 Bound Exciton and the Excitonic Molecule in CdS (I): Excited States of the I_2 Bound Exciton
(b) Vol. **94**, No. 1, p. 199–204 (1979).
See also 20.3, p. 332; 22.4.1, p. 451.

P. PETELENZ
*Mixing of Frenkel Excitons and Ionic Excited States in a Linear Crystal of Donor–Acceptor Complex
(b) Vol. **94**, No. 2, p. 371–379 (1979).
See also *13.4*, p. *173*; 22.9, p. 511.

W. Schmid
*Determination of Thermodynamical Parameters of the EHD in Silicon from Luminescence Data
(b) Vol. **94**, No. 2, p. 413–419 (1979).
See also 6, p. 71; *20.3*, p. *333*; 22.1.2, p. 422.

Ch. Uihlein and S. Feierabend
*Magnetoreflectance of Exciton-Polaritons in Cubic II-VI Compounds (II): Semiclassical Treatment
(b) Vol. **94**, No. 2, p. 421–428 (1979).
See also 20.1, p. 316; 22.4, p. 449.

H. J. Fischbeck and R. Strehlow
*Single-Site Theory of Wannier Excitons in Disordered Semiconductors (I): General Considerations
(b) Vol. **94**, No. 2, p. 457–468 (1979).
See also 2, p. 46; 22, p. 402.

V. May, K. Henneberger, and F. Henneberger
*Renormalization of Polaritons Due to Virtual Formation of Biexcitons at High Densities of Excitation
(b) Vol. **94**, No. 2, p. 611–620 (1979).
See also 20.1, p. 317; 20.2, p. 325; 22.5, p. 461.

N. Magnea, D. Bensahel, J. L. Pautrat, and J. C. Pfister
Optical Identification of Substitutional Acceptors in Refined ZnTe
(b) Vol. **94**, No. 2, p. 627–639 (1979).
See also *13.4*, p. *174*; *20.3*, p. *333*; 22.4.3, p. 455.

P. Kleinert
*The Influence of Contact Excitons on the Electroabsorption in Disordered Semiconductors
(b) Vol. **94**, No. 2, p. K145–K147 (1979).
See also 2, p. 46; 14.3.4, p. 235; *20.1*, p. *317*; 22, p. 403.

L. G. Suslina, A. G. Plyukhin, O. Goede, and D. Hennig
*Compositional Fluctuation-Induced Broadening of Bound-Exciton Lines in II–VI Semiconductor Mixed Crystals
(b) Vol. **94**, No. 2, p. K185–K189 (1979).
See also *20.1*, p. *317*; 22.4.4, p. 459.

V. A. Benderskii, V. Kh. Brikenshtein, A. G. Burshtein, A. G. Lavrushko, A. G. Prikhozhenko, and P. G. Filippov
*Non-Linear Fluorescence Quenching in Molecular Crystals (II): Recombination of Free Excitons
(b) Vol. **95**, No. 1, p. 47–57 (1979).
See also *20.3*, p. *334*; 22.9, p. 512.

A. A. O'Connel-Bronin and V. G. Plenkhanov
*Resonant Raman Scattering in Crystals with Self-Trapping Excitons
(b) Vol. **95**, No. 1, p. 75–84 (1979).
See also *6*, p. *72*; 20.3, p. 334; 22.5.2, p. 470.

I. Broser and M. Rosenzweig
*A Quantitative Study of Excitonic Polariton Reflectance in CdS (II): B-Exciton and k-Linear Term
(b) Vol. **95**, No. 1, p. 141–154 (1979).
See also *20.1*, p. *318*; 22.4.1, p. 451.

S. W. Koch and H. Haug
*Treatment of the Electron–Hole Droplet Nucleation in the Fokker-Planck Approximation
(b) Vol. **95**, No. 1, p. 155–161 (1979).
See also 20.1, p. 318; 20.3, p. 334; 22.1.1, p. 412.

A. S. Rao and R. J. Kearney
Logarithmic Derivative Reflectance Spectra of BaO and SrO
(b) Vol. **95**, No. 1, p. 243–250 (1979).
See also *20.1*, p. *319*; 22.6, p. 483.

J. D'hertoghe and G. Jacobs
*Self-Trapped Exciton and F Centre Formation by Picosecond Laser Pulses in Alkali Bromides and Iodides
(b) Vol. **95**, No. 1, p. 291–300 (1979).
See also 20.3, p. 334; 22.5.2, p. 470.

K. B. Tolpygo and S. M. Zubkova
*On the Indirect Exciton Energy Levels in Crystals of Diamond and Sphalerite Type with Spin–Orbit Interaction
(b) Vol. **95**, No. 2, p. 341–348 (1979).
See also 13.1, p. 152; 22.1.1, p. 412; 22.1.2, p. 424; 22.2, p. 431.

K. Bärner, R. Stearns, and R. Braunstein
*Evidence for Excitonic Absorption in Metallic $MnAs_{0.88}P_{0.12}$
(b) Vol. **95**, No. 2, p. 483–489 (1979).
See also *20.1*, p. *319*; 22.8, p. 496.

R. Schwabe, F. Thuselt, H. Weinert, and R. Bindemann
*Electron–Hole Liquid in GaP. The Influence of the Isoelectronic Impurity Nitrogen
(b) Vol. **95**, No. 2, p. 571–583 (1979).
See also *20.3*, p. *335*; 22.2.2, p. 438.

R. Kink and M. Selg
*Polariton Effects in Crystalline Xenon
(b) Vol. **96**, No. 1, p. 101–109 (1979).
See also 20.1, p. 320; 20.3, p. 336; 23, p. 516.

H. J. Mattausch and Ch. Uihlein
*Fine Structure of P Excitons in CuBr
(b) Vol. **96**, No. 1, p. 189–200 (1979).
See also 20.1, p. 321; 22.5, p. 462.

V. I. Sugakov and V. I. Tovstenko
*Exciton States in Semi-Infinite Molecular Crystals with a Point Defect near the Surface
(b) Vol. **96**, No. 1, p. 307–312 (1979).
See also 13.4, p. 178.

K. Hübner and F. Bechstedt
*Temperature Dependence of Atomic Core Levels in Solids (I): Separation of the Temperature Dependence of Conduction-Band and Core-Exciton Energies
(b) Vol. **96**, No. 1, p. **351–357** (1979).
See also *13.1*, p. *153*; 22.1.2, p. 425; 22.2.2, p. 438; 22.6, p. 484.

W. Stössel and H. J. Wagner
*The Influence of Spatial Dispersion and of the Angle of Incidence on Thermomodulation Spectra of the $A_{n=1}$-Exciton in CdS
(b) Vol. **96**, No. 1, p. **369–378** (1979).
See also *20.1*, p. *321*; 22.4.1, p. 451.

A. Kuroiwa, H. Saito, and S. Shionoya
*Exciton Reflection Spectra under High-Intensity Excitation in CdS and CdSe
(b) Vol. **96**, No. 1, p. **483–486** (1979).
See also *20.1*, p. *321*; 20.3, p. 336; 22.4.1, p. 451; 22.4.2, p. 454.

J. L. Jansons, Z. A. Rachko, and J. A. Valbis
*Fine Structure of the Exciton Luminescence in MgO Single Crystals
(b) Vol. **96**, No. 1, p. **K29–K31** (1979).
See also *20.3*, p. *336*; 22.6, p. 484.

A. Twardowski, M. Nawrocki, and J. Ginter
*Excitonic Magnetoabsorption in $Cd_{1-x}Mn_xTe$ Mixed Crystals
(b) Vol. **96**, No. 2, p. **497–506** (1979).
See also 18, p. 264; 20.1, p. 322; 22.4.4, p. 459.

K. Hübner and F. Bechstedt
*Temperature Dependence of Atomic Core Levels in Solids (II): Chemical Shift and Relaxation Energy of Core Levels and Their Dependence on Temperature
(b) Vol. **96**, No. 2, p. **595–603** (1979).
See also *13.1*, p. *154*; 22.1.2, p. 426; 22.2.2, p. 438; 22.6, p. 485.

J. Puls, I. Rückmann, and J. Voigt
Excitation Spectroscopy of the I_2 Bound Exciton and the Excitonic Molecule in CdS (II): Relation between I_2-Line and EM-Lines
(b) Vol. **96**, No. 2, p. **641–646** (1979).
See also 20.3, p. 336; 22.4.1, p. 452.

O. Goede, D. Hennig, and L. John
*Disorder Effects on Free Excitons in $CdS_{1-x}Se_x$ Mixed Crystals
(b) Vol. **96**, No. 2, p. **671–681** (1979).
See also *20.1*, p. *323*; 22.4.4, p. 459.

V. D. Iskra
*Phonon-Assisted Interband Optical Absorption in Disordered Semiconductors
(b) Vol. **96**, No. 2, p. **843–856** (1979).
See also *20.1*, p. *323*; 22, p. 407.

J. Mazzaschi, J. Barrau, M. Brousseau, J. Collet, and H. Maaref
*Luminescence Line Shape of Free Excitons in Natural Diamond
(b) Vol. **96**, No. 2, p. **K61–K64** (1979).
See also *20.3*, p. *337*; 22.1, p. 408.

I. Egri and A. Stahl
*Real Space Wave Equation for Exciton-Polaritons of Wannier Type
(b) Vol. **96**, No. 2, p. K83 – K86 (1979).
See also 20, p. 303; 22, p. 407.

A. Coret and A. Haydar
*Photoconductivity in the 1S Excitonic Line of Cu_2O. Shape of the Response
(a) Vol. **56**, No. 2, p. 531 – 534 (1979).
See also *16*, p. *255*; 22.6, p. 485.

13.5.2 Plasmons

Original Papers and Short Notes

J. T. DEVREESE, F. BROSENS, and L. F. LEMMENS
*Dynamical Exchange Effects on the Plasmon Dispersion in Aluminium
(b) Vol. **91**, No. 1, p. 349–354 (1979).
See also 14.4, p. 238; 21, p. 338.

M. FUKUI, C. V. Y. SO, and R. NORMANDIN
*Lifetimes of Surface Plasmons in Thin Silver Films
(b) Vol. **91**, No. 1, p. K61–K64 (1979).
See also 21.6, p. 387.

W. GASSER and R. HÖFLING
*Calculation of the Dielectric Function of the One-Dimensional Hubbard Model
(b) Vol. **92**, No. 1, p. 91–97 (1979).
See also 14, p. 197; 18.1, p. 266.

S. GUHA and S. GHOSH
*Longitudinal Phonon–Plasmon Interaction in Inhomogeneous Semiconductors
(b) Vol. **92**, No. 2, p. K95–K97 (1979).
See also *6*, p. *67*; 22, p. 397.

D. N. TRIPATHY, S. S. MANDAL, and B. K. RAO
*On the Study of Plasmon Dispersion beyond k_c
(b) Vol. **92**, No. 2, p. K113–K116 (1979).
See also 14.4, p. 239; 21, p. 342.

I. M. DYKMAN and L. A. PROKHNITSKII
*Eigenoscillations of Hot Electrons in an Inhomogeneous Semiconductor under Classically Strong Magnetic Field
(b) Vol. **93**, No. 1, p. 119–125 (1979).
See also 14.3.4, p. 233; 22, p. 398.

W. BRAUNE, W. SCHNEIDER, J. LEBECH, and K. SAERMARK
*Magnetoplasma Waves in Bismuth and Semimetallic BiSb Alloys at Microwave Frequencies 9 to 100 GHz
(b) Vol. **93**, No. 1, p. 269–278 (1979).
See also 13.5.4, p. 195; 19, p. 291; 21.7, p. 391.

YA. M. KSENDZOV
*The Electron Structure of 3d Metals
(b) Vol. **93**, No. 1, p. 415–425 (1979).
See also *13.1*, p. *147*; 21.1, p. 358; 21.1.1, p. 372.

E. GERLACH and H. KLUTTIG
Electron Scattering by a Layer of Dislocation Lines in a Semiconductor
(b) Vol. **93**, No. 1, p. 443–446 (1979).
See also *10.2*, p. *107*; *14.3*, p. *212*; 22, p. 398.

A. F. Barabanov and A. M. Tsvelik
*On the Renormalization of Collective Excitations in Intermediate Valence Metals
(b) Vol. **93**, No. 2, p. 817–822 (1979).
See also 7, p. 78; 21.4, p. 383.

H. Stolz and R. Zimmermann
Correlated Pairs and a Mass Action Law in Two-Component Fermi Systems. Excitons in an Electron–Hole Plasma
(b) Vol. **94**, No. 1, p. 135–146 (1979).
See also *13.5.1*, p. *186*; 22, p. 401.

C. Salustri
*Plasmons in Three-Component Systems
(b) Vol. **94**, No. 2, p. 517–524 (1979).
See also 21, p. 347; 21.1, p. 362.

G. D. Priftis, J. Boviatsis, A. Vradis, and K. Kountouris
*On the X-Ray Plasmon Scattering from Small Particles of Graphite at $k = 0$
(b) Vol. **95**, No. 1, p. 301–305 (1979).
See also 22.1, p. 408.

W. Gasser and R. Höfling
*Local-Field Corrections to the Plasmon Frequency of the Half-Filled Band Hubbard Model
(b) Vol. **95**, No. 2, p. K127–K130 (1979).
See also 14, p. 198; 18.1, p. 268.

13.5.3 Polarons

Original Papers and Short Notes

HJ. BERNHARDT
Defect Electron Centres in Lead Molybdate Crystals
(a) Vol. **52**, No. 1, p. 223–229 (1979).
See also 10.2, p. 105; *20.1*, p. *308*; *20.3*, p. *328*; 22.8.1, p. 499.

R. MAŃKA
*The Large Polaron Phase Transition in the Strong Coupling Limit
(b) Vol. **93**, No. 1, p. 53–58 (1979).
See also 18.1, p. 267.

G. BEHNKE and H. BÜTTNER
Erratum to: The Exciton Dispersion in Polar Semiconductors
(b) Vol. **93**, No. 1, p. 447 (1979).
See also *13.5.1*, p. *185*; 22.5, p. 461.

P. KUIVALAINEN, J. SINKKONEN, K. KASKI, and T. STUBB
*Bound Magnetic Polaron in Magnetic Semiconductors
(b) Vol. **94**, No. 1, p. 181–190 (1979).
See also 14.3, p. 213; *18.2*, p. *276*; 22, p. 401.

Y. KAYANUMA
*Anomalous Reduction of the Two Phonon Pinning Interaction in Magneto-Optics
(b) Vol. **94**, No. 1, p. K53–K56 (1979).
See also *20.1*, p. *315*; 22.5.1, p. 463.

R. MAŃKA
*The First-Order Phase Transition for the Molecular Polaron
(b) Vol. **94**, No. 1, p. K67–K70 (1979).

HJ. BERNHARDT
*On the Polaron Character of the Coloration Centres in $LiNbO_3$ Crystals
(a) Vol. **54**, No. 2, p. 597–603 (1979).
See also *10.2*, p. *111*; 20.1, p. 317; 20.3, p. 334; 22.8.2, p. 506.

SH. N. GIFEISMAN and YU. G. MALKOCH
The Theory of Magneto-Optical Effects in Shallow F-Centres
(b) Vol. **96**, No. 1, p. 73–90 (1979).
See also 13.4, p. 177; *18*, p. *263*; *20*, p. *302*; 22, p. 405.

Z. ŠIMŠA
Optical Properties of Charge Carriers in Magnetite and Mn Ferrites
(b) Vol. **96**, No. 2, p. 581–587 (1979).
See also *20.1*, p. *322*; 22.8.2, p. 508.

Y. LÉPINE and D. MATZ
*Mean Field Theory of a Single Fröhlich Polaron. Possible Existence of Phase Transitions
(b) Vol. **96**, No. 2, p. 797–806 (1979).
See also 22, p. 407.

13.5.4 Magnons

Review Article

W. Nolting
*Theory of Ferromagnetic Semiconductors
(b) Vol. **96**, No. 1, p. 11–54 (1979).
See also 18.2, p. 270; 22, p. 393.

Original Papers and Short Notes

E. Bałszczak, B. Westwański, and K. Skrobiś
*Elementary Excitations in Ferrimagnetic and Antiquadrupolar Biquadratic Systems ($S = 3/2$)
(b) Vol. **91**, No. 1, p. 283–290 (1979).
See also 18.2, p. 270; 18.3, p. 283.

E. Kolley and W. Kolley
*Renormalization of the Spin Wave Stiffness in Ni-Based Alloys
(b) Vol. **91**, No. 2, p. K173–K177 (1979).
See also 18.2, p. 271; 21.1, p. 355.

C. E. Patton
*Spin-Wave Instability Theory in Cubic Single Crystal Magnetic Insulators (I): General Theory
(b) Vol. **92**, No. 1, p. 211–220 (1979).
See also 18.2, p. 272; 22, p. 395.

E. Kolley and W. Kolley
*Spin Wave Stiffness Constant of Narrow-Band Ferromagnetic Alloys with Random Hopping
(b) Vol. **92**, No. 2, p. 473–481 (1979).
See also 18.2, p. 273; 21, p. 342.

C. E. Patton
*Spin-Wave Instability Theory in Cubic Single Crystal Magnetic Insulators (II): Applications
(b) Vol. **93**, No. 1, p. 63–69 (1979).
See also 18.2, p. 273; 22, p. 398.

W. Braune, W. Schneider, J. Lebech, and K. Saermark
Magnetoplasma Waves in Bismuth and Semimetallic BiSb Alloys at Microwave Frequencies 9 to 100 GHz
(b) Vol. **93**, No. 1, p. 269–278 (1979).
See also *13.5.2*, p. *191*; 19, p. 291; 21.7, p. 391.

Diep-The-Hung, J. C. S. Levy, and O. Nagai
Effects of Surface Spin Waves and Surface Anisotropy in Magnetic Thin Films at Finite Temperatures
(b) Vol. **93**, No. 1, p. 351–361 (1979).
See also *18.2.1*, p. *281*; *18.4*, p. *287*.

14. Electrical Properties. Transport Phenomena

Original Papers and Short Notes

C. Benoit, M. Galtier, and A. Montaner
*A Model for TTF-TCNQ Conductivity
(b) Vol. **91**, No. 1, p. 269–274 (1979).
See also *20*, p. *298*; 22.9, p. 509.

D. Chakravorty, A. R. Haranahalli, and D. Kumar
*Electrical Properties of V_2O_5–P_2O_5 Glasses Containing Metallic Particles
(a) Vol. **51**, No. 1, p. 275–280 (1979).
See also 22.8.1, p. 498.

W. Gasser and R. Höfling
Calculation of the Dielectric Function of the One-Dimensional Hubbard Model
(b) Vol. **92**, No. 1, p. 91–97 (1979).
See also *13.5.2*, p. *191*; 18.1, p. 266.

A. V. Gaisanyuk, K. I. Kugel, and V. A. Petrov
*Some Peculiarities of the Hall Constant in Disordered Materials: the ZrC–C System as an Example
(a) Vol. **52**, No. 1, p. K81–K83 (1979).
See also *1.2*, p. *17*; 22.3, p. 444.

S. B. Patil, H. V. Keer, and D. K. Chakrabarty
*Structural, Electrical, and Magnetic Properties in the System $Ba_xLa_{1-x}CoO_3$
(a) Vol. **52**, No. 2, p. 681–686 (1979).
See also 1.1, p. 12; *18.2*, p. *273*; 22.8, p. 493.

M. Grünewald and P. Thomas
*Percolation and Effective Medium Theory of Conductivity and Thermopower in the Hopping Transport Regime
(b) Vol. **93**, No. 1, p. K17–K20 (1979).
See also 2, p. 43; *15*, p. *250*; 22, p. 398.

A. M. Berezhkovskii
Certain Features of Semiconductor Behaviour in Strong Parallel Magnetic and Electric Fields (IV): Zener Tunnelling
(b) Vol. **94**, No. 1, p. 87–94 (1979).
See also *13*, p. *139*; 22, p. 400; 22.2.3, p. 440.

G. Mihály, A. Jánossy, J. Kürti, L. Forró, and G. Grüner
*Nonlinear Transport in $Qn(TCNQ)_2$
(b) Vol. **94**, No. 1, p. 287–296 (1979).
See also 22.9, p. 511.

S. Oseroff, R. Calvo, J. Stankiewicz, Z. Fisk, and D. C. Johnston
*Magnetic and Electric Properties of EuB_6
(b) Vol. **94**, No. 2, p. K133–K135 (1979).
See also *18.2*, p. *277*; 19, p. 294; 21.7, p. 392.

14.1 Metals. Semi-Metals

Original Papers and Short Notes

E. Heiner
An Information Theoretical Method (ITM) Applied to the Metal–Insulator Transition of the Neutral Hubbard Model
(b) Vol. **91**, No. 1, p. 87–98 (1979).
See also *13.5*, p. *181*; 14.3, p. 209.

Y. Waseda and W. A. Miller
The Structure of Liquid Mercury over a Wide Range of Temperature (–30 to 290 °C)
(b) Vol. **91**, No. 1, p. 141–146 (1979).
See also *2*, p. *40*; 15, p. 249; 21, p. 338.

K. Elk, J. Richter, and V. Christoph
*Influence of Electron Correlation on the Electrical Resistivity of Disordered Transition Metal Alloys
(b) Vol. **91**, No. 1, p. K39–K42 (1979).
See also 13, p. 136; 21, p. 339.

I. O. Bashkin, E. G. Ponyatovskii, and M. E. Kost
On the Possibility of Hydrogen Ordering in Rare-Earth Hydrides
(b) Vol. **91**, No. 2, p. 401–405 (1979).
See also *1.2*, p. *15*; 14.3, p. 210; 21.4, p. 382.

F. Goedsche, R. Richter, and G. Vojta
*Force–Force Correlation Function Approach to the Electrical Resistivity of Disordered Transition Metal Alloys
(b) Vol. **91**, No. 2, p. 581–591 (1979).
See also 13, p. 136; 21.1, p. 355.

V. Christoph
*Temperature Dependence of the Electrical Resistivity of Disordered Metallic Alloys
(b) Vol. **91**, No. 2, p. 593–603 (1979).
See also 21, p. 339.

J. Zieliński
*The Effects of Magnetic Field and Direct Interaction between Impurities in the Anderson Model. Resistivity
(b) Vol. **91**, No. 2, p. 615–625 (1979).
See also *18.1*, p. *265*; 21.1, p. 355; 21.6, p. 387.

J. Heinrichs
The Virial Theorem for the Jellium Model of Metal Surfaces
(b) Vol. **92**, No. 1, p. 185–192 (1979).
See also *13.3*, p. *158*.

O. A. Troitskii, V. I. Spitsyn, N. V. Sokolov, and V. G. Ryzhkov
*Application of High-Density Current in Plastic Working of Metals
(a) Vol. **52**, No. 1, p. 85–93 (1979).
See also 21, p. 341.

W. Schüle and R. Scholz
Properties of Vacancies and Divancancies in F.C.C. Metals
(b) Vol. **93**, No. 2, p. K119—K123 (1979).
See also 9, p. 87; *10.1*, p. *96*; 21, p. **344**; 21.1, p. 359; 21.6, p. 388.

R. Malmhäll, S. M. Bhagat, K. V. Rao, and G. Bäckström
*Transport Properties of Amorphous Ferromagnets — Hall and Ohmic Resistivities
(a) Vol. **53**, No. 2, p. 641—651 (1979).
See also *2*, p. *44*; 18.2, p. 276; 21.1, p. 360; 21.1.1, p. 374.

W. Kleinert and R. Franke
*Change of Electrical Resistivity and Stress Amplitude during Push–Pull Fatigue of Polycrystalline Nickel
(a) Vol. **53**, No. 2, p. K177—K179 (1979).
See also *10.1*, p. *97*; 21.1, p. 361.

H. Schmidt and E. Mann
Anisotropy of the Electron–Phonon Interaction in Copper (II): Scattering and Transport Properties
(b) Vol. **94**, No. 1, p. 95—105 (1979).
See also *6*, p. *70*; *13.1*, p. *149*; 21.1, p. 361.

W. Schiller, V. Christoph, and R. Seliger
*Force-Correlation Function Method Applied to the Hall Coefficient of Paramagnetic Transition Metals
(b) Vol. **94**, No. 1, p. 215—220 (1979).
See also 21.1, p. 361.

H. Sato
*Sondheimer Oscillation in Aluminium Single Crystals
(b) Vol. **94**, No. 1, p. 309—315 (1979).
See also 13.2, p. 156; 21, p. 345.

Y. D. Yao
*Temperature-Dependent Resistivity in Ni-Rich Alloys
(b) Vol. **94**, No. 2, p. K137—K140 (1979).
See also 18.1, p. 268; 18.2, p. 277; 21.1, p. 363.

J. Zieliński
*Electrical Resistivity in the Anderson Model
(b) Vol. **94**, No. 2, p. K141—K143 (1979).
See also 21, p. 347.

C. Radomsky and H. Löffler
*On the Extra-Resistivity of Coherent Precipitates
(a) Vol. **54**, No. 2, p. 615—620 (1979).
See also 10.1, p. 98; 21.1, p. 363.

A. Möbius, F. Goedsche, D. Elefant, K.-H. Berthel, and G. Vojta
*Deviations from Matthiessen's Rule in the Low-Temperature Resistivity of Iron
(b) Vol. **95**, No. 1, p. 203—214 (1979).
See also 21.1.1, p. 375.

H. RAUH
*Untersuchungen zum elektrischen Widerstand von Versetzungen in Metallen (I): Allgemeine Theorie
(Investigation of the Electrical Resistance of Dislocations in Metals (I): General Theory)
(b) Vol. **96**, No. 2, p. 627–640 (1979).
See also *10.1*, p. *101*; 13.2, p. 157; 21, p. 353.

C. MALINOWSKA-ADAMSKA and L. WOJTCZAK
*Pseudoharmonic Effect of Phonons in Electrical Resistivity
(b) Vol. **96**, No. 2, p. 697–702 (1979).
See also *6*, p. *76*; 21.1, p. 368; 21.6, p. 390.

T. OHTANI, K. KOSUGE, and S. KACHI
*Anomalous Thermal Hysteresis in Metal–Semiconductor Transition of $(Ni_{1-x}Fe_x)_{1-y}S$ $(0 \leqq x \leqq 0.30, 0 \leqq y \leqq 0.035)$
(b) Vol. **96**, No. 2, p. K69–K72 (1979).
See also *14.3*, p. *217*; 21.1, p. 368; 21.1.1, p. 377.

K. H. J. BUSCHOW and N. M. BEEKMANS
On the Thermal Stability and the Electrical Resistivity of Amorphous Th–Fe Alloys
(a) Vol. **56**, No. 2, p. 505–511 (1979).
See also *2*, p. *50*; 21.1.1, p. 377; 21.4, p. 384.

C. RADOMSKY and H. LÖFFLER
*On the Resistivity Behaviour of Spinodally Decomposing Al–Zn Alloys
(a) Vol. **56**, No. 2, p. 711–715 (1979).
See also 10.1, p. 102; 21.1, p. 368.

14.2 Superconductivity. Superconducting Materials and Devices

Original Papers and Short Notes

V. S. Postnikov, V. V. Postnikov, and V. M. Fedorov
*Superconductivity and Phonon Softening in Instable Systems
(b) Vol. 92, No. 1, p. 347–351 (1979).
See also *6*, p. *66*.

A. Kiendl
*Kondensation der supraleitenden Phase in massivem Blei
(Condensation of the Superconducting Phase in Bulk Lead)
(a) Vol. 52, No. 1, p. 131–138 (1979).

F. Dettmann and P. Pertsch
*Nb–NbO_x–PbIn Tunnel Junctions with Ultra-Thin Niobium Electrodes
(a) Vol. 52, No. 1, p. K17–K20 (1979).
See also 21, p. 341.

W. Zahn and F. Dettmann
*Critical Currents of Nb–NbO_x–Pb Tunnel Junctions with High Normal Resistance
(a) Vol. 52, No. 1, p. K57–K60 (1979).
See also 21, p. 342.

A. S. Sidorenko, E. V. Minenko, N. Ya. Fogel, R. I. Shekhter, and I. O. Kulik
*Critical Field Anisotropy in Inhomogeneous Superconducting Films
(a) Vol. 52, No. 1, p. K89–K91 (1979).
See also 21, p. 342.

C. Sułkowski
*Superconducting Transition Temperature and Electronic Properties of Indium Alloys
(b) Vol. 92, No. 2, p. K109–K111 (1979).
See also 21, p. 342.

M. Fähnle
*Irradiation Effects in Nb_3Sn — New Theoretical Results
(b) Vol. 92, No. 2, p. K127–K129 (1979).
See also *11*, p. *121*; 21, p. 342.

R. Tidecks and G. von Minnigerode
*The Influence of High-Frequency Radiation on the *U–I* Characteristics of Superconducting Tin Whiskers
(a) Vol. 52, No. 2, p. 421–426 (1979).
See also 21.7, p. 391.

A. J. Drehman and W. L. Johnson
*Effects of Thermal Relaxation on an Amorphous Superconducting Zr–Rh Alloy
(a) Vol. 52, No. 2, p. 499–507 (1979).
See also 2, p. 42; 21, p. 343.

C. S. Pande
Transmission Electron Microscopy of Radiation Damage in Nb_3Sn
(a) Vol. 52, No. 2, p. 687–696 (1979).
See also *1.4*, p. *28*; *11*, p. *121*.

V. P. Ledovskoi, N. A. Potsar, V. L. Laska, and E. Seidowski
*Superconducting Properties of Nb Films Deposited in a Quasi-Closed Volume
(a) Vol. 54, No. 1, p. K37—K40 (1979).
See also 21, p. 346.

A. E. Karkin, V. E. Arkhipov, V. A. Marchenko, and B. N. Goshchitskii
*Electrical Resistivity of V_3Si and Nb_3Sn under Neutron Radiation
(a) Vol. 54, No. 1, p. K53—K58 (1979).
See also 11, p. 123; 21, p. 346.

E. Schachinger, H. W. Weber, and H. Zlöbl
*Dynamic Destruction of Superconductivity in Pb-n% Sb Alloys
(a) Vol. 54, No. 2, p. 557—563 (1979).
See also 18, p. 263; 21, p. 347.

M. V. Sadovskii and Yu. N. Skryabin
*Superconductivity in Spin-Glasses
(b) Vol. 95, No. 1, p. 59—64 (1979).
See also 18.2, p. 278; 21, p. 348.

J. Schumann and D. Elefant
*Tunneling Investigations on V_3Si
(b) Vol. 95, No. 1, p. 91—97 (1979).
See also 21.1, p. 364.

A. Handstein, B. Pietrass, and G. Behr
Magnetic Susceptibility of the A-15 Compound System $(V_{1-x}Cr_x)_3Si$
(b) Vol. 95, No. 1, p. 131—135 (1979).
See also *18.1*, p. *268*; 21.1, p. 364.

V. G. Chudinov, N. V. Moseev, B. N. Goshchitskii, and V. I. Protasov
*On the Influence of Irradiation Type on the Formation of Radiation Defects in Nb-Based Superconductors with A-15 Structure
(a) Vol. 55, No. 1, p. 109—112 (1979).
See also *11*, p. *123*; 21, p. 348.

Yu. D. Chistyakov, S. V. Potapov, D. N. Gulidov, and H. G. Schneider
*The Effect of Residual Mechanical Stresses and Strains on the Superconductivity of Hetero-epitaxial Layers of Nb on α-Al_2O_3
(a) Vol. 55, No. 1, p. 147—152 (1979).
See also 1.5, p. 36; 22.6, p. 483.

V. L. Laska, V. P. Ledovskoi, N. A. Potsar, E. Seidowski, and A. Kh. Chizhov
*Quasi-Closed Volume as the Method of Preparing the Complex Superconducting Films
(a) Vol. 55, No. 1, p. K107—K108 (1979).
See also *3*, p. *53*; 21, p. 349.

W. Zacharko and R. Kubiak
The Anisotropy of the Upper Critical Fields in the Tetragonal Superconductor In_3Sn
(b) Vol. 96, No. 1, p. K21—K24 (1979).
See also 21, p. 352.

14.3 Semiconductors

Review Articles

O. A. Golikova
*Boron and Boron-Based Semiconductors
(a) Vol. 51, No. 1, p. 11–40 (1979).
See also *1*, p. *9*; 13.4, p. 163; 14.3.4, p. 231; 15, p. 249; 20.1, p. 304; 22, p. 393; 22.1.3, p. 428; 22.2, p. 430.

R. S. Brazis, J. K. Furdyna, and J. K. Požela
Microwave Effects in Narrow-Gap Semiconductors (I)
(a) Vol. 53, No. 1, p. 11–41 (1979).
See also *13*, p. *136*; 13.1, p. 143; 18.1, p. 265; *20*, p. *298*; 22.4.3, p. 455; 22.4.4, p. 458; 22.7, p. 488.

M. Averous
Symmetry-Induced Zero-Gap Semiconductors
(b) Vol. 95, No. 1, p. 9–28 (1979).
See also *13.1*, p. *143*; 13.4, p. 163; 22.1.3, p. 428; 22.4, p. 448.

Original Papers and Short Notes

E. Heiner
An Information Theoretical Method (ITM) Applied to the Metal–Insulator Transition of the Neutral Hubbard Model
(b) Vol. 91, No. 1, p. 87–98 (1979).
See also *13.5*, p. *181*; 14.1, p. 199.

K. Szlenk, Z. Dziuba, and R. R. Gałązka
*Contribution of Resonant States to the Conductivity of HgTe
(b) Vol. 91, No. 1, p. 255–261 (1979).
See also 13.4, p. 163; 22.4.3, p. 455.

B. G. Arnaudov, D. S. Domanevskii, and I. Y. Yanchev
*Hopping Conductivity in Heavily Doped Strongly Compensated GaAs
(b) Vol. 91, No. 1, p. 311–318 (1979).
See also 22.2.1, p. 432.

T. Satow, O. Uemura, and M. Sanzyō
*The Electricl Conductivity of Bismuth Chalcogenides in the Liquid State
(b) Vol. 91, No. 1, p. 325–330 (1979).
See also *2*, p. *40*; 22, p. 393.

C. K. Sarkar and D. Chattopadhyay
*Effect of Band Non-Parabolicity on the Microwave Magnetoconductivity in a Large Quantizing Magnetic Field
(b) Vol. 91, No. 1, p. K9–K11 (1979).
See also *13.1*, p. *144*; 22, p. 393.

R. M. Mehra, Radhey Shyam, and P. C. Mathur
*Negative Magnetoresistance of an Amorphous Se–Te–Ge Sample
(a) Vol. 51, No. 2, p. K165 – K169 (1979).
See also 22.1.1, p. 410; 22.1.3, p. 428.

M. Samoć and A. Samoć
Thermally Stimulated Current with the Participation of Two Trapping Levels
(a) Vol. 51, No. 2, p. K195 – K199 (1979).
See also *13.4*, p. *165*; 22.9, p. 510.

L. Gouskov and A. Gouskov
*Resistivity Anisotropy in the Layer Plane of GaTe
(a) Vol. 51, No. 2, p. K213 – K215 (1979).
See also 22.3, p. 444.

T. D. Dzhafarov, E. A. Skoryatina, E. S. Guds, and I. E. Maoronchuk
The Diffusion and Electric Activity of Co in Epitaxial GaAs
(a) Vol. 51, No. 2, p. K221 – K225 (1979).
See also *9*, p. *85*; 22.2.1, p. 433.

K. Takita, K. Onabe, and S. Tanaka
*Anomalous Magnetoresistance and Band Crossing in Uniaxially Compressed HgTe
(b) Vol. 92, No. 1, p. 297 – 306 (1979).
See also *13.1*, p. *145*; 22.4.3, p. 455.

J. C. Bourgoin, J. Krynicki, and B. Blanchard
*Boron Concentration and Impurity-to-Band Activation Energy in Diamond
(a) Vol. 52, No. 1, p. 293 – 298 (1979).
See also *13.4*, p *167*; 22.1, p. 408.

S. P. Singh and H. Kuzmany
*Amplification of Bleustein-Gulyaev Waves in a Piezoelectric Coated with a Semiconducting Film
(a) Vol. 52, No. 1, p. K93 – K96 (1979).
See also *7*, p. *78*.

Yu. D. Arbuzov and V. M. Evdokimov
*On the Static Conductivity of a Heavily Doped Semiconductor
(b) Vol. 92, No. 2, p. 579 – 583 (1979).
See also 22, p. 397.

J. C. Bourgoin, G. Frossati, A. Ravex, D. Thoulouze, M. Vandorpe, and B. Waksmann
*Conducting at Low Temperature in Ion-Implanted Silion
(b) Vol. 92, No. 2, p. 585 – 594 (1979).
See also 11, p. 120; 22.1.2, p. 418.

J. P. Farges and A. Brau
*Some Comments on the Hall Effect and the Thermopower in $TEA(TCNQ)_2$
(b) Vol. 92, No. 2, p. K131 – K133 (1979).
See also *15*, p. *249*; 22.9, p. 510.

L. Borisova and S. Dimitrova
*Kinetic Coefficients in Solid Solutions of the $(Ag_{x/2}Pb_{1-x}Sb_{x/2}Te$ Type
(a) Vol. **53**, No. 1, p. 403–407 (1979).
See also 15, p. 250; 22, p. 399.

L. D. Borisova
Thermoelectric Properties of Impurity Doped PbTe
(a) Vol. **53**, No. 1, p. K19–K22 (1979).
See also *15*, p. *250*; 22, p. 399.

D. V. Gitsu, F. M. Muntyanu, M. I. Onu, and A. I. Russu
*Magnetoresistance Anisotropy of $Bi_{0.91}Sb_{0.09}$ Alloy in Strong Magnetic Field
(b) Vol. **93**, No. 2, p. K127–K129 (1979).
See also 18, p. 262; 22.7, p. 489.

Yen-chu Wang
*A Simple Analytical Model for Electron Transport in GaAs
(a) Vol. **53**, No. 2, p. K113–K117 (1979).
See also 22.2.1, p. 434.

G. Weill, I. A. Smirnov, and V. N. Gurin
*Electrical Transport Properties of EuB_6 under Pressure. New Experimental Data
(a) Vol. **53**, No. 2, p. K119–K123 (1979).
See also 22, p. 400.

A. I. Merkulov, S. I. Radautsan, and V. E. Tezlevan
*Effect of Annealing on the Electrical Conductivity of In-Doped $CdCr_2Se_4$ Single Crystals
(a) Vol. **53**, No. 2, p. K129–K131 (1979).
See also 22.8, p. 494.

T. R. Jervis
*Temperature Dependence of the Electron Mobility in GaAs
(a) Vol. **53**, No. 2, p. K199–K202 (1979).
See also 22.2.1, p. 434.

V. N. Lugovoi
*Bistability and Hysteresis Phenomena in an Optical Parametric Oscillator
(b) Vol. **94**, No. 1, p. 79–86 (1979).
See also *20*, p. *300*.

M. Grünewald and P. Thomas
*A Hopping Model for Activated Charge Transport in Amorphous Silicon
(b) Vol. **94**, No. 1, p. 125–133 (1979).
See also 2, p. 45; 15, p. 250; 22.1.2, p. 420.

P. Kuivalainen, J. Sinkkonen, K. Kaski, and T. Stubb
Bound Magnetic Polaron in Magnetic Semiconductors
(b) Vol. **94**, No. 1, p. 181–190 (1979).
See also *13.5.3*, p. *193*; *18.2*, p. *276*; 22, p. 401.

G. Röpke and V. Christoph
*Derivation of DC Hopping Conductivity Formula Using the Correlation Function Method
(b) Vol. 95, No. 1, p. K15–K20 (1979).
See also 2, p. 47; 22, p. 404.

M. D. Varentsov
*Quantum Current Oscillations in γ-Irradiated n-Si with Layer Inhomogeneity
(b) Vol. 95, No. 1, p. K47–K49 (1979).
See also 11, p. 123; 22.1.2, p. 423.

B. R. Nag and M. Deb Roy
*Monte Carlo Calculation of the Low-Field Electron Mobility
(b) Vol. 95, No. 1, p. K65–K67 (1979).
See also 22.2.3, p. 441.

J. P. Farges, H. Grassi, A. Brau, and P. Dupuis
*Phase Transition in the TCNQ Complex Salt Formed from the Cation (Methyl-1-N-Ethyl Benzimidazolium)$^+$?
(a) Vol. 55, No. 1, p. 89–98 (1979).
See also 22.9, p. 512.

D. Bensahel, M. Dupuy, and J. C. Pfister
Electrical, SEM, and TEM Studies in ZnTe (I): Impurity Segregation during Short Annealings and Quenching
(a) Vol. 55, No. 1, p. 211–221 (1979).
See also *1.4*, p. *30*; *10.2*, p. *113*; 22.4.3, p. 456.

C. de Blasi, S. Galassini, C. Manfredotti, G. Micocci, L. Ruggiero, and A. Tepore
*Thermally Stimulated Currents in n-GaS
(a) Vol. 55, No. 1, p. 291–296 (1979).
See also 13.4, p. 176; 22.3, p. 445.

P. Byszewski, K. Szlenk, J. Kossut, and R. R. Gałązka
*Energy Levels at Γ-Point in $Hg_{1-x}Mn_xTe$ in Intense Magnetic Fields
(b) Vol. 95, No. 2, p. 359–367 (1979).
See also *13.4*, p. *176*; 22.4.4, p. 459.

K. Szlenk
*Temperature Dependence of Electron Concentration in Intrinsic-Like HgTe
(b) Vol. 95, No. 2, p. 445–452 (1979).
See also *13.4*, p. *176*; 22.4.3, p. 456.

E. I. Terukov, W. Reichelt, D. Ihle, and H. Oppermann
*Isotope Effect on the Verwey Transition Temperature of Magnetite
(b) Vol. 95, No. 2, p. 491–495 (1979).
See also 6, p. 74; 13.4, p. 176; 22.6, p. 484.

J. P. Estienne and J. L. Farvacque
*Experimental Evidence for Piezoelectrical Free-Carrier Scattering by Screw Dislocations in Tellurium
(b) Vol. 95, No. 2, p. 593–600 (1979)
See also *10.2*, p. *113*; 22.1.3, p. 429.

Y. Ohmura, T. Shigematsu, and R. Nakayama
Effects of Irradiation Time and Subsequent Heat Treatment on Nd:YAG Laser Annealing of High-Dose As-Implanted Silicon
(a) Vol. **56**, No. 1, p. 225–230 (1979).
See also *11*, p. *125*; 22.1.2, p. 425.

L. Tichý, M. Frumar, and J. Klikorka
*Some Electrical Properties of $GeBi_2Te_4$ Single Crystals
(a) Vol. **56**, No. 1, p. 323–326 (1979).
See also 13.1, p. 153; *15*, p. *251*; 22.8, p. 496.

T. I. Kolchenko, V. M. Lomako, and I. G. Tarutin
Introduction of Defect Clusters in GaAs by High-Energy γ-Rays
(a) Vol. **56**, No. 1, p. 387–393 (1979).
See also *11*, p. *125*; 22.2.1, p. 435.

S. Acharya and O. N. Srivastava
Electronic Bandgap Measurements of SnS_2 Polytypes
(a) Vol. **56**, No. 1, p. K1–K4 (1979).
See also 1.1, p. 14; *13.1*, p. *154*; 22, p. 406.

V. B. Sulimov
*Balance Equation for Non-Equilibrium Electrons in the Mobility Gap of a Disordered Semiconductor in the Presence of Light
(b) Vol. **96**, No. 2, p. 507–515 (1979).
See also *2*, p. *49*; 13.4, p. 179; 22, p. 406.

P. Kuivalainen, K. Kaski, S. Eränen, T. Stubb, and H. Stubb
*AC Conductivity and Dielectric Anomaly in Annealed EuSe
(b) Vol. **96**, No. 2, p. 703–712 (1979).
See also 14.4, p. 243; 22, p. 406.

J. Vindevoghel and M. Vindevoghel
*Contribution of Interband Transitions to the Conductivity of p-Si at Submillimeter Wavelengths
(b) Vol. **96**, No. 2, p. 717–725 (1979).
See also *13.1*, p. *154*; 20.1, p. 323; 22.1.2, p. 426.

O. V. Emelyanenko, G. N. Ivanova, T. S. Lagunova, D. D. Nedeoglo, G. M. Shmelev, and A. V. Simashkevich
*Scattering Mechanisms of Electrons in ZnSe Crystals with High Mobility
(b) Vol. **96**, No. 2, p. 823–833 (1979).
See also 13.5.3, p. 194; 22.4.2, p. 454.

T. Ohtani, K. Kosuge, and S. Kachi
*Anomalous Thermal Hysteresis in Metal–Semiconductor Transition of $(Ni_{1-x}Fe_x)_{1-y}S$ $(0 \leqq x \leqq 0.30, 0 \leqq y \leqq 0.035$
(b) Vol. **96**, No. 2, p. K69–K72 (1979).
See also *14.1*, p. *203*; 21.1, p. 368; 21.1.1, p. 377.

14.3.1 Films

Original Papers and Short Notes

N. Saito, H. Fujiyasu, and S. Yamada
*Doping Effects of Sb and Ga on Evaporated Amorphous Ge
(a) Vol. **51**, No. 1, p. 235–242 (1979).
See also *2*, p. *40*; 15, p. 249; 22.1.1, p. 410.

M. Andrási, G. Forgács, and A. Lörinczy
*The Effect of Potassium Contamination of Tungsten on Filament-Metallized Devices
(a) Vol. **51**, No. 2, p. 573–577 (1979).
See also 14.3.2, p. 223; 22.1.2, p. 416; 22.6, p. 478.

V. V. Mitin, Yu. A. Tkhorik, and Yu. M. Shvarts
*Anisotropy of the Transverse Magnetoresistance in p-Ge Films on GaAs
(a) Vol. **51**, No. 2, p. K133–K135 (1979).
See also 22.1.1, p. 410.

V. A. Gritsenko, E. E. Meerson, K. P. Mogilnikov, and S. P. Sinitsa
*High-Field Conductivity of Amorphous Insulator Films
(a) Vol. **52**, No. 1, p. 47–57 (1979).
See also *2*, p. *42*; 14.3.4, p, 232; 22, p. 396.

P. L. Li and N. C. Halder
*Frequency Response Characteristics of Amorphous Germanium Thin Films
(a) Vol. **52**, No. 1, p. 365–370 (1979).
See also 2, p. 42; 22.1.1, p. 411.

J. Świątek
*Thermally Stimulated Currents in Thin Polycrystalline p-Polyphenyl Layers
(a) Vol. **52**, No. 1, p. K69–K71 (1979).
See also 13.4, p. 167; 22.9, p. 510.

H. Neumann, D. Peters, B. Schumann, and G. Kühn
*Electrical Properties of $CuGaTe_2$ Epitaxial Layers
(a) Vol. **52**, No. 2, p. 559–564 (1979).
See also 13.4, p. 168; 22.8, p. 493.

W. Paul and B. Fromm
*Some Remarks on the AC Behaviour of Thin Sputtered Amorphous GeSe Films
(a) Vol. **52**, No. 2, p. K209–K212 (1979).
See also *2*, p. *43*; 22.7, p. 489.

T. V. Rao and K. L. Chopra
*Electron Transport Properties of Cu-Doped PVC Films
(a) Vol. **53**, No. 1, p. 43–50 (1979).
See also 22.9, p. 511.

I. P. Lisovskii, V. G. Litovchenko, and R. O. Litvinov
*Effect of UV Illumination on the Electrical Properties of MOS Layer Structures
(a) Vol. **53**, No. 1, p. 253–262 (1979).
See also *13.3*, p. *159*; 22.1.2, p. 419.

CHUN CHIANG
*A Theory of Threshold Switching in Amorphous Thin Films
(a) Vol. **54**, No. 2, p. 735–738 (1979).
See also *2*, p. *46*.

H. NEUMANN, E. BUTTER, P. KRAMER, H.-G. ERNST, and W. KOHL
*Simultaneous Doping of GaP with Sn and Zn
(a) Vol. **55**, No. 1, p. 99–104 (1979).
See also *13.4*, p. *176*; 22.2.2, p. 438.

J. CHUTIA and K. BARUA
*Electrical Conduction in Polyacrylonitrile Films
(a) Vol. **55**, No. 1, p. K13–K18 (1979).
See also 22.9, p. 512.

A. L. DAWAR, K. V. KRISHNA, O. P. TANEJA, and P. C. MATHUR
*Negative Magnetoresistance of Two-Phase Polycrystalline p-Type InSb Films
(a) Vol. **55**, No. 1, p. K19–K22 (1979).
See also 22.2.3, p. 441.

S. RADHAKRISHNAN and P. C. MEHENDRU
*Electronic Conduction in Dye-Sensitized Selenium Films
(a) Vol. **55**, No. 2, p. 635–638 (1979).
See also 22.1.3, p. 429.

M. TAKEUCHI
*Photoconductive Properties of Sputtered TiO_2 Films
(a) Vol. **55**, No. 2, p. 653–659 (1979).
See also *16*, p. *254*; 22.6, p. 484.

R. V. KONAKOVA, G. D. MELNIKOV, YU. A. TKHORIK, and M. YU. FILATOV
*Influence of Structural Defects and Concentration Inhomogeneities in GaAs Epitaxial Layers on Reverse *I–U* Characteristics of Schottky Barrier Diodes
(a) Vol. **55**, No. 2, p. K131–K133 (1979).
See also 1.5, p. 36; *14.3.3*, p. *230*; 22.2.1, p. 435.

J. TOUŠKOVÁ, J. TOUŠEK, E. KLIER, and R. KUŽEL
*Preparation and Basic Electrical Properties of CdTe Thick Films
(a) Vol. **56**, No. 1, p. 315–322 (1979).
See also 1.5, p. 37; 22.4.3, p. 456.

C. A. HOGARTH and M. Y. NADEEM
*The Effect of Electroforming on the Alternating Current Behaviour of Thin Amorphous Insulating Films
(a) Vol. **56**, No. 1, p. K37–K40 (1979).
See also *2*, p. *49*; 22.6.1, p. 487.

M. SUZUKI, K. MARUYAMA, M. KUMEDA, and T. SHIMIZU
*Properties of a-$Si_{100-x}Al_x$ Prepared by High Pressure RF Sputtering
(a) Vol. **56**, No. 1, p. K49–K52 (1979).
See also *2*, p. *49*; 19, p. 297; 22.1.2, p. 426.

14.3.2 Surfaces and Interfaces

Original Papers and Short Notes

A. K. Walton and J. C. Dutt
*Evidence of Surface Conduction on Magnetically Frozen-out InSb
(b) Vol. **91**, No. 1, p. K57–K60 (1979).
See also 22.2.3, p. 439.

D. V. Morgan and J. Frey
*Comments on the Modification of Schottky Barrier Height by Interfacial Oxides
(a) Vol. **51**, No. 1, p. K29–K33 (1979).
See also *14.3.4*, p. *231*; 22.1.2, p. 415; 22.2.1, p. 432; 22.2.2, p. 437.

P. A. Tove, M. P. Ali, M. M. Ibrahim, and H. Norde
*Electrical Properties of Junctions between Ge Films and Monocrystalline Silicon
(a) Vol. **51**, No. 2, p. 491–496 (1979).
See also *14.3.3*, p. *226*; 22.1.1, p. 410; 22.1.2, p. 416.

M. Andrási, G. Forgács, and A. Lörinczy
The Effect of Potassium Contamination of Tungsten on Filament-Metallized Devices
(a) Vol. **51**, No. 2, p. 573–577 (1979).
See also *14.3.1*, p. *219*; 22.1.2, p. 416; 22.6, p. 478.

V. G. Litovchenko, I. P. Lisovskii, R. O. Litvinov, H. Flietner, and W. Füssel
Specific Memory Effect in MOS Structures
(a) Vol. **51**, No. 2, p. K161–K164 (1979).
See also 11, p. 120; *13.3*, p. *158*; 22.1.2, p. 417.

R. Piccolo
A Note on the Photomagnetoelectric Effect
(a) Vol. **51**, No. 2, p. K217–K219 (1979).
See also *16*, p. *252*; 22, p. 395.

N. Lieske and R. Hezel
*Surface States on Si(111) 7 × 7 Studied by Low-Energy Electron Loss Spectroscopy
(b) Vol. **92**, No. 1, p. 159–170 (1979).
See also *13.3*, p. *158*; 22.1.2, p. 417.

H. Köhler and M. Roos
*The Influence of the Valley Splitting and Lineshape in the Density of States on the Quantum Oscillations of n-Type (100) Inverted Silicon MOSFET Surfaces
(b) Vol. **92**, No. 2, p. 489–494 (1979).
See also *13*, p. *137*; 22.1.2, p. 418.

J. Gallardo and D. Mattis
*Green's Theorem Calculation of Energy of Exciton near Hard Walls
(b) Vol. **93**, No. 1, p. 229–235 (1979).
See also *13.5.1*, p. *185*; 20.1, p. 311.

S. M. Lindsay
*Laser Light Spot Mapping of Depletion in Power Semiconductor Devices
(a) Vol. **53**, No. 1, p. 311–320 (1979).
See also 14.3.3, p. 227; 22.1.2, p. 419.

14.3.3 Devices. Juntions
(Contact Problems See 14.3.4)

Review Article

A. Rose
A Global View of Solar Energy in Rational Units
(a) Vol. **56**, No. 1, p. 11–26 (1979).
See also *16*, p. *252*.

Original Papers and Short Notes

H. Köhler and M. Roos
*Quantitative Determination of the Valley Splitting in n-Type Inverted Silicon (100) MOSFET Surfaces
(b) Vol. **91**, No. 1, p. 233–241 (1979).
See also *13.2*, p. *156*; 18, p. 261; 22.1.2, p. 415.

J. Niewodniczańska-Zawadzka and J. Grembowicz
*Phonon-Assisted Tunneling in HgCdTe–Insulator–Metal and HgCdSe–Insulator–Metal Junctions
(b) Vol. **91**, No. 1, p. K13–K15 (1979).
See also *6*, p. *64*; 22.4.4, p. 458.

A. Yu. Leiderman and P. M. Karageorgy-Alkalaev
*On the Theory of Current–Voltage Characteristics of Semiconductor Diode Structures with Strong Carrier Accumulation
(a) Vol. **51**, No. 1, p. 63–70 (1979).
See also 14.3.4, p. 231; 22.2.1, p. 432.

O. Breitenstein, B. Rheinländer, and R. Bindemann
*DLTS Capacitance Studies on Deep Level Centres in VPE GaP:N
(a) Vol. **51**, No. 1, p. 79–86 (1979).
See also *13.4*, p. *164*; 22.2.2, p. 437.

S. G. Christov and C. M. Vodenicharov
*Band Structure Effects on the Current–Voltage Characteristics of Me–GeS–Me Junctions in the Field Emission Region
(a) Vol. **51**, No. 1, p. 201–207 (1979).
See also 13.1, p. 144; 22, p. 394.

B. Joseph
*The Time Dependence of the Flatband Voltage Shift of MOS Capacitors Caused by the Injection of Minority Carriers from Si into SiO_2
(a) Vol. **51**, No. 1, p. K9–K11 (1979).
See also *14.3.4*, p. *231*; 22.1.2, p. 415; 22.6, p. 477.

A. J. Harris and H. Thomas
*On the Electroluminescent Efficiency of SiC Diode Types
(a) Vol. **51**, No. 1, p. K49–K51 (1979).
See also *20.3*, p. *327*; 22.3, p. 444.

H. F. KAPPERT, G. SIXT, and G. H. SCHWUTTKE
Minority Carrier Lifetime in Silicon after Ar^+ and Si^+ Implantation
(a) Vol. **52**, No. 2, p. 463—474 (1979).
See also *11*, p. *121*; 22.1.2, p. 418.

H. KRAUSE
*Tunnel Hopping Current and Trap Filling in Insulating Layers
(a) Vol. **52**, No. 2, p. 565—575 (1979).
See also *13.4*, p. *168*; 22.6, p. 479.

H. PFLEIDERER
*Magnetodiode Model with Gradation of Lifetime and Mobility
(a) Vol. **53**, No. 1, p. 187—194 (1979).
See also 14.3.4, p. 233; 22.1.2, p. 419.

G. KADEN and H. REIMER
*MOS-Strukturen mit sperrseitig vorgespanntem externem pn-Übergang bei linearer Gate-spannungsansteuerung (I): Impulsstromverhalten
(MOS Structures with Reserve-Biased External p–n Junction for Linear Gate Voltage Selection(I): Impulse-Current Behaviour)
(a) Vol. **53**, No. 1, p. 195—206 (1979).
See also *13.3*, p. *159*; 14.3.4, p. 234; 22, p. 399.

R. MACH, YU. V. BOCHKOV, B. SELLE, and A. N. GEORGOBIANI
*Physical Properties of Schottky Diodes on ZnS Single Crystals
(a) Vol. **53**, No. 1, p. 263—270 (1979).
See also *16*, p. *253*; 22.4.1, p. 450.

P. LEMASSON and J. GAUTRON
*Current Photoresponse of an Electrolyte–Zinc Selenide Junction
(a) Vol. **53**, No. 1, p. 303—309 (1979).
See also *16*, p. *253*; 22.4.2, p. 453.

S. M. LINDSAY
Laser Light Spot Mapping of Depletion in Power Semiconductor Devices
(a) Vol. **53**, No. 1, p. 311—320 (1979).
See also *14.3.2*, p. *223*; 22.1.2, p. 419.

G. JONES and R. A. COLLINS
*Threshold and Memory Switching in Amorphous Selenium Thin Films
(a) Vol. **53**, No. 1, p. 339—350 (1979).
See also 2, p. 43; *14.3.1*, p. *220*; 22.1.3, p. 428.

A. MÖSCHWITZER
*A Simple Threshold Model for Depletion MOST
(a) Vol. **53**, No. 1, p. K43—K45 (1979).
See also *14.3.4*, p. *234*; 22.1.2, p. 419.

S. A. HAMID
*Pyroelectric Infrared Radiation Detector Using $Ba(NO_2)_2 \cdot H_2O$
(a) Vol. **53**, No. 1, p. K75—K77 (1979).
See also 22.8, p. 494.

Hang H. Lai, Bong Ho, and G. E. Stillman
*Wave Interpretation of *nl* Product in Gunn Instability
(a) Vol. **54**, No. 2, p. 493–498 (1979).
See also 14.3.4, p. 235.

N. L. Dmitruk and O. Yu. Borkovskaya
*Photosensitivity Losses in Ideal and Real Schottky Barriers
(a) Vol. **54**, No. 2, p. 523–527 (1979).
See also *16*, p. *253*; 22.2.1, p. 435.

X. Aymerich-Humet and F. Serra-Mestres
*Zero Bias Anomaly Due to Elastic Tunneling in MIM Structures
(a) Vol. **54**, No. 2, p. 655–663 (1979).
See also 13.4, p. 174; 22.6, p. 482.

L. M. Lambert
Theory of ISC Measurements for p-Type Schottky Diodes with Deep Acceptor Impurities
(a) Vol. **54**, No. 2, p. 665–673 (1979).
See also *13.4*, p. *175*; 22, p. 403.

A. K. Owczarek
*Potential and Carrier Concentration Distributions in MOS Transistors at Punch-Through
(a) Vol. **54**, No. 2, p. 761–771 (1979).
See also *14.3.4*, p. *235*; 22.1.2, p. 423; 22.6, p. 482.

H. Köhler and M. Roos
*Evidence for a Landé-Factor $g = 2$ in n-Type Inverted Silicon MOSFET Surfaces
(b) Vol. **95**, No. 1, p. 107–115 (1979).
See also *13.2*, p. *157*; 18, p. 263; 22.1.2, p. 423.

K. Hess, J. F. Detry, and C. T. Sah
*Trapping Noise in Charge Coupled Devices
(a) Vol. **55**, No. 1, p. 243–249 (1979).
See also *13.3*, p. *161*.

N. D. Stojadinović and R. S. Popović
*A New Method for Elimination of Emitter Edge Dislocations of Silicon Planar NPN Transistors
(a) Vol. **55**, No. 1, p. 307–313 (1979).
See also *10.2*, p. *113*; 22.1.2, p. 424.

R. D. Gould and C. A. Hogarth
*The Time Dependence of Circulating and Emission Currents in Evaporated Thin Film Sandwich Structures of Au–SiO_x–Au
(a) Vol. **55**, No. 1, p. 325–332 (1979).
See also 17, p. 258; 22.6, p. 483.

N. D. Stojadinović
*Emitter Diffusion-Induced Stress Effect on Common-Emitter Current Gain of Silicon Planar Transistors
(a) Vol. **55**, No. 1, p. K89–K93 (1979).
See also 22.1.2, p. 424.

14.3.4 High-Field Phenomena, Space-Charge Effects, Inhomogeneities, Injected Carriers (Electroluminescence See 20.3; Junctions See 14.3.3)

Review Article

O. A. Golikova
Boron and Boron-Based Semiconductors
(a) Vol. **51**, No. 1, p. 11–40 (1979).
See also *1*, p. *9*; 13.4, p. 163; *14.3*, p. *209*; 15, p. 249; 20.1, p. 304; 22, p. 393; 22.1.3, p. 428; 22.2, p. 430.

Original Papers and Short Notes

O. Breitenstein and K. Unger
*Distribution of the Perturbation Potential and the Density of States in High-Doped Semiconductors
(b) Vol. **91**, No. 1, p. 243–250 (1979).
See also *13.1*, p. *143*; 22, p. 393.

A. Yu. Leiderman and P. M. Karageorgy-Alkalaev
On the Theory of Current–Voltage Characteristics of Semiconductor Diode Structures with Strong Carrier Accumulation
(a) Vol. **51**, No. 1, p. 63–70 (1979).
See also *14.3.3*, p. *225*; 22.2.1, p. 432.

A. Pincevičius, A. Sakalas, and J. Viščakas
*Double Injection into Structures with a Low-Conductivity Region (I): General Considerations
(a) Vol. **51**, No. 1, p. 303–312 (1979).
See also 22, p. 394.

B. Joseph
*The Time Dependence of the Flatband Voltage Shift of MOS Capacitors Caused by the Injection of Minority Carriers from Si into SiO_2
(a) Vol. **51**, No. 1, p. K9–K11 (1979).
See also *14.3.3*, p. *225*; 22.1.2, p. 415; 22.6, p. 477.

D. V. Morgan and J. Frey
*Comments on the Modification of Schottky Barrier Height by Interfacial Oxides
(a) Vol. **51**, No. 1, p. K29–K33 (1979).
See also *14.3.2*, p. *223*; 22.1.2, p. 415; 22.2.1, p. 432; 22.2.2, p. 437.

S. P. S. Arya and H. P. Singh
*I–U Characteristics of Al–Al_2O_3–Hg Systems
(a) Vol. **51**, No. 1, p. K67–K70 (1979).
See also *14.3.3*, p. *226*; 22.6, p. 477.

M. Asche, H. Kostial, and O. G. Sarbey
*Ionization of Low Donor Levels and Recombination of Hot Electrons in n-Si at Low Temperatures
(b) Vol. **91**, No. 2, p. 521–530 (1979).
See also *13.4*, p. *164*; 22.1.2, p. 416.

R. MURRAY and B. L. EVANS
The Electro-Absorption Edge Spectra of 2H-MoS_2
(b) Vol. **92**, No. 2, p. 551–556 (1979).
See also *13.5.1*, p. *184*; 20.1, p. 309; 22, p. 396.

Ľ. HRIVNÁK and B. ĎUROVOVÁ
*High-Field Capture of Electrons by a Screened Coulomb Attractive Center
(b) Vol. **92**, No. 2, p. K99–K101 (1979).
See also *13.4*, p. *168*; 22, p. 397.

S. GUHA, P. K. SEN, and S. GHOSH
*Parametric Instability of Acoustic Waves in Transversely Magnetised Piezoelectric Semiconductors
(a) Vol. **52**, No. 2, p. 407–414 (1979).
See also *7*, p. *78*; 22.2.3, p. 439.

L. A. DEMBEREL, A. S. POPOV, and D. B. KUSHEV
*Deep Levels in Fe-Doped GaP
(a) Vol. **52**, No. 2, p. 653–656 (1979).
See also *13.4*, p. *168*; 20.1, p. 310; 20.3, p. 329; 22.2.2, p. 437.

G. OELGART
*Avalanche Multiplication of Electron-Beam Excited Carriers in GaP p–n Junction
(a) Vol. **52**, No. 2, p. K205–K208 (1979).
See also 22.2.2, p. 437.

I. M. DYKMAN and L. A. PROKHNITSKII
Eigenoscillations of Hot Electrons in an Inhomogeneous Semiconductor under Classically Strong Magnetic Field
(b) Vol. **93**, No. 1, p. 119–125 (1979).
See also *13.5.2*, p. *191*; 22, p. 398.

M. ASCHE and H. KOSTIAL
*Peculiarities of Current–Voltage Characteristics along ⟨110⟩ in n-Si
(b) Vol. **93**, No. 1, p. K89–K92 (1979).
See also 22.1.2, p. 419.

A. SZADKOWSKI
*Field Effect Studies of the Electron Transport in β-Rhombohedral Boron
(a) Vol. **53**, No. 1, p. 95–104 (1979).
See also *13.4*, p. *170*; 22.1, p. 408.

J. GODLEWSKI and J. KALINOWSKI
*Photo-Enhanced Currents in Organic Insulators
(a) Vol. **53**, No. 1, p. 161–170 (1979).
See also *16*, p. *253*; 22.9, p. 511.

H. PFLEIDERER
*Magnetodiode Model with Gradation of Lifetime and Mobility
(a) Vol. **53**, No. 1, p. 187–194 (1979).
See also *14.3.3*, p. *227*; 22.1.2, p. 419.

S. P. S. BADWAL and H. J. DE BRUIN
*Electrode Kinetics at the Pt/Yttria-Stabilized Zirconia Interface by Complex Impedance Dispersion Analysis
(a) Vol. 54, No. 1, p. 261–270 (1979).
See also *13.3*, p. *160*; 21.6, p. 388; 22.6, p. 481.

R. S. POPOVIĆ
Analytical Approximations of Non-Degenerate Schottky Diode Current–Voltage Characteristics
(a) Vol. 54, No. 1, p. 337–348 (1979).
See also *14.3.3*, p. *228*; 22, p. 402.

V. A. ZHUKOV, V. V. SAVYALOV, B. V. KOBA, V. L. LITVINOV, V. P. SADIKOV, and N. A. UKHIN
*Electrical Properties of Silicon, Irradiated by Neutrons and γ-Quanta, in Strong Electric Fields
(a) Vol. 54, No. 1, p. K49–K52 (1979).
See also *11*, p. *122*; 22.1.2, p. 421.

P. KLEINERT
The Influence of Contact Excitons on the Electroabsorption in Disordered Semiconductors
(b) Vol. 94, No. 2, p. K145–K147 (1979).
See also 2, p. 46; *13.5.1*, p. *187*; *20.1*, p. *317*; 22, p. 403.

HANG H. LAI, BONG HO, and G. E. STILLMAN
Wave Interpretation of *nl* Product in Gunn Instability
(a) Vol. 54, No. 2, p. 493–498 (1979).
See also *14.3.3*, p. *229*.

V. F. SYNOROV, V. D. LINNIK, and B. I. SYSOEV
*The Energy Spectrum of Discrete Centres in the High Inhomogeneous Electric Field of the Space Charge Region of a Semiconductor
(a) Vol. 54, No. 2, p. 517–521 (1979).
See also *13.4*, p. *174*; 22, p. 403.

B. T. KOLOMIETS, V. CH. SHPUNT, and E. A. SMORGONSKAYA
*Characteristic Features of the Thermal Switching Process in Chalcogenide Vitreous Semiconductors
(a) Vol. 54, No. 2, p. 585–591 (1979).
See also *2*, p. *46*; 22, p. 403.

A. K. OWCZAREK
*Potential and Carrier Concentration Distributions in MOS Transistors at Punch-Through
(a) Vol. 54, No. 2, p. 761–771 (1979).
See also *14.3.3*, p. *229*; 22.1.2, p. 423; 22.6, p. 482.

Z. N. CHIGOGIDZE and N. P. KHUCHUA
*Study of Electrical Breakdown Peculiarities in High-Resistivity Compensated p-InSb
(a) Vol. 54, No. 2, p. 773–781 (1979).
See also *13.4*, p. *175*; 22.2.3, p. 441.

D. P. BHATTACHARYA and Z. S. KACHLISHVILI
*Nonradiative Multiphonon Carrier Capture by Deep Neutral Traps
(b) Vol. 95, No. 1, p. 117–130 (1979).
See also *13.4*, p. *175*; 22, p. 404.

J. Godlewski and J. Kalinowski
*Trapping Effects on Emission-Limited Current Flow in Narrow-Band Insulators
(a) Vol. **56**, No. 1, p. 293–303 (1979).
See also 22.9, p. 513.

J. Godlewski and J. Kalinowski
*On Action Spectra of Photocurrent in Organic Insulators
(a) Vol. **56**, No. 1, p. K57–K60 (1979).
See also *16*, p. *254*; 22.9, p. 513.

K. Seeger and H. Vana
Hot-Carrier Infrared Absorption in n-Type Germanium
(b) Vol. **96**, No. 2, p. 605–610 (1979).
See also *13*, p. *142*; *20.1*, p. *322*; 22.1.1, p. 413.

Doan Nhat Quang
*On the Theory of Electroreflectance for Heavily Doped Degenerate Semiconductors
(b) Vol. **96**, No. 2, p. K73–K76 (1979).
See also *20.1*, p. *323*; 22, p. 407.

D. Haack and H. G. Wagemann
*Electronic Conductivity in Highly Doped SiO_2 by Means of Ion Implantation
(a) Vol. **56**, No. 2, p. 581–591 (1979).
See also *13.4*, p. *180*; 22.6, p. 485.

R. R. Sharma and R. C. Sharma
*Submillimeter Harmonic Generation in Semiconductor Waveguides
(a) Vol. **56**, No. 2, p. 743–747 (1979).
See also 22.1.1, p. 413; 22.2.3, p. 441.

V. Šmíd, A. Tříska, and V. Janda
The Influence of Different Contact Materials on the Behaviour of the Metal–Amorphous Ge Junctions
(a) Vol. **56**, No. 2, p. K101–K104 (1979).
See also *2*, p. *50*; *14.3.2*, p. *224*; 22.1.1, p. 414.

14.4 Dielectrics

Original Papers and Short Notes

M. E. van Hulle and W. Maenhout-van der Vorst
*Influence of Pb^{2+} Ions on the Ionic Conductivity and the Space Charge Characteristics of Cubic Silver Halide Microcrystals
(a) Vol. 52, No. 1, p. 277–282 (1979).
See also *10.2*, p. *105*; 22.5.1, p. 463.

K. Hübner
Frequency- and Wavenumber-Dependent Dielectric Function in Terms of Bond Parameters
(b) Vol. 92, No. 2, p. K73–K76 (1979).
See also *4*, p. *56*; *13.1*, p. *146*; 22.6, p. 479.

S. C. Goyal and V. Kumar
*Strain Derivatives of the Static Dielectric Constant of Silver Halides
(b) Vol. 92, No. 2, p. K77–K80 (1979).
See also *12.2*, p. *134*; 22.5.1, p. 463.

D. N. Tripathy, S. S. Mandal, and B. K. Rao
*On the Study of Plasmon Dispersion beyond k_c
(b) Vol. 92, No. 2, p. K113–K116 (1979).
See also *13.5.2*, p. *191*; 21, p. 342.

K. Hübner
Chemical Bond and Related Properties of SiO_2 (VI): Electronic Structures of SiO_x
(a) Vol. 52, No. 2, p. 541–547 (1979).
See also *4*, p. *56*; *13.1*, p. *147*; 22.6, p. 479.

H. Fiedler and H.-R. Beilich
*Influence of the Substrate Temperature on the Dielectric Relaxation of Amorphous Selenium Layers in the Glass Transition Range
(a) Vol. 52, No. 2, p. K157–K159 (1979).
See also *2*, p. *43*; 22.1.3, p. 428.

R. Muccillo and L. L. Campos
*Thermally Stimulated Depolarization Currents in Ceramic ThO_2
(a) Vol. 52, No. 2, p. K183–K187 (1979).
See also *10.2*, p. *106*; 22.6, p. 479.

R. Bindemann, E. Hempel, and K. Kreher
Determination of the Dielectric Constant of GaP from S–C Donor–Acceptor Pair Spectra
(a) Vol. 52, No. 2, p. K201–K203 (1979).
See also *13.4*, p. *169*; *20.3*, p. *329*; 22.2.2, p. 437.

R. A. Bartels, J. C. Koo, and M. L. Thomas
*Temperature ad Pressure Dependence of the Dielectric Constants of CaO and SrO
(a) Vol. 52, No. 2, p. K213–K216 (1979).
See also 22.6, p. 479.

O. Litzman and V. Holý
*Some Special Models of the Spatial Theory of Dispersion
(b) Vol. 93, No. 1, p. 135–146 (1979).
See also *20.1*, p. *311*.

Y. A. BADR and R. KAMEL
*Low-Temperature Structure Changes in Nitrate Crystals
(a) Vol. 53, No. 2, p. K161–K164 (1979).
See also *1.2*, p. *18*; 22.8.1, p. 500.

F. BECHSTEDT and R. ENDERLEIN
Electronic Polarization (Relaxation) Effects in the Core Level Spectra of Semiconductors (I): General Theory of Electronic Polarization (Relaxation) in Semiconductors
(b) Vol. 94, No. 1, p. 239–248 (1979).
See also *13.1*, p. *150*; 22.1.2, p. 420; 22.2, p. 430.

S. NOOR MOHAMMAD
*On the Screening of Impurity Ions in Semiconductors with Spatially Variable Dielectric Constants
(b) Vol. 94, No. 1, p. K5–K10 (1979).
See also 13.4, p. 172; 22, p. 401.

N. RIHON
Field Emission Microscopy from Glass-Coated Tips. Observation of Fowler-Nordheim Emission Regime and of Periodic Current Oscillations
(a) Vol. 54, No. 1, p. 189–194 (1979).
See also *17.1*, p. *260*.

D. L. KIRK
*The Relationship between the Defect Structure and the Intrinsic Ionic Conductivity of Monocrystalline Sodium Chloride
(a) Vol. 54, No. 1, p. 251–259 (1979).
See also *10.2*, p. *110*; 22.5.2, p. 469.

P. K. REDDY and S. R. JAWALEKAR
*Dielectric Properties of Tantalum Oxynitride Films
(a) Vol. 54, No. 1, p. K63–K66 (1979).
See also 22, p. 402.

A. V. TURIK, N. B. SHEVCHENKO, V. G. GAVRILYACHENKO, and E. G. FESENKO
*On the Nature of Dielectric Permittivity of $PbTiO_3$ Single Crystals
(b) Vol. 94, No. 2, p. 525–528 (1979).
See also 6, p. 72; 22.8.1, p. 501.

C. ANCE
*Energie dispersive dans les semiconducteurs
(Dispersion Energy in Semiconductors)
(b) Vol. 94, No. 2, p. 557–568 (1979).
See also *13.1*, p. *151*; 22, p. 402; 22.2.4, p. 442.

G. DUGGAN, G. J. MORGAN, and A. LETTINGTON
*The Effective Screening Function in Si, Ge, and GaAs and the Effects of Exchange
(b) Vol. 94, No. 2, p. 659–665 (1979).
See also *13*, p. *140*; 22.1.1, p. 412; 22.1.2, p. 422; 22.2.1, p. 432.

E. LILLEY and J. E. STRUTT
*Bulk and Grain Boundary Ionic Conductivity in Polycrystalline β''-Alumina
(a) Vol. 54, No. 2, p. 639–650 (1979).
See also *10.2*, p. *111*; 22.6, p. 482.

M. Połomska, M. Malinowski, and H. H. Otto
*Dielectric and Electric Study of $Pb_5(Ge, Si)_3\ O_{11}$ Single Crystals
(a) Vol. 56, No. 1, p. 335–339 (1979).
See also 22.8.1, p. 503.

J. B. Wolfenstine, R. F. Fies, and T. G. Stoebe
*Lattice Defect Equilibria in KCl: Eu^{++}
(a) Vol. 56, No. 1, p. 355–364 (1979).
See also *10.2*, p. *116*; 20.1, p. 322; 22.5.2, p. 472.

B. Lalevic, M. Gvishi, and M. Shoga
Schottky Barrier Effects on the Electrical Properties of the Bi–NbO_2–Bi System
(a) Vol. 56, No. 1, p. 379–386 (1979).
See also *14.3.3*, p. *230*; 22.6, p. 485.

J.-L. Goudard, Amir A. Lakhani, and N. K. Hota
*Dielectric Constant of High and Low Temperature Phases of K-TCNQ
(a) Vol. 56, No. 1, p. K53–K56 (1979).
See also 22.9, p. 513.

F. Callens and W. Maenhout-van der Vorst
*Theoretical Treatment of the Activation Energy ΔE for Ionic Conductivity in AgX Microcrystals
(a) Vol. 56, No. 1, p. K71–K74 (1979).
See also *10.2*, p. *116*; 22.5.1, p. 464.

P. Kuivalainen, K. Kaski, S. Eränen, T. Stubb, and H. Stubb
AC Conductivity and Dielectric Anomaly in Annealed EuSe
(b) Vol. 96, No. 2, p. 703–712 (1979).
See also *14.3*, p. *217*; 22, p. 406.

W. Jamroz and J. Karniewicz
*Three-Dimensional Model of Nonlinear Magneto-Electric Susceptibilities
(b) Vol. 96, No. 2, p. K87–K90 (1979).
See also *20*, p. *303*; 22.2.1, p. 436; 22.4.3, p. 457.

V. F. Antyushin, B. I. Sysoev, and V. F. Synorov
*Identification of Alkali Metal Ions in Silicon Dioxide Films
(a) Vol. 56, No. 2, p. K91–K95 (1979).
See also *10.2*, p. *117*; 22.6, p. 486.

14.4.1 Ferroelectrics

Review Article

Original Papers and Short Notes

E. Castellucci and V. Schettino
Angular Dispersion of the Internal Phonon Modes in Ferroelectric Sodium Nitrite
(b) Vol. **91**, No. 2, p. 641–645 (1979).
See also *6*, p. *64*; *20.1*, p. *305*.

A. M. Balagurov, I. D. Dutt, Z. Gheorghiu, B. N. Savenko, and L. A. Shuvalov
*A Method for Observing the Ferroelectric Domain Structure by Using a Time-of-Flight Neutron Diffractometer
(a) Vol. **51**, No. 2, p. 367–374 (1979).
See also 22.8.1, p. 498.

Z. Czapla, T. Lis, and L. Sobczyk
*Ferroelectric Properties of NH_4HSeO_4 Crystals
(a) Vol. **51**, No. 2, p. 609–612 (1979).
See also *1.2*, p. *16*; 22.8.1, p. 499.

H. Shimizu, H. Yamada, N. Yasuda, and S. Fujimoto
*Effect of Hydrostatic Pressure on the Ferroelectric Phase Transition in NH_4IO_3
(a) Vol. **51**, No. 2, p. K121–K124 (1979).
See also 1.2, p. 16; 22.8.1, p. 499.

G. Sorge and A. Almeida
*Elastic Compliances and Ultrasonic Attenuation in $(NH_4)_2BeF_4$ in the Temperature Range from 170 to 190 K
(a) Vol. **51**, No. 2, p. K181–K184 (1979).
See also 1.2, p. 16; 22.8, p. 492.

V. N. Kashcheev
*Thermal Conductivity of Low-Dimensional Displacive-Type Ferroelectrics
(b) Vol. **92**, No. 1, p. K5–K7 (1979).
See also *8*, p. *81*.

Sh. Sh. Bashkirov and V. N. Lebedev
Isomer Shift and Quadrupolar Splitting of the Mössbauer Spectrum of Ferroelectrics
(b) Vol. **92**, No. 2, p. 397–402 (1979).
See also *5*, p. *60*; 22.8.1, p. 499.

I. H. Ismailzade, A. Yu. Kudzin, and L. Ya. Sadovskaya
*X-Ray Diffractional and Optical Investigations of Ferroelectric $SrTeO_3$
(a) Vol. **52**, No. 2, p. K105–K109 (1979).
See also 1.1, p. 12; 22.8.1, p. 499.

H. Pykacz
*Notes on the First-Order Ferroelectric Phase Transitions in an Electric Field
(a) Vol. **52**, No. 2, p. K179–K182 (1979).
See also 22.8.1, p. 499.

W. Meyer, P. Würfel, R. Munser, and G. Müller-Vogt
Kinetics of Fixation of Phase Holograms in $LiNbO_3$
(a) Vol. **53**, No. 1, p. 171–180 (1979).
See also *16*, p. *253*; 20.1, p. 312; 22.8.1, p. 500.

H. Pykacz
*Investigation of a Phase Transition in $NaH_3(SeO_3)_2$ by Measurement of the Pyroelectric Current
(a) Vol. **53**, No. 1, p. 367–372 (1979).
See also 22.8, p. 493.

K. Hołderna-Matuszkiewicz
*On the Dielectric Properties of Rochelle Salt under Hydrostatic Pressure and near the Upper Curie Point
(a) Vol. **53**, No. 1, p. K85–K89 (1979).
See also *14.4*, p. *240*; 22.8, p. 494.

M. P. Ozolinsh
*Pulse Response of the Electrooptic Effect in PMN Single Crystals
(a) Vol. **53**, No. 2, p. K169–K171 (1979).
See also *20.1*, p. *314*; 22.8.1, p. 500.

Č. Koňák
Changes of Optical Properties at an Incommensurate–Commensurate Phase Transition in $(NH_4)_2BeF_4$ Crystals
(a) Vol. **54**, No. 1, p. 99–104 (1979).
See also *1.2*, p. *18*; 20.1, p. 316; 22.8.1, p. 500.

V. N. Kashcheev
*Central Peak in Ising Model with Transverse Field
(b) Vol. **94**, No. 2, p. 731–737 (1979).

M. M. Nadoliisky, S. D. Toshev, and T. K. Vasileva
*Electric Conductivity and Dielectric Properties of Ferroelectric PZT Ceramics Doped with Cr_2O_3
(a) Vol. **54**, No. 2, p. K145–K150 (1979).
See also 22.8.1, p. 501.

G. Völkel, W. Brunner, W. Windsch, and D. Steudel
*Electron Spin-Echo (ESE) Investigations of Critical Paramagnetic Relaxation Anomalies in X-Irradiated Triglycine Sulphate (TGS:X)
(b) Vol. **95**, No. 1, p. 99–105 (1979).
See also *19*, p. *295*; 22.9, p. 512.

H.-G. Unruh, W. Eller, and G. Kirf
*Spectroscopic and Dielectric Investigations of K_2SeO_4
(a) Vol. **55**, No. 1, p. 173–180 (1979).
See also 6, p. 73; 20.1, p. 319; 22.8.1, p. 502.

T. Suski and W. Müller-Lierheim
*Anharmonic Effects in Ferroelectric $Pb_5Ge_3O_{11}$
(b) Vol. **95**, No. 2, p. 349–357 (1979).
See also *6*, p. *73*; 8, p. 83; 22.8.1, p. 502.

A. V. Turik and N. B. Shevchenko
*Dielectric Spectrum of $BaTiO_3$ Single Crystals
(b) Vol. **95**, No. 2, p. 585–592 (1979).
See also 22.8.1, p. 502.

A. E. KRUMINS and P. GUENTER
Photovoltaic Effect and Photoconductivity in Reduced Potassium Niobate Crystals
(a) Vol. 55, No. 2, p. K185 – K189 (1979).
See also *16*, p. *254*; 22.8.1, p. 503.

M. FÄHNLE, B. BARBARA, and H. KRONMÜLLER
Thermally Activated Domain Wall Motion and Coercive Field in Materials with Extremely Narrow Domain Walls
(b) Vol. 96, No. 1, p. 343 – 350 (1979).
See also *1.4*, p. *32*; 18, p. 264; 21, p. 351.

A. JAŚKIEWICZ and J. PRZESŁAWSKI
*Exponential Switching of the TGS Crystal Composed of Regular, Partially Stabilized Domains
(a) Vol. 56, No. 1, p. 365 – 370 (1979).

R. BLINC, M. BURGAR, J. SLAK, V. RUTAR, and F. MILIA
^{13}C NMR Study of the Incommensurate Phase Transition in $[N(CH_3)_4]_2ZnCl_4$
(a) Vol. 56, No. 1, p. K65 – K69 (1979).
See also *1.2*, p. *21*; *19*, p. *297*; 22.9, p. 514.

G. K. ROY and A. N. DAS
*Mode Damping by Electrostrictive Interaction in a Uniaxial Ferroelectric
(b) Vol. 96, No. 2, p. 727 – 734 (1979).
See also *6*, p. *76*.

W. J. KULIŃSKI, S. K. HOFFMANN, and A. B. WIĘCKOWSKI
*EPR Studies of VO^{2+} Ions in Rochelle Salt Crystal
(b) Vol. 96, No. 2, p. 745 – 755 (1979).
See also *19*, p. *297*.

I. SAVATINOVA, E. ANACHKOVA, J. ROHLEDER, and B. JAKUBOVSKI
C≡N Vibrations in Potassium Ferrocyanide Trihydrate
(b) Vol. 96, No. 2, p. K97 – K101 (1979).
See also *6*, p. *77*; *20.1*, p. *323*; 22.8, p. 497.

P. E. TOMASZEWSKI and A. PIETRASZKO
Thermal Expansion of Lithium Ammonium Sulphate
(a) Vol. 56, No. 2, p. 467 – 472 (1979).
See also *1.2*, p. *21*; 22.8.1, p. 503.

G. LUTHER and H.-G. UNRUH
*Dielectric Properties of Ferrielectric Ammonium Sulphate in the Microwave Region
(a) Vol. 56, No. 2, p. 597 – 606 (1979).
See also 22.8.1, p. 504.

G. KLEER, H. SCHMITT, and A. KLÖPPERPIEPER
Investigation of Stoichiometry Defects in Ammonium Rubidium Sulphate Single Crystals by Electron Probe Microanalysis
(a) Vol. 56, No. 2, p. K75 – K77 (1979).
See also *1*, p. *10*; 22.8.1, p. 504.

15. Thermoelectric and Thermomagnetic Properties

Review Article

O. A. Golikova
Boron and Boron-Based Semiconductors
(a) Vol. **51**, No. 1, p. 11–40 (1979).
See also *1*, p. *9*; 13.4, p. 163; *14.3*, p. *209*; 14.3.4, p. 231; 20.1, p. 304; 22, p. 393; 22.1.3, p. 428; 22.2, p. 430.

Original Papers and Short Notes

D. Ihle and Th. Eifrig
*Thermoelectric Power near the Atomic Limit of the Hubbard Model
(b) Vol. **91**, No. 1, p. 135–140 (1979).
See also 13.1, p. 143.

Y. Waseda and W. A. Miller
The Structure of Liquid Mercury over a Wide Range of Temperature (–30 to 290 °C)
(b) Vol. **91**, No. 1, p. 141–146 (1979).
See also *2*, p. *40*; 14.1, p. 199; 21, p. 338.

D. Ihle
*On the Thermoelectric Power of Superionic Conductors
(b) Vol. **91**, No. 1, p. K49–K52 (1979).
See also 22, p. 393.

N. Saito, H. Fujiyasu, and S. Yamada
Doping Effects of Sb and Ga on Evaporated Amorphous Ge
(a) Vol. **51**, No. 1, p. 235–242 (1979).
See also *2*, p. *40*; *14.3.1*, p. *219*; 22.1.1, p. 410.

L. G. Khvostantsev and N. A. Nikolaev
*Behaviour of the Thermopower of Cerium under Pressure up to 7500 MPa
(a) Vol. **51**, No. 1, p. K57–K61 (1979).
See also 21.4, p. 382.

E. Shatkovskii, A. Tamaševičius, and E. Matulionis
*Photomagnetoelectric Effect in Strongly Excited $Cd_xHg_{1-x}Te$
(a) Vol. **51**, No. 1, p. K97–K99 (1979).
See also *16*, p. *252*; 22.4.4, p. 458.

H. Süssmann und W. Heiliger
*Zur Anisotropie der thermoelektrischen Eigenschaften von p-$Sb_xBi_{2-x}Te_3$
(On the Anisotropy of the Thermoelectric Properties of p-$Sb_xBi_{2-x}Te_3$)
(a) Vol. **51**, No. 2, p. K201–K206 (1979).
See also 22.7.1, p. 490.

J. P. Farges and A. Brau
*Some Comments on the Hall Effect and the Thermopower in $TEA(TCNQ)_2$
(b) Vol. **92**, No. 2, p. K131–K133 (1979).
See also *14.3*, p. *211*; 22.9, p. 510.

B. VISWANATHAN, A. RAMANAN, and T. K. VARADARAJAN
*Magnetic Properties of $HoFe_2O_4$
(a) Vol. 55, No. 1, p. K87–K88 (1979).
See also *18.3*, p. *284*; 22.8, p. 496.

K. N. KHANNA and P. K. SHARMA
A Pseudopotential Study of the Electronic Properties of Simple Metals
(b) Vol. 96, No. 1, p. 111–116 (1979).
See also *2*, p. *48*; *14.1*, p. *202*; 21.1, p. 366.

M. A. SHAIKH
Electrical and Thermoelectric Properties of Dibenzotetrathiofulvalinium–Tetracyanoquinodimethane
(b) Vol. 96, No. 1, p. K7–K9 (1979).
See also *14.3*, p. *216*; 22.9, p. 513.

L. TICHÝ, M. FRUMAR, and J. KLIKORKA
*Some Electrical Properties of $GeBi_2Te_4$ Single Crystals
(a) Vol. 56, No. 1, p. 323–326 (1979).
See also 13.1, p. 153; *14.3*, p. *217*; 22.8, p. 496.

W. KUBITZKI and B. HOFFMANN
*Application of an AC Thermopower Measurement to Electronic Ceramics
(a) Vol. 56, No. 2, p. 573–579 (1979).
See also *14.3*, p. *218*; 22.8.1, p. 503.

16. Photoconductivity. Photovoltaic Effects

Review Article

Original Papers and Short Notes

F. Fischer
A Simple Model for Photo-Induced Polarization
(a) Vol. **52**, No. 1, p. 189 – 194 (1979).
See also *10.2*, p. *105*; 13.4, p. 167; 14.4, p. 238; 22.5.2, p. 466.

P. R. Herczfeld and M. J. Fisher
*Utilization of Noise Theory and Measurements in Modelling Electronic Processes and Transport Phenomena in Photoconductors
(a) Vol. **52**, No. 1, p. 357 – 363 (1979).
See also 22.4.1, p. 450; 22.4.2, p. 453.

R. Baubinas, F. Senulis, and J. Vaitkus
*Temperature Dependence of the Optical Ionization Energy of Deep Local Levels in CdSe Single Crystals
(a) Vol. **52**, No. 2, p. K143 – K146 (1979).
See also *13.4*, p. *169*; 22.4.2, p. 453.

J. Godlewski and J. Kalinowski
*Photo-Enhanced Currents in Organic Insulators
(a) Vol. **53**, No. 1, p. 161 – 170 (1979).
See also *14.3.4*, p. *233*; 22.9, p. 511.

W. Meyer, P. Würfel, R. Munser, and G. Müller-Vogt
*Kinetics of Fixation of Phase Holograms in $LiNbO_3$
(a) Vol. **53**, No. 1, p. 171 – 180 (1979).
See also 14.4.1, p. 245; 20.1, p. 312; 22.8.1, p. 500.

R. Mach, Yu. V. Bochkov, B. Selle, and A. N. Georgobiani
*Physical Properties of Schottky Diodes on ZnS Single Crystals
(a) Vol. **53**, No. 1, p. 263 – 270 (1979).
See also *14.3.3*, p. *227*; 22.4.1, p. 450.

P. Lemasson and J. Gautron
*Current Photoresponse of an Electrolyte–Zinc Selenide Junction
(a) Vol. **53**, No. 1, p. 303 – 309 (1979).
See also *14.3.3*, p. *227*; 22.4.2, p. 453.

V. M. Fridkin, R. Nitsche, N. Korchagina, N. A. Kosonogov, R. Magomadov, A. I. Rodin, and K. A. Verkhovskaya
*Photoferroelastic Phenomena in Sb_5O_7I Crystals
(a) Vol. **54**, No. 1, p. 231 – 237 (1979).
See also 12.2, p. 135; 22.8, p. 495.

N. L. Dmitruk and O. Yu. Borkovskaya
*Photosensitivity Losses in Ideal and Real Schottky Barriers
(a) Vol. **54**, No. 2, p. 523 – 527 (1979).
See also *14.3.3*, p. *229*; 22.2.1, p. 435.

D. N. Popov and K. I. Terzijsky
*Electron Trapping of α_A- and F_A-Centers in KCl:Na
(b) Vol. **95**, No. 1, p. K43 – K46 (1979).
See also *10.2*, p. *112*; 13.4, p. 175; 22.5.2, p. 470.

J. A. Kaladė, E. A. Montrimas, and J. P. Varaneckas
*Radiographic Contrast of Selenium Electroradiographic Layers
(a) Vol. 55, No. 1, p. 261—269 (1979).
See also 22.1.3, p. 429.

T. X. Hoai, L. V. Hong, and N. V. Vuong
*On the Recombination Barrier in Highly Compensated GaAs
(a) Vol. 55, No. 1, p. K99—K102 (1979).
See also *14.3.4*, p. *236*; 22.2.1, p. 435.

M. Takeuchi
*Photoconductive Properties of Sputtered TiO_2 Films
(a) Vol. 55, No. 2, p. 653—659 (1979).
See also *14.3.1*, p. *221*; 22.6, p. 484.

A. E. Krumins and P. Guenter
*Photovoltaic Effect and Photoconductivity in Reduced Potassium Niobate Crystals
(a) Vol. 55, No. 2, p. K185—K189 (1979).
See also 14.4.1, p. 247; 22.8.1, p. 503.

J. D. Wiley, S. Pennington, and E. Schönherr
*Anisotropy of the Intrinsic Photoconductivity of GeS
(b) Vol. 96, No. 1, p. K43—K46 (1979).
See also 22, p. 406.

D. Valentovič, J. Červenák, S. Luby, M. L. Aldea, and T. Boţilă
Some Non-Equilibrium Phenomena in Sputtered CdTe Thin Films
(a) Vol. 56, No. 1, p. 341—347 (1979).
See also *1.5*, p. *37*; 13.4, p. 179; 22.4.3, p. 457.

J. Godlewski and J. Kalinowski
*On Action Spectra of Photocurrent in Organic Insulators
(a) Vol. 56, No. 1, p. K57—K60 (1979).
See also *14.3.4*, p. *237*; 22.9, p. 513.

N. N. Sirota and E. Z. Katsnelson
*Photogalvanomagnetic Effects in Ferrites
(a) Vol. 56, No. 1, p. K61—K64 (1979).
See also 18.3, p. 285; 22.8.2, p. 508.

W. Pickin
*The Kinetics of Electron Transitions for Donors with Excited States
(b) Vol. 96, No. 2, p. 617—626 (1979).
See also *13.4*, p. *179*; 22.1.1, p. 413.

R. Basset and P. Viktorovitch
*Spectral Response of Metal–Amorphous Silicon Barriers and Hole Drift Mobility
(a) Vol. 56, No. 2, p. 495—499 (1979).
See also 2, p. 50; *14.3.3*, p. *230*; 22.1.2, p. 426.

A. Coret and A. Haydar
*Photoconductivity in the 1S Excitonic Line of Cu_2O. Shape of the Response
(a) Vol. **56**, No. 2, p. 531–534 (1979).
See also *13.5.1*, p. *190*; 22.6, p. 485.

17. Emission of Electrons and Ions

Original Papers and Short Notes

B. A. Orłowski and J. M. Langer
Determination of the CdF_2 Valence Band Structure by Photoemission Measurement
(b) Vol. **91**, No. 1, p. K53 – K56 (1979).
See also *13.1*, p. *144*; 13.5.1, p. 183; 22.5, p. 460.

V. Siegel, W. Rasp, and H.-H. Kirchner
*Adsorption-Induced Exoelectron Maximum of BeO Ceramics
(a) Vol. **51**, No. 1, p. 123 – 130 (1979).
See also 22.6, p. 477.

H. Nanto and M. Kawanishi
*Effect of Reanodization on Exoelectron Emission of Anodic Ta_2O_5 Films
(a) Vol. **51**, No. 1, p. 191 – 196 (1979).
See also 22.6, p. 477.

U. Mizutani, T. Kondow, and M. Uda
Experimental Observation of the Fermi Surface–Brillouin Zone Interaction in Pure Zinc by Means of X-Ray Photoelectron Spectroscopy
(b) Vol. **91**, No. 2, p. 693 – 696 (1979).
See also *13.1*, p. *144*; 21.1, p. 355.

H. Kirihata and J. Akutsu
*Formation of Exoelectron Emission Centers by Slow Electron Impact
(a) Vol. **51**, No. 2, p. K109 – K112 (1979).
See also 13.3, p. 158; 21.1.1, p. 371.

H. Kirihata
*Effects of Gas Pressure and Sample Temperature during Slow Electron Excitation on Exoelectron Emission
(a) Vol. **52**, No. 1, p. K97 – K99 (1979).
See also 13.3, p. 159; 21.1.1, p. 372.

G. Rudlof, J. Becherer, and H. Glaefeke
*A Remark on Determining Activation Energies by Fractional Glow Technique (FGT) in Case of Complex Glow Spectra – A Procedure of Improving This Method
(a) Vol. **52**, No. 2, p. K137 – K141 (1979).
See also *13.4*, p. *169*; 22.6, p. 479.

A. Szász and Á. Tichy-Rács
*Photoemission Investigation of Ni near the Curie Temperature
(b) Vol. **93**, No. 1, p. K13 – K15 (1979).
See also 18.1, p. 267; 18.2, p. 274; 21.1, p. 358.

U. Berg, R. Fuhrmann, and O. Brümmer
*Angular-Dependent X-Ray Photoemission Study on the Surface of GaAs (111) and (111) Single Crystals
(a) Vol. **53**, No. 1, p. K1 – K4 (1979).
See also *1.6*, p. *38*; 22.2.1, p. 433.

B. RAUSCHENBACH and G. BLASEK
Determination of Very Low Diffusion Coefficients in Glass with Means of Secondary-Ion Mass Spectrometry
(a) Vol. **53**, No. 1, p. K11–K13 (1979).
See also *2*, p. *44*; *9*, p. *87*; 22, p. 399.

A. BERNDTSSON, R. NYHOLM, N. MÅRTENSSON, R. NILSSON, and J. HEDMAN
*ESCA Studies of Rhenium and Osmium
(b) Vol. **93**, No. 2, p. K103–K105 (1979).
See also *13.1*, p. *149*; 21, p. 344.

G. M. MLADENOV and M. BRAUN
*Ion Bombardment Induced Line Radiation from Sputtered Beryllium
(a) Vol. **53**, No. 2, p. 631–640 (1979).
See also 21.3, p. 381.

V. T. CHEREPIN, M. A. VASILEV, and O. A. YAKUBTSOV
Appearance Potential Spectroscopy Study of the Oxidation of CoNi Alloy Surfaces
(a) Vol. **53**, No. 2, p. K207–K210 (1979).
See also *13.3*, p. *159*; 21.1, p. 361.

S. ARNOLD, M. POPE, and T. K. T. HSIEH
*Double Quantum External Photoelectric Effect in Crystalline Tetracene
(b) Vol. **94**, No. 1, p. 263–272 (1979).
See also 13.4, p. 172; 22.9, p. 511.

G. M. RENFRO and H. J. FISCHBECK
*Optically Stimulated Exoelectron Emission from Potassium Chloride Single Crystals
(a) Vol. **54**, No. 1, p. 85–92 (1979).
See also 10.2, p. 109; 22.5.2, p. 468.

K. BERNDT, U. MARX, and O. BRÜMMER
Photoelectron Spectra of Core Electrons from Palladium–Silver Alloys
(b) Vol. **94**, No. 2, p. 541–548 (1979).
See also *13*, p. *139*; 21, p. 347; 21.6, p. 388.

J. F. MCGILP and A. W. PARKE
*Angle-Resolved UPS of GaS, GaSe, and $GaS_{0.6}Se_{0.4}$
(b) Vol. **94**, No. 2, p. 685–689 (1979).
See also *13.1*, p. *151*; 22.3, p. 445; 22.3.1, p. 447.

M. BÜCHEL and H. LÜTH
Defect-Induced Surface States on Cleaved GaAs (110) Surfaces Measured in UV Photoemission
(b) Vol. **94**, No. 2, p. K155–K158 (1979).
See also *13.3*, p. *160*; 22.2.1, p. 435.

V. KORTOV, V. ISAKOV, A. GAPRINDOSHVILY, H.-J. FITTING, H. GLAEFEKE und W. WILD
*Untersuchung des Austritts von Exoelektronen aus geladenen Isolatorschichten mit Hilfe des Monte-Carlo-Verfahrens
(Study of the Exoelectron Emission from Charged Insulating Layers by Means of the Monte Carlo Method)
(a) Vol. **54**, No. 2, p. 633–638 (1979).
See also 22.6, p. 482.

L. Wolber

*Simultaneous Measurement of Thermally Stimulated Exoelectron Emission and Surface Charging of $BaSO_4$
(a) Vol. 56, No. 1, p. 207–212 (1979).
See also 13.3, p. 162; 22.8.1, p. 503.

H.-J. Hunger and L. Küchler

*Measurements of the Electron Backscattering Coefficient for Quantitative EPMA in the Energy Range of 4 to 40 keV
(a) Vol. 56, No. 1, p. K45–K48 (1979).
See also 21, p. 352.

17.1 Field Emission Microscope Investigations

Original Papers and Short Notes

N. Rihon
*Field Emission Microscopy from Glass-Coated Tips. Observation of Fowler-Nordheim Emission Regime and of Periodic Current Oscillations
(a) Vol. **54**, No. 1, p. 189–194 (1979).
See also 14.4, p. 241.

Ch. Kleint
Stationary Stochastic Adparticle Flip-Flop as a Precursor to Surface Diffusion
(a) Vol. **55**, No. 2, p. 447–456 (1979).
See also *1.6*, p. *39*; *9*, p. *88*; 22.9, p. 513.

F. Okuyama
*A Novel Technique for Preparing Tungsten Field Cathodes with [001] Orientation
(a) Vol. **55**, No. 2, p. 793–800 (1979).
See also 21, p. 351.

F. Okuyama
Oriented Structure Transformation from W_2C to W as Revealed by Field Electron and Transmission Electron Microscopy
(a) Vol. **56**, No. 1, p. 285–292 (1979).
See also *1.2*, p. *21*; 21, p. 352; 22, p. 406.

18. Magnetic Properties

Original Papers and Short Notes

H. Köhler and M. Roos
Quantitative Determination of the Valley Splitting in n-Type Inverted Silicon (100) MOSFET Surfaces
(b) Vol. **91**, No. 1, p. 233–241 (1979).
See also *13.2*, p. *156*; *14.3.3*, p. *225*; 22.1.2, p. 415.

J. Schoenes, E. Kaldis, W. Thöni, and P. Wachter
*Optical, Magnetic, and Magnetooptical Properties of the Europium Silicate Glass $Eu_{0.14}Si_{0.31}O_{0.55}$
(a) Vol. **51**, No. 1, p. 173–181 (1979).
See also 2, p. 40; *20.1*, p. *305*; 22.6, p. 477.

N. E. Alekseevskii, G. Wolf, K. Bohmhammel, N. M. Dobrovolskii, and C. Hohlfeld
*Singularities in the Magnetic and Superconducting Properties of $YbFe_xMo_6S_8$
(a) Vol. **51**, No. 2, p. 399–405 (1979).
See also 8, p. 80; *14.2*, p. *204*; 22.8, p. 492 .

K. Sokalski
*Linear Ising Chain in the Molecular Field
(b) Vol. **92**, No. 1, p. K15–K17 (1979).

T. A. Abramovskaya and M. D. Zviadadze
*On Effects of the Dipole Reservoir in Radiofrequency Discrete Saturation
(a) Vol. **52**, No. 1, p. 111–120 (1979).
See also *13*, p. *137*; 19, p. 291.

S. Mager, E. Wieser, T. Zemčík, O. Schneeweiss, P. N. Stetsenko, and V. V. Surikov
*Investigations of the Magnetic Moments in $(Fe_{1-x}Mn_x)_3Al$ Alloys
(a) Vol. **52**, No. 1, p. 249–258 (1979).
See also 21.1, p. 357; 21.1.1, p. 371.

J. Morkowski and A. M. Oleś
On the Origin of Localized Magnetic States in Transition Metal Alloys (I): Wolff Model
(b) Vol. **92**, No. 2, p. 525–536 (1979).
See also *13.4*, p. *167*; 21.1, p. 357.

A. M. Berezhkovskii and A. A. Ovchinnikov
Certain Features of Semiconductor Behaviour in Strong Parallel Magnetic and Electric Fields (I): Effect of Electric Field on the Magnetic Moment of a Specimen
(b) Vol. **92**, No. 2, p. 573–578 (1979).
See also *13*, p. *138*; 22, p. 396.

L. Palermo and X. A. da Silva
Spin Polarization of a Magnetic Electron Gas Induced by a van Vleck Ion
(b) Vol. **92**, No. 2, p. 619–626 (1979).
See also *13*, p. *138*.

A. M. Oleś and J. Morkowski
*On the Origin of Localized Magnetic States in Transition Metal Alloys (II): Doubly Degenerate Orbitals
(b) Vol. **93**, No. 1, p. 243–253 (1979).
See also *13.4*, p. *169*; 21.1, p. 358.

C. ALDRICH and R. L. GREENE
*Hydrogen-Like Systems in Arbitrary Magnetic Fields. A Variational Approach
(b) Vol. **93**, No. 1, p. 343—350 (1979).
See also *13*, p. *138*.

R. C. THOMPSON, V. K. GUJRAL, H. J. WAGNER, and C. F. SCHWERDTFEGER
*Magnetic and Electric Properties of $Mn(TCNQ)_2$ and Its Hydrates
(a) Vol. **53**, No. 1, p. 181—186 (1979).
See also *14.3*, p. *212*; 19, p. 292; 22.9, p. 511.

H. U. HABERMEIER, W. KLEIN, R. AOKI, and H. KRONMÜLLER
*Magneto-Optical Investigation of Flux Penetration in Plastically Deformed Niobium Monocrystals
(a) Vol. **53**, No. 1, p. 225—235 (1979).
See also 10.1, p. 95; *20*, p. *300*; 21, p. 344.

K. H. J. BUSCHOW, U. GOEBEL, and E. DORMANN
Valence Instabilities in $CeSn_3$ and $YbAl_3$
(b) Vol. **93**, No. 2, p. 607—615 (1979).
See also *13*, p. *139*; 19, p. 292; 22.7, p. 489.

E. RÖBER, K. HACKSTEIN, H. COUFAL, and S. SOTIER
*Magnetic Sucseptibility of Liquid $Na_{1-x}Bi_x$ Alloys
(b) Vol. **93**, No. 2, p. K99—K102 (1979).
See also *2*, p. *44*; 22, p. 400.

D. V. GITSU, F. M. MUNTYANU, M. I. ONU, and A. I. RUSSU
Magnetoresistance Anisotropy of $Bi_{0.91}Sb_{0.09}$ Alloy in Strong Magnetic Field
(b) Vol. **93**, No. 2, p. K127—K129 (1979).
See also *14.3*, p. *213*; 22.7, p. 489.

A. V. GOLTSEV, I. G. LANG, and S. T. PAVLOV
Electron–Phonon and Electron Raman Scattering in Polar Semiconductors in High Magnetic Fields
(b) Vol. **94**, No. 1, p. 37—50 (1979).
See also *6*, p. *70*; *20.1*, p. *314*; 22, p. 400.

G. BASTARD, C. RIGAUX, Y. COUDER, H. THOME, and A. MYCIELSKI
*Effect of Exchange on the Impurity Magnetoabsorption in Zero-Gap $Hg_{1-k}Mn_kTe$ Mixed Crystals
(b) Vol. **94**, No. 1, p. 205—210 (1979).
See also 13.4, p. 172; *20.1*, p. *314*; 22.4.4, p. 458.

V. K. ARORA and H. N. SPECTOR
*Quantum-Limit Magnetoresistance for Impurity Scattering
(b) Vol. **94**, No. 1, p. 323—336 (1979).
See also 6, p. 70; 13.4, p. 172; *14.3*, p. *214*; 22.2.3, p. 440.

M. ROOS and H. KÖHLER
Investigation of the Reproducibility of Electronic Properties in n-Type Inverted Silicon MOSFET Surfaces
(b) Vol. **94**, No. 2, p. 469—475 (1979).
See also *13.3*, p. *160*; *14.3.3*, p. *228*; 22.1.2, p. 422.

G. I. Georgiev and M. P. Sandalski
*One-Dimensional Ising Model with an Infinite Radius of Interaction in a Transverse Magnetic Field. Some Thermodynamic Properties of the Rigorously Solvable Part
(b) Vol. **94**, No. 2, p. 477–486 (1979).
See also 8, p. 82; *13*, p. *139*.

P. Vorderwisch, S. Hautecler, and J. B. Suck
*Inelastic Neutron Scattering by Cerium Hydrides (IV): Crystal Field
(b) Vol. **94**, No. 2, p. 569–576 (1979).
See also *13*, p. *140*; 21.4, p. 384.

E. Schachinger, H. W. Weber, and H. Zlöbl
Dynamic Destruction of Superconductivity in Pb–n% Sb Alloys
(a) Vol. **54**, No. 2, p. 557–563 (1979).
See also *14.2*, p. *207*; 21, p. 347.

H. Köhler and M. Roos
*Evidence for a Landé-Factor $g = 2$ in n-Type Inverted Silicon MOSFET Surfaces
(b) Vol. **95**, No. 1, p. 107–115 (1979).
See also *13.2*, p. *157*; *14.3.3*, p. *229*; 22.1.2, p. 423.

A. Singh and K. N. Shrivastava
Optical-Acoustic Two-Phonon Relaxation in Spin Systems
(b) Vol. **95**, No. 1, p. 273–277 (1979).
See also *6*, p. *73*.

J. Kötzler and W. Scheithe
*Determination of Crystalline Field Interaction in $GdCl_3$
(b) Vol. **95**, No. 1, p. K1–K4 (1979).
See also *13*, p. *140*; 22.5, p. 462.

A. Freudenhammer and D. Deutges
On the $1/T$-Behaviour of the Specific Heat of a RKKY Ising Spin Glass
(b) Vol. **95**, No. 1, p. K55–K59 (1979).
See also 2, p. 47; *8*, p. *83*; 21, p. 348.

N. Shamir, M. Melamud, and H. Shaked
Crystalline Electric Field Studies of Uranium Compounds Using Neutron Inelastic Scattering at the NRCN
(a) Vol. **55**, No. 1, p. K49–K52 (1979).
See also *13*, p. *140*; 21.5, p. 386; 22.1.2, p. 424; 22.5, p. 462; 22.6, p. 483.

Y. Obi, H. Morita, and H. Fujimori
*Exchange Anisotropy in Ni–Mn–Si–B Amorphous Alloys
(a) Vol. **55**, No. 1, p. K67–K70 (1979).
See also *2*, p. *48*; 21.1, p. 365.

M. Kiwi, E. Pestana, and R. Ramírez
Magnitude of Localized Magnetic Moments in Metals
(b) Vol. **95**, No. 2, p. 497–502 (1979).
See also *13.4*, p. *176*; 21, p. 350.

Sh. N. Gifeisman and Yu. G. Malkoch
*The Theory of Magneto-Optical Effects in Shallow F-Centres
(b) Vol. **96**, No. 1, p. 73–90 (1979).
See also 13.4, p. 177; 13.5.3, p. 193; *20*, p. *302*; 22, p. 405.

18.1 Paramagnetic Properties

Review Articles

R. S. Brazis, J. K. Furdyna, and J. K. Poželа
*Microwave Effects in Narrow-Gap Semiconductors (I)
(a) Vol. **53**, No. 1, p. 11–41 (1979).
See also *13*, p. *136*; 13.1, p. 143; 14.3, p. 209; *20*, p. *298*; 22.4.3, p. 455; 22.4.4, p. 458; 22.7, p. 488.

R. S. Brazis, J. K. Furdyna, and J. K. Poželа
Microwave Effects in Narrow-Gap Semiconductors (II)
(a) Vol. **54**, No. 1, p. 11–27 (1979).
See also 13, p. 136; *19*, p. *289*; 20, p. 298; 22.2.1, p. 432; 22.2.3, p. 439; 22.2.4, p. 442; 22.4.3, p. 455; 22.4.4, p. 458.

Original Papers and Short Notes

J. Zieliński and B. Kozarzewski
*Second-Order Self-Energy for the Anderson Model. Analytical Results
(b) Vol. **91**, No. 1, p. 355–360 (1979).
See also 21, p. 338.

J. Ilarraz and A. del Moral
*Magnetic Ordering Temperatures in Rare Earth–Fe_2 Laves Phases Determined by DTA
(a) Vol. **51**, No. 1, p. K41–K43 (1979).
See also *18.3*, p. *283*; 21.1.1, p. 370; 21.4, p. 382.

N. V. Fedoseeva, T. A. Velikanova, and A. G. Zvegintsev
*High-Pressure Cubic Phase of $RbMnCl_3$ – Magnetic Properties
(a) Vol. **51**, No. 1, p. K93–K96 (1979).
See also *18.4*, p. *286*; 22.8, p. 491.

G. Schanz, V. Müller, and E. J. Unterhorst
*Nuclear-Acoustic-Resonance Determination of the Core Polarization Contribution to the Knight Shift and Hyperfine Fields in Single-Crystal Tantalum
(b) Vol. **91**, No. 2, p. 421–425 (1979).
See also *19*, p. *290*; 21, p. 339.

J. Zieliński
*The Effects of Magnetic Field and Direct Interaction between Impurities in the Anderson Model. Resistivity
(b) Vol. **91**, No. 2, p. 615–625 (1979).
See also *14.1*, p. *199*; 21.1, p. 355; 21.6, p. 387.

M. V. Medvedev
*Short-Range Magnetic Order in a Spin Glass with Nearst-Neighboure Random Exchange Bonds
(b) Vol. **91**, No. 2, p. 713–724 (1979).
See also *13*, p. *137*.

C. Bansal
*Local Environment Effects in Ni–Cu and Ni–Au Alloys
(a) Vol. **52**, No. 2, p. K119–K123 (1979).
See also *18.2*, p. *273*; 21.1, p. 358; 21.6, p. 387.

R. Mańka
The Large Polaron Phase Transition in the Strong Coupling Limit
(b) Vol. **93**, No. 1, p. 53–58 (1979).
See also *13.5.3*, p. *193*.

Yu. B. Gaididei and A. S. Trofimov
Theory of Exciton Migration in a Critical Region
(b) Vol. **93**, No. 1, p. 97–103 (1979).
See also *13.5.1*, p. *185*; 18.2, p. 273.

A. Szász and Á. Tichy-Rács
Photoemission Investigation of Ni near the Curie Temperature
(b) Vol. **93**, No. 1, p. K13–K15 (1979).
See also *17*, p. *256*; 18.2, p. 274; 21.1, p. 358.

J. Roessler, R. Ferrer, and B. Glass
*On the Susceptibility of a Mott Insulator
(b) Vol. **93**, No. 1, p. K53–K57 (1979).

W. Wegener, J. Vanderhaeghen, D. Scheerlinck, S. Hautecler, E. Legrand, and G. Will
*Magnetic Interactions in α-$MnSO_4$ and β-$MnSeO_4$
(b) Vol. **93**, No. 2, p. 597–605 (1979).
See also 18.4, p. 287; 22.8.1, p. 500.

K. H. J. Buschow and A. M. van der Kraan
*Magnetic Properties of Amorphous $Th_{1-x}Fe_x$ Alloys
(a) Vol. **53**, No. 2, p. 665–669 (1979).
See also 2, p. 44; 5, p. 61; *18.2*, p. *276*; 21.1.1, p. 374; 21.4, p. 383.

E. Török and G. Hausch
Bulk Modulus Anomaly of Fe–Ni and Fe–Pt Invar Alloys
(a) Vol. **53**, No. 2, p. K147–K151 (1979).
See also *12.1*, p. *130*; *18.2*, p. *276*; 21.1, p. 360; 21.1.1, p. 374; 21.6, p. 388.

M. Bünger and R. J. Jelitto
*On the Phase Diagram of Itinerant Magnets in the Presence of Unmagnetic Electron Bands
(b) Vol. **94**, No. 1, p. 191–198 (1979).
See also 13.1, p. 149; *18.2*, p. *276*.

H. R. Khan, U. Köbler, K. Lüders, Ch. J. Raub, and Z. Szücz
*Magnetic Susceptibility of V_xMo_{1-x} Solid Solutions
(b) Vol. **94**, No. 1, p. K27–K29 (1979).
See also 21.1, p. 361.

I. Pop, E. Rus, M. Coldea, and O. Pop
*Knight Shift and Magnetic Susceptibilities of Intermetallic Compounds $PrCu_4$ and $PrCu_5$
(a) Vol. **54**, No. 1, p. 365–368 (1979).
See also *19*, p. *293*; 21.1, p. 362; 21.4, p. 384.

K. Elk

On the Susceptibility of Strongly Correlated Electrons
(b) Vol. **96**, No. 1, p. 251 – 257 (1979).
See also 13, p. 141.

J. Pebler, K. Schmidt, and G. Weiser

Magnetic Susceptibility and Electrical Conductivity Measurements of $V_{1-x}Fe_xO_{2-x}F_x$ in the Concentration Range $0 < x < 0.20$
(a) Vol. **56**, No. 2, p. 457 – 465 (1979).
See also *1.2*, p. *21*; 13.4, p. 180; 14.3, p. 218; 22.6.1, p. 487.

18.2 Ferromagnetic Properties

Review Articles

R. CISZEWSKI
*New Effects of Magnetic Critical Phenomena in Iron
(a) Vol. **52**, No. 1, p. 11–28 (1979).
See also *1*, p. *9*; 21.1.1, p. 370.

W. NOLTING
Theory of Ferromagnetic Semiconductors
(b) Vol. **96**, No. 1, p. 11–54 (1979).
See also *13.5.4*, p. *195*; 22, p. 393.

Original Papers and Short Notes

H. KRONMÜLLER and K. VETTER
Isotope Effects of Diatomic TiH and TiD Complexes in FeTi Crystals
(b) Vol. **91**, No. 1, p. 201–210 (1979).
See also *10.1*, p. *92*; 21.1.1, p. 370.

E. BAŁSZCZAK, B. WESTWAŃSKI, and K. SKROBIŚ
Elementary Excitations in Ferrimagnetic and Antiquadrupolar Biquadratic Systems ($S = 3/2$)
(b) Vol. **91**, No. 1, p. 283–290 (1979).
See also *13.5.4*, p. *195*; 18.3, p. 283.

R. CISZEWSKI and J. MILCZAREK
*Temperature Shift of the Maximum of Neutron Critical Scattering from an Iron Polycrystal. A Test for the Dynamic Scaling Theory
(b) Vol. **91**, No. 1, p. 299–304 (1979).

J. D. STEPHENSON, V. KELHÄ, M. TILLI, and T. TUOMI
*Quasi-Lifetime Topography of Magnetic Domain Movements in Grain Oriented (100) [001] Fe–3 wt% Si Polycrystals Using White Synchrotron Radiation. The Influence of an External Magnetic Field
(a) Vol. **51**, No. 1, p. 93–105 (1979).
See also *1.4*, p. *26*; 21.1.1, p. 370.

J. L. PORTESEIL
*On the Model of the Potential Function $V(x)$. Motion of a Bloch Wall in an Imperfect Crystal
(a) Vol. **51**, No. 1, p. 107–116 (1979).

M. A. SIGAL
*Ferromagnetic Resonance Absorption in a Thin Uniaxial Platelet with Stripe Domain Structure Magnetized along the Easy Axis
(a) Vol. **51**, No. 1, p. 151–161 (1979).
See also *19*, p. *289*; 22.8.2, p. 505.

S. Niziol, R. Fruchart, and J. P. Senateur
*Magnetic Properties of CoMnSi and $CoMnSi_{0.85}Ge_{0.15}$
(a) Vol. **51**, No. 1, p. K23 – K27 (1979).
See also *18.4*, p. *286*; 22.8, p. 491.

V. I. Malyutin
*On the Determination of the Energy of Charged Walls in Magnetic Domain Structures
(a) Vol. **51**, No. 1, p. K53 – K56 (1979).

J. Leciejewicz, H. Ptasiewicz-Bąk, and A. Zygmunt
*Magnetic Phase Transition in UPd_2Si_2
(a) Vol. **51**, No. 1, p. K71 – K73 (1979).
See also *1.1*, p. *11*; 18.4, p. 286; 22.8, p. 491.

N. I. Kulikov
Band Structure and Electronic Properties of Transition Metal Hydrides
(b) Vol. **91**, No. 2, p. 753 – 762 (1979).
See also *13.1*, p. *145*; 14.2, p. 204; 20.1, p. 306; 21.1, p. 355.

V. M. Laptev and Yu. N. Skryabin
*Critical Behaviour of the Random Spin Models with a Coupling to a Nonfluctuating Parameter
(b) Vol. **91**, No. 2, p. K143 – K147 (1979).
See also *18.4*, p. *286*.

E. Kolley and W. Kolley
*Renormalization of the Spin Wave Stiffness in Ni-Based Alloys
(b) Vol. **91**, No. 2, p. K173 – K177 (1979).
See also *13.5.4*, p. *195*; 21.1, p. 355.

Th. Kleinefeld and H. H. Mende
*Calculations for Optimizing the Magnification of the Magneto-Optic Kerr Effect by Interference Layers on Iron Whiskers
(a) Vol. **51**, No. 2, p. 487 – 490 (1979).
See also *20*, p. *299*; 21.1.1, p. 371.

G. A. Jones and I. B. Puchalska
*The Birefringent Effects of Magnetic Colloid Applied to the Study of Magnetic Domain Structures
(a) Vol. **51**, No. 2, p. 549 – 558 (1979).
See also 22.8.2, p. 505.

A. V. Deryagin, N. V. Kudrevatykh, R. Z. Levitin, and Yu. F. Popov
*Magnetostriction of Single Crystals of Intermetallic R_2Co_{17} (R = Ho, Er, Tm, Lu) Compounds
(a) Vol. **51**, No. 2, p. K125 – K127 (1979).
See also 21.1, p. 356; 21.4, p. 382.

Yu. P. Lobastov and V. I. Pushkarskii
*The Influence of Internal Field Non-Uniformity in Ferromagnetic Samples on the Magnetization Process in the High Magnetic Field Region
(a) Vol. **51**, No. 2, p. K155 – K159 (1979).
See also 21.1.1, p. 371.

E. Kolley and W. Kolley
Spin Wave Stiffness Constant of Narrow-Band Ferromagnetic Alloys with Random Hopping
(b) Vol. 92, No. 2, p. 473–481 (1979).
See also *13.5.4*, p. *195*; 21, p. 342.

M. A. Gusmão and C. Scherer
*Reaction Field Approximation for the Ordered Phase of Heisenberg Systems
(b) Vol. 92, No. 2, p. 595–600 (1979).
See also *18.4*, p. *286*; 22.5, p. 461.

D. Kapor and A. Ivić
*Erratum to: The Coefficients in a Boson Series Expansion of Spin 1/2 Operators
(b) Vol. 92, No. 2, p. K151 (1979).
See also *13*, p. *138*.

R. Gemperle, P. Novotný, and A. Meňovský
*Bitter Figure Observations on U_3As_4 at Low Temperatures
(a) Vol. 52, No. 2, p. 587 – 596 (1979).
See also 22, p. 397.

S. B. Patil, H. V. Keer, and D. K. Chakrabarty
*Structural, Electrical, and Magnetic Properties in the System $Ba_xLa_{1-x}CoO_3$
(a) Vol. 52, No. 2, p. 681 – 686 (1979).
See also 1.1, p. 12; *14*, p. *197*; 22.8, p. 493.

V. E. Antonov, I. T. Belash, B. K. Ponomarev, E. G. Ponyatovskii, and V. G. Thiessen
*Magnetic Properties of Hydrogen Solid Solutions in Fe–Ni–Mn Alloys
(a) Vol. 52, No. 2, p. 703 – 710 (1979).
See also *18.4*, p. *287*; 21.1, p. 358; 21.1.1, p. 372.

C. Bansal
*Local Environment Effects in Ni–Cu and Ni–Au Alloys
(a) Vol. 52, No. 2, p. K119 – K123 (1979).
See also *18.1*, p. *267*; 21.1, p. 358; 21.6, p. 387.

C. E. Patton
Spin-Wave Instability Theory in Cubic Single Crystal Magnetic Insulators (II): Applications
(b) Vol. 93, No. 1, p. 63 – 69 (1979).
See also *13.5.4*, p. *195*; 22, p. 398.

E. Szczepaniak
*On Singlet–Triplet Ferromagnet Dynamics
(b) Vol. 93, No. 1, p. 89 – 95 (1979).
See also *13.5*, p. *181*.

Yu. B. Gaididei and A. S. Trofimov
Theory of Exciton Migration in a Critical Region
(b) Vol. 93, No. 1, p. 97 – 103 (1979).
See also *13.5.1*, p. *185*; 18.1, p. 267.

F. P. Livi, L. Amaral, and S. G. Fries
*The Magnetic Hyperfine Field at Hf Sites in the (Zr, Hf) Fe_2 Laves Pseudo-Binary Compound
(a) Vol. **53**, No. 1, p. 379 – 382 (1979).
See also 21, p. 344; 21.1.1, p. 373.

K. Hoselitz
*Crystallisation of Magnetic Fe–Si–B Metallic Glasses
(a) Vol. **53**, No. 1, p. K23 – K26 (1979).
See also *2*, p. *44*; 21.1.1, p. 373.

H. E. Khodenkov
*The Motion of the Head-On Domain Wall in the Ferromagnet: New Exact Solution
(a) Vol. **53**, No. 1, p. K103 – K105 (1979).

A. Kobayashi and K. Suzuki
*Erratum to: High Field Transport Properties of a Magnetic Semiconductor
(a) Vol. **53**, No. 1, p. K107 (1979).
See also *14.3.4*, p. *234*; 22.8, p. 494.

M. A. Continentino and N. Rivier
*Magnetic Excitations in Amorphous Ferromagnets
(b) Vol. **93**, No. 2, p. 721 – 733 (1979).
See also *2*, p. *44*.

Kh. I. Pushkarov and D. I. Pushkarov
Solitary Bound State of Magnons and Lattice Deformation in an Anharmonic One-Dimensional Ferromagnet
(b) Vol. **93**, No. 2, p. 735 – 742 (1979).
See also *13.5.4*, p. *196*.

M. Iliev and H. Pink
*Thermoreflectance of $CdCr_2S_4$ near the Curie Point
(b) Vol. **93**, No. 2, p. 799 – 804 (1979).
See also *20.1*, p. *313*; 22.8, p. 494.

L. Biegała
*Negative Magnetocaloric Effect in Uniaxial Ferromagnets
(b) Vol. **93**, No. 2, p. K125 – K126 (1979).

J. Katriel
*Explicit Expressions for the Coefficients in Boson Series Expansions of Arbitrary Spin Operators
(b) Vol. **93**, No. 2, p. K177 – K179 (1979).
See also *13*, p. *139*.

T. Tuomi, J. D. Stephenson, M. Tilli, and V. Kelhä
*Application of Synchrotron Topography to the Study of Magnetic Domain Movements in (100) Fe–3 wt% Si Produced by a Mechanical Stress
(a) Vol. **53**, No. 2, p. 571 – 576 (1979).
See also *1.4*, p. *29*; 12.1, p. 130; 21.1.1, p. 374.

S. F. Dubinin, S. G. Teplouchov, and S. K. Sidorov
Atomic Structure of the Invar Fe–Ni Alloys
(a) Vol. 54, No. 1, p. 239–244 (1979).
See also *1.3*, p. *24*; 21.1, p. 362; 21.1.1, p. 374.

R. Krishnan, M. Tessier, and J. Sztern
*Magnetic Properties of Amorphous $Fe_{38}Ni_{38}B_{24}$ Alloy
(a) Vol. 54, No. 1, p. K77–K79 (1979).
See also *2*, p. *46*; 21.1.1, p. 375.

A. Richter, G. Röpke, and F. Goedsche
Functional Integral Approach for a Doubly Degenerate Hubbard Model
(b) Vol. 94, No. 2, p. 437–450 (1979).
See also *13.1*, p. *151*; 18.1, p. 268; 18.4, p. 287.

B. K. Chakrabarti, B. P. Barua, and S. K. Sinha
*Critical Temperature of a Compressible Ising Magnet
(b) Vol. 94, No. 2, p. 621–625 (1979).
See also 6, p. 72; 13, p. 140.

J. Spałek and Z. Tarnawski
*Static Magnetic Properties of Doped Magnetic Semiconductors. Mean Field Approach
(b) Vol. 94, No. 2, p. 711–720 (1979).
See also 18.3, p. 284; 18.4, p. 287; 22, p. 403.

A. I. Mitsek, D. I. Sirota, and N. P. Kolmakova
*Phenomenological Theory of Electromagnetic and Sound Energy Attenuation in Many-Domain Magnetics
(b) Vol. 94, No. 2, p. 739–747 (1979).
See also 7, p. 79; *18.4*, p. *287*; 20, p. 301; 21.1, p. 362; 22.5, p. 462; 22.6, p. 482.

S. Oseroff, R. Calvo, J. Stankiewicz, Z. Fisk, and D. C. Johnston
*Magnetic and Electric Properties of EuB_6
(b) Vol. 94, No. 2, p. K133–K135 (1979).
See also *14*, p. *197*; 19, p. 294; 21.7, p. 392.

Y. D. Yao
Temperature-Dependent Resistivity in Ni-Rich Alloys
(b) Vol. 94, No. 2, p. K137–K140 (1979).
See also *14.1*, p. *201*; 18.1, p. 268; 21.1, p. 363.

A. Amamou
Mössbauer Effect and Short-Range Order in Fe–P–B and Fe–P–C Amorphous Alloys
(a) Vol. 54, No. 2, p. 565–572 (1979).
See also *2*, p. *46*; *5*, p. *61*; 21.1.1, p. 375.

A. S. Ermolenko, A. Z. Menshikov, and Yu. A. Dorofeev
*Magnetic Heterogeneities in $Y(Co_{1-x}Ni_x)_5$ Alloys and Their Relation to Coercive Force
(a) Vol. 54, No. 2, p. K113–K116 (1979).
See also 21.1, p. 363; 21.4, p. 384.

M. V. SADOVSKII and YU. N. SKRYABIN
Superconductivity in Spin-Glasses
(b) Vol. 95, No. 1, p. 59–64 (1979).
See also *14.2*, p. *207*; 21, p. 348.

J. M. YELLUP and B. A. PARKER
*The Determination of Compositions in Non-Homogeneous Ferromagnetic Materials by Curie Temperature Measurement
(a) Vol. 55, No. 1, p. 137–145 (1979).
See also 1.3, p. 24; 21.1, p. 364.

L. M. SANDLER, V. P. POPOV, and YA. V. NAGLYUK
*Magnetic Anisotropy Induced by Rolling Single Crystals of Ferromagnetic Alloys (I): Face-Centered Cubic Single Crystals
(a) Vol. 55, No. 1, p. 271–277 (1979).
See also *10.1*, p. *98*; 21.1, p. 364; 21.1.1, p. 375.

J. ECHIGOYA and S. HAYASHI
*Anomaly of Dislocation Motion in Fe–35%Ni Invar Alloys
(a) Vol. 55, No. 1, p. 279–284 (1979).
See also *10.1*, p. *98*; 21.1, p. 364; 21.1.1, p. 375.

R. HAUG, G. KAPPEL, and A. JAÉGLÉ
Electrical Resistivity Studies of the System Mn_5Ge_3–Mn_5Si_3
(a) Vol. 55, No. 1, p. 285–290 (1979).
See also *14.1*, p. *202*; 18.4, p. 288; 21.1, p. 364.

A. ZIĘBA, S. HANEDA, Y. YAMAGUCHI, and H. WATANABE
*Latent Ferromagnetism in $MnAs_{0.96}P_{0.04}$
(a) Vol. 55, No. 1, p. K27–K29 (1979).
See also 22.2.4, p. 443.

R. RICHTER, M. WOLF, and F. GOEDSCHE
On the Spin-Disorder Resistivity of Amorphous Ferromagnets
(b) Vol. 95, No. 2, p. 473–482 (1979).
See also *2*, p. *48*; *14.1*, p. *202*; 21, p. 349.

A. KÜHNEL and S. TRIMPER
*A Fundamental Breakdown of an Approximation to the Heisenberg Ferromagnet
(b) Vol. 95, No. 2, p. K93–K97 (1979).

J. JASENEK
*Solution of a Transient Phenomenon in the Spin System Using the Eigen-Coordinate System of the Magnetization Vector $\boldsymbol{M}$
(b) Vol. 95, No. 2, p. K131–K134 (1979).
See also 22.8.2, p. 507.

H. R. HILZINGER, H. HILLMANN, and A. MAGER
*Magnetostriction Measurements on Co-Base Amorphous Alloys
(a) Vol. 55, No. 2, p. 763–769 (1979).
See also *2*, p. *48*; 12.1, p. 131; 21.1, p. 366; 21.1.1, p. 376.

Y. D. YAO and S. ARAJS
*Overlapping between Magnetic and α–γ Transition in the FePd System
(a) Vol. **55**, No. 2, p. K201–K205 (1979).
See also 14.1, p. 202; *18.1*, p. *268*; 21.1.1, p. 376.

F. GOEDSCHE, A. MÖBIUS, and A. RICHTER
On the Low-Temperature Resistivity of Ferromagnetic Transition Metal Alloys
(b) Vol. **96**, No. 1, p. 279–286 (1979).
See also *14.1*, p. *202*; 21.1, p. 366.

E. JÄGER
*Faraday Effect in Inhomogeneously Magnetized Solids
(b) Vol. **96**, No. 1, p. K11–K16 (1979).
See also *20.1*, p. *321*; 21, p. 352.

A. I. MERKULOV, YU. M. YAKOVLEV, S. I. RADAUTSAN, V. E. TEZLEVAN, and T. N. LYANGUZOVA
*The Influence of Acceptor States on the Exchange Interaction in $CdCr_2Se_4$ Single Crystals
(b) Vol. **96**, No. 1, p. K57–K59 (1979).
See also *18.4*, p. *288*; 22.8.2, p. 507.

L. M. SANDLER, V. P. POPOV, and YA. V. NAGLYUK
*Magnetic Anisotropy Induced by Rolling Single Crystals of Ferromagnetic Alloys (II): Body-Centered Cubic Single Crystals
(a) Vol. **56**, No. 1, p. 55–60 (1979).
See also *10.1*, p. *100*; 21.1, p. 366; 21.1.1, p. 376.

L. H. DUNLEAVY and H. J. BLYTHE
*Magnetic Disaccommodation between 30 and 300 K in Low-Temperature Neutron-Irradiated Iron
(a) Vol. **56**, No. 1, p. 237–246 (1979).
See also 10.1, p. 101; *11*, p. *125*; 21.1.1, p. 376.

L. TAKÁCS
Electronic Structure of Transition Metal–Metalloid Crystalline and Glassy Alloys
(a) Vol. **56**, No. 1, p. 371–377 (1979).
See also *2*, p. *48*; 5, p. 62; *13.1*, p. *154*; 21.1, p. 367; 21.1.1, p. 377.

V. E. RODE
*Role of the Exchange Interaction between Ferro- and Antiferromagnetic Components in Invar Alloys
(a) Vol. **56**, No. 1, p. 407–417 (1979).
See also *18.4*, p. *288*; 21.1, p. 367; 21.1.1, p. 377.

M. KOLÁŘ
*A Contribution to the Theoretical Study of the Magnetic Surface Anisotropy of Transition Metals
(b) Vol. **96**, No. 2, p. 683–690 (1979).
See also 13.3, p. 162; 21.1, p. 368.

C. A. MOYER, J. R. KELLY, and S. ARAJS
*Model for Local Moments in a Spin Density Wave Host
(b) Vol. **96**, No. 2, p. 775–783 (1979).
See also 13.4, p. 179; *18.4*, p. *288*; 21.1, p. 368.

18.2.1 Ferromagnetic Films

Original Papers and Short Notes

S. Kuriki
*Film Growth and Magnetic Anisotropy of Thin Ni Electrodeposits on (001) Cu Films
(a) Vol. **51**, No. 1, p. 227–233 (1979).
See also *1.5*, p. *34*; 21.1, p. 354.

M. S. Vijayaraghavan, V. Sivaramakrishnan, and K. V. S. Rama Rao
*Magneto-Optical Properties of Cu-Substituted Manganese–Antimony Films
(a) Vol. **51**, No. 2, p. 543–547 (1979).
See also *20*, p. *299*; 22.7, p. 489.

H. Hoffmann, A. J. Owen, and F. Schröpf
Electron Microscopy of Evaporated and Sputtered Gd/Co and Ho/Co Films
(a) Vol. **52**, No. 1, p. 161–174 (1979).
See also *1.5*, p. *34*; 2, p. 42; 21.1, p. 356; 21.4, p. 382.

D. Hafner and H. Hoffmann
*Microscopic and Macroscopic Inhomogeneity of Magnetization and Anisotropy in Amorphous Rare Earth/Transition Metal Films
(a) Vol. **52**, No. 2, p. 549–558 (1979).
See also 2, p. 43; *20.1*, p. *310*; 21.1, p. 357; 21.4, p. 383.

Diep-The-Hung, J. C. S. Levy, and O. Nagai
*Effects of Surface Spin Waves and Surface Anisotropy in Magnetic Thin Films at Finite Temperatures
(b) Vol. **93**, No. 1, p. 351–361 (1979).
See also 13.5.4, p. 195; *18.4*, p. *287*.

J. Šimšová, I. Tomáš, P. Görnert, M. Nevřiva, and M. Maryško
Preparation and Properties of Co-Substituted (001) YIG Films with Rectangular Bubble Domains
(a) Vol. **53**, No. 1, p. 297–301 (1979).
See also *3*, p. *52*; 22.8.2, p. 506.

W. Maciejewski and A. R. Ferchmin
*On Surface and Bulk Magnetic Phases in Diluted Ising Films with Various Surface Orientations
(b) Vol. **93**, No. 2, p. K173–K176 (1979).
See also 21, p. 344.

J. Białoń and B. Wysłocki
*Investigation of Domain Walls in Thin Permalloy Films by the Low-Angle Electron Deflection Method
(a) Vol. **53**, No. 2, p. 455–459 (1979).
See also 21.1, p. 360; 21.1.1, p. 373.

K. Twarowski and H. K. Lachowicz
*Magnetostriction of Amorphous $Gd_{1-x}Co_x$ RF Sputtered Thin Films
(a) Vol. **53**, No. 2, p. 599–603 (1979).
See also *2*, p. *44*; 18.3, p. 284; 21.1, p. 360; 21.4, p. 383.

18.3 Ferrimagnetic Properties

Original Papers and Short Notes

E. Bałszczak, B. Westwański, and K. Skrobiś
Elementary Excitations in Ferrimagnetic and Antiquadrupolar Biquadratic Systems ($S = 3/2$)
(b) Vol. **91**, No. 1, p. 283 – 290 (1979).
See also *13.5.4*, p. *195*; 18.2, p. 270.

M. Bałanda and S. Niziol
*Spin Reorientation in HoIG Investigated by Means of Faraday Effect
(b) Vol. **91**, No. 1, p. 291 – 298 (1979).
See also 22.8.2, p. 505.

Z. M. Stadnick and W. Zarek
*On the Exchange Integrals in Nickel Ferrite
(b) Vol. **91**, No. 1, p. K83 – K85 (1979).
See also 22.8.2, p. 505.

J. Ilarraz and A. Del Moral
*Magnetic Ordering Temperatures in Rare Earth–Fe_2 Laves Phases Determined by DTA
(a) Vol. **51**, No. 1, p. K41 – K43 (1979).
See also *18.1*, p. *265*; 21.1.1, p. 370; 21.4, p. 382.

S. Prasad, S. B. Singh, R. M. Mehra, and P. K. Singh
*A New Expression for Diagonal Components of the Permeability Tensor and Resonance in Partially Magnetised Ferrites
(a) Vol. **51**, No. 1, p. K75 – K78 (1979).
See also 22.8.2, p. 505.

Z. Jirák, S. Vratislav, and J. Zajiček
*The Magnetic Structure of $Pr_{0.9}Ca_{0.1}MnO_3$
(a) Vol. **52**, No. 1, p. K39 – K43 (1979).
See also 22.8.1, p. 499.

A. Stępniak and J. Skwarcz
*Optical and Magnetic-Optical Properties of $Sm_{0.55}Tb_{0.45}FeO_3$
(a) Vol. **52**, No. 1, p. K77 – K79 (1979).
See also *20.1*, p. *309*; 22.8.2, p. 505.

J. M. Alberdi, J. M. Barandiarán, and J. Ilarraz
*Temperature Dependence of Coercivity in Some Rare-Earth–Fe_2 Intermetallic Compounds
(a) Vol. **52**, No. 1, p. K101 – K103 (1979).
See also *18.2*, p. *272*; 21.1.1, p. 372; 21.4, p. 383.

M. Popescu and C. Ghizdeanu
*Cation Distribution in Cobalt Ferrite-Aluminates
(a) Vol. **52**, No. 2, p. K169 – K172 (1979).
See also *1.1*, p. *12*; 22.8.2, p. 505.

F. Walz, H. Deusch, and H. Kronmüller
*Investigation of Magnetic After-Effects in Magnetite near the Verwey Temperature
(a) Vol. **53**, No. 2, p. 519–528 (1979).
See also 10.2, p. 109; 22.6, p. 480.

K. Twarowski and H. K. Lachowicz
Magnetostriction of Amorphous $Gd_{1-x}Co_x$ RF Sputtered Thin Films
(a) Vol. **53**, No. 2, p. 599–603 (1979).
See also *2*, p. *44*; *18.2.1*, p. *281*; 21.1, p. 360; 21.4, p. 383.

E. Błaszczak and B. Westwański
Zero-Point Effect in Ferrimagnetic and Antiquadrupolar Biquadratic Systems ($S = 3/2$)
(b) Vol. **94**, No. 1, p. 29–35 (1979).
See also *13.5.4*, p. *196*; 18.2, p. 276.

E. Kratochvílová, V. Roskovec, and M. Nekvasil
*Magnetocrystalline Anisotropy of Neodymium-Substituted Yttrium–Iron Garnet
(a) Vol. **54**, No. 1, p. 105–110 (1979).
See also 22.8.2, p. 506.

H. Szymczak and N. Tsuya
*Phenomenological Theory of Magnetostriction and Growth-Induced Anisotropy in Garnet Films
(a) Vol. **54**, No. 1, p. 117–120 (1979).
See also *1*, p. *9*; 22.8.2, p. 506.

B. Kuhlow
*Lichtbeugung an magnetischen Bubble-Bereichsgittern in substituiertem Yttrium–Eisengranat
(Light Diffraction by Magnetic Bubble Domain Gratings in Substituted Yttrium–Iron Garnet)
(a) Vol. **54**, No. 1, p. 281–289 (1979).
See also 22.8.2, p. 506.

E. Błaszczak and B. Westwański
Phase Transitions in Two-Sublattice Biquadratic Systems ($S = 3/2$)
(b) Vol. **94**, No. 2, p. 405–411 (1979).
See also *1.2*, p. *19*.

J. Spałek and Z. Tarnawski
Static Magnetic Properties of Doped Magnetic Semiconductors. Mean Field Approach
(b) Vol. **94**, No. 2, p. 711–720 (1979).
See also *18.2*, p. *277*; 18.4, p. 287; 22, p. 403.

B. Viswanathan, A. Ramanan, and T. K. Varadarajan
*Magnetic Properties of $HoFe_2O_4$
(a) Vol. **55**, No. 1, p. K87–K88 (1979).
See also *15*, p. *251*; 22.8, p. 496.

A. A. Kazakov and N. A. Litvinenko
*Energy Spectrum, Magnetization, and Critical Field of the Uniaxial Two-Sublattice Ferrimagnet by the Green Function Method
(b) Vol. **95**, No. 2, p. 333–340 (1979).
See also *13.5*, p. *182*.

A. V. Mitin and V. A. Tarasov
*Excitation of Magnetoacoustic Surface Waves by Bloch Walls
(b) Vol. **95**, No. 2, p. 461 – 472 (1979).
See also *7*, p. *79*.

G. Srinivasan
*Domain Wall Resonance in Zn and Cu Substituted Magnetite
(a) Vol. **55**, No. 2, p. K149 – K152 (1979).
See also 22.8.2, p. 507.

H. B. von Basel and K. A. Hempel
*High Orientation of Coprecipitated Barium Ferrite by Uniaxial Hot-Pressing
(a) Vol. **55**, No. 2, p. K183 – K184 (1979).
See also 22.8.2, p. 507.

G. N. Falileev
Calculation of the Magnetization–Temperature Dependence of Rare-Earth and Yttrium Iron Garnet Solid Solutions
(b) Vol. **96**, No. 1, p. 429 – 435 (1979).
See also 22.8.2, p. 507.

N. N. Sirota and E. Z. Katsnelson
Photogalvanomagnetic Effects in Ferrites
(a) Vol. **56**, No. 1, p. K61 – K64 (1979).
See also *16*, p. *254*; 22.8.2, p. 508.

S. G. Stoyanov and I. S. Gorban
Diffuse Reflectance and Thermoreflectance Spectra of Pure and Ga-Diluted YIG near the Curie Point
(a) Vol. **56**, No. 2, p. 655 – 661 (1979).
See also *13.1*, p. *155*; 20.1, p. 324; 22.8.2, p. 508.

18.4 Antiferromagnetic Properties

Original Papers and Short Notes

H. Bartholin, D. Florence, and O. Vogt
*Magnetic Phase Diagrams of $Ce_x(La_{0.76}Y_{0.24})_{1-x}Sb$ Compounds
(a) Vol. **52**, No. 2, p. 647–652 (1979).
See also 22.7, p. 489.

V. E. Antonov, I. T. Belash, B. K. Ponomarev, E. G. Ponyatovskii, and V. G. Thiessen
*Magnetic Properties of Hydrogen Solid Solutions in Fe–Ni–Mn Alloys
(a) Vol. **52**, No. 2, p. 703–710 (1979).
See also *18.2*, p. *273*; 21.1, p. 358; 21.1.1, p. 372.

B. G. S. Doman
*An Improved Molecular Field Approach to the Triangular Ising Antiferromagnet
(b) Vol. **93**, No. 1, p. 147–151 (1979).
See also 1.2, p. 17; *18.2*, p. *274*.

Diep-The-Hung, J. C. S. Levy, and O. Nagai
*Effects of Surface Spin Waves and Surface Anisotropy in Magnetic Thin Films at Finite Temperatures
(b) Vol. **93**, No. 1, p. 351–361 (1979).
See also 13.5.4, p. 195; *18.2.1*, p. *281*.

W. Wegener, J. Vanderhaeghen, D. Scheerlinck, S. Hautecler, E. Legrand, and G. Will
Magnetic Interactions in α-$MnSO_4$ and β-$MnSeO_4$
(b) Vol. **93**, No. 2, p. 597–605 (1979).
See also *18.1*, p. *267*; 22.8.1, p. 500.

J. Hönigschmid and G. Will
*A Neutron Diffraction Investigation of the H–T Phase Diagram of α-Fe_2O_3
(a) Vol. **53**, No. 2, p. 557–563 (1979).
See also 22.6, p. 481.

A. Richter, G. Röpke, and F. Goedsche
Functional Integral Approach for a Doubly Degenerate Hubbard Model
(b) Vol. **94**, No. 2, p. 437–450 (1979).
See also *13.1*, p. *151*; 18.1, p. 268; 18.2, p. 277.

J. Spałek and Z. Tarnawski
Static Magnetic Properties of Doped Magnetic Semiconductors. Mean Field Approach
(b) Vol. **94**, No. 2, p. 711–720 (1979).
See also *18.2*, p. *277*; 18.3, p. 284; 22, p. 403.

A. I. Mitsek, D. I. Sirota, and N. P. Kolmakova
*Phenomenological Theory of Electromagnetic and Sound Energy Attenuation in Many-Domain Magnetics
(b) Vol. **94**, No. 2, p. 739–747 (1979).
See also 7, p. 79; *18.2*, p. *277*; 20, p. 301; 21.1, p. 362; 22.5, p. 462; 22.6, p. 482.

R. Rai, R. Morehouse, and E. Dibaj
Dynamic Shift of the ESR Field in the One-Dimensional Magnetic Copper Dipyridine Dichloride
(b) Vol. **94**, No. 2, p. K159–K164 (1979).
See also 18.1, p. 268; *19*, p. *294*; 22.8, p. 495.

19. Magnetic Resonance

Review Article

R. S. Brazis, J. K. Furdyna, and J. K. Požela
*Microwave Effects in Narrow-Gap Semiconductors (II)
(a) Vol. **54**, No. 1, p. 11–27 (1979).
See also 13, p. 136; 18.1, p. 265; 20, p. 298; 22.2.1, p. 432; 22.2.3, p. 439; 22.2.4, p. 442; 22.4.3, p. 455; 22.4.4, p. 458.

Original Papers and Short Notes

H.-J. Gollnest, G. Kuka, W. Braune, and R. Herrmann
*Temperature Dependence of the Relaxation Time of the Holes in Bismuth
(b) Vol. **91**, No. 1, p. 177–183 (1979).
See also 21.7, p. 391.

D. Ploumbidis and T. Sutter
*Temperature and Concentration Dependence of the Knight Shift of ^{115}In in Liquid In–Hg Alloys
(b) Vol. **91**, No. 1, p. 185–188 (1979).
See also 2, p. 40; 21, p. 338.

S. A. Marshall, T. Marshall, S. A. Zager, and D. N. Zimmerman
*Electron Spin Resonance Absorption Spectrum of Divalent Manganese in Single Crystal Calcite
(b) Vol. **91**, No. 1, p. 275–282 (1979).
See also *13.4*, p. *163*; 22.8.1, p. 498.

V. K. Jain
*Electron Spin Resonance of VO^{2+} Ions in Antiferroelectric ADP Crystals
(b) Vol. **91**, No. 1, p. K35–K37 (1979).
See also 14.4.1, p. 244; 22.8.1, p. 498.

G. Schwachheim, S. F. Machado, and C. Tsallis
*Thermal Behaviour of the ESR Relaxation Time in Slightly Dirty Superconductors
(b) Vol. **91**, No. 1, p. K43–K47 (1979).
See also *14.2*, p. *204*.

T. Grochulski, K. Leibler, A. Sienkiewicz, and R. R. Gałązka
*EPR Investigations of Ordering Effects in $Hg_{1-x}Mn_xTe$
(b) Vol. **91**, No. 1, p. K73–K76 (1979).
See also 22.4.4, p. 458.

M. A. Sigal
*Ferromagnetic Resonance Absorption in a Thin Uniaxial Platelet with Stripe Domain Structure Magnetized along the Easy Axis
(a) Vol. **51**, No. 1, p. 151–161 (1979).
See also *18.2*, p. *270*; 22.8.2, p. 505.

J.-J. Pilloud and C. Jaccard
Dynamic Properties of $[X_2^-]$ Centres in Cesium Halides
(b) Vol. 92, No. 1, p. 233–239 (1979).
See also *13.4*, p. *166*; *20.3*, p. *328*; 22.5.2, p. 466.

Zh. Egemberdiev, V. Seeman, S. Zazubovich, and N. Jaanson
Polarized Luminescence and EPR Study of Sn^+ Centres in KCl: $SnCl_2$ Crystals
(b) Vol. 92, No. 1, p. 335–345 (1979).
See also *10.2*, p. *105*; 20.1, p. 308; *20.3*, p. *328*; 22.5.4, p. 475.

T. A. Abramovskaya and M. D. Zviadadze
On Effects of the Dipole Reservoir in Radiofrequency Discrete Saturation
(a) Vol. 52, No. 1, p. 111–120 (1979).
See also *13*, p. *137*; *18*, p. *261*.

H. Sachs and G. Lehmann
*Correlation of Zero-Field Splittings and Site Distortions (I): $MnCl_4^{2-}$ in Approximately Trigonal Symmetry
(b) Vol. 92, No. 2, p. 417–424 (1979).
See also 22.9, p. 510.

A. Goltzené and C. Schwab
Impurity Scattering Effect on the Cyclotron Resonance of Carriers in Cu_2O
(b) Vol. 92, No. 2, p. 483–487 (1979).
See also *13.1*, p. *146*; 22.6, p. 479.

M. Bogdan, A. Nicula, and D. Lupu
Spin-Lattice Relaxation Time and Quadrupole Coupling of Deuterium in $PdD_{0.70}$
(a) Vol. 52, No. 2, p. 615–619 (1979).
See also *9*, p. *86*; 21, p. 343.

G. D. Sootha, G. K. Padam, and S. K. Gupta
*ESR Study of Oxygen Radicals Formed in Cadmium Sulphide
(a) Vol. 52, No. 2, p. K125–K128 (1979).
See also 22.4.1, p. 450.

J. Isoya and J. A. Weil
*Uncompensated Titanium (3+) Center in α-Quartz
(a) Vol. 52, No. 2, p. K193–K196 (1979).
See also *10.2*, p. *107*; 11, p. 121; 22.6, p. 479.

P. Mialhe
*Approximate Technique in ESR Transition Probability Treatments
(b) Vol. 93, No. 1, p. 189–195 (1979).
See also 22.4, p. 448; 22.5, p. 461; 22.6, p. 479.

W. Braune, W. Schneider, J. Lebech, and K. Saermark
Magnetoplasma Waves in Bismuth and Semimetallic BiSb Alloys at Microwave Frequencies 9 to 100 GHz
(b) Vol. 93, No. 1, p. 269–278 (1979).
See also *13.5.2*, p. *191*; 13.5.4, p. 195; 21.7, p. 391.

D. Baisa, A. Bondar, A. Gordon, and S. Maltsev
*NQR Study of Phase Transitions in Proustite at High Pressure
(b) Vol. **93**, No. 2, p. 805–809 (1979).
See also *1.2*, p. *18*; 22.8, p. 494.

U. Steigenberger and M. von Ortenberg
*Cyclotron Resonance in Uniaxially Stressed Tellurium
(b) Vol. **93**, No. 2, p. K139–K141 (1979).
See also 12.2, p. 134; *13.1*, p. *149*; 22.1.3, p. 429.

A. Watterich and R. Voszka
Effect of X-Irradiation on Divalent Metal Aggregates in NaCl Crystals at RT
(b) Vol. **93**, No. 2, p. K161–K165 (1979).
See also *10.2*, p. *108*; *11*, p. *122*; 22.5.2, p. 468.

I. C. Baianu, K. Rubinson, and J. Patterson
The Observation of Structural Relaxation in FeNiPB Glass by Ferromagnetic Resonance and X-Ray Scattering
(a) Vol. **53**, No. 2, p. K133–K138 (1979).
See also 1.3, p. 24; *2*, p. *44*; *18.2*, p. *276*; 21.1, p. 360; 21.1.1, p. 374.

O. J. Żogał
*^{139}La NMR Study of Sub-Stoichiometric Lanthanum Trihydride
(a) Vol. **53**, No. 2, p. K203–K205 (1979).
See also 21.4, p. 384.

M. D. Glinchuk, M. F. Deigen, and V. S. Rozhko
Local Fields and Electric Dipole Moment of Localized Impurities in Crystals
(b) Vol. **94**, No. 1, p. 63–68 (1979).
See also *13.4*, p. *171*; 22, p. 400.

V. Lupei and A. Lupei
*On the Nature of U^{5+} Centres in LiF
(b) Vol. **94**, No. 1, p. 301–307 (1979).
See also *13.4*, p. *172*; 22.5.2, p. 468.

J. Englich and P. Čížek
*N^{14} NMR in Non-Stoichiometric Cubic Manganese Nitrides
(b) Vol. **94**, No. 1, p. K15–K17 (1979).
See also 22, p. 402.

J. G. Kliava, Z. A. Konstants, J. J. Purāns et A. I. Dimante
*RPE de Mn^{2+} dans les polycristaux de basse symmétrie. Le métaphosphate de magnésium (EPR of Mn^{2+} in Polycrystals of Low Symmetry. Magnesium Metaphosphate)
(a) Vol. **54**, No. 1, p. 61–66 (1979).
See also 22.8.1, p. 500.

I. Pop, E. Rus, M. Coldea, and O. Pop
*Knight Shift and Magnetic Susceptibilities of Intermetallic Compounds $PrCu_4$ and $PrCu_5$
(a) Vol. **54**, No. 2, p. 365–368 (1979).
See also *18.1*, p. *267*; 21.1, p. 362; 21.4, p. 384.

J. Witters, A. Stesmans, and J. van Meijel
*Conductivity Electron Spin Resonance and Inhomogeneous Broadening Due to Local Moments
(b) Vol. 94, No. 2, p. 353–357 (1979).
See also 21, p. 346.

M. Friedrich and W. Karthe
*Structure and Kinetics of Fe^{3+}-Doped $PbMoO_4$
(b) Vol. 94, No. 2, p. 451–456 (1979).
See also *13.4*, p. *173*; 22.8.1, p. 501.

Ch. Hoentzsch and J. M. Spaeth
*Dynamical Superhyperfine Interaction of Atomic Hydrogen in KCl and RbCl
(b) Vol. 94, No. 2, p. 497–503 (1979).
See also *13.4*, p. *173*; 22.5.2, p. 469.

V. K. Arora and H. N. Spector
*Quantum-Limit Cyclotron Resonance Linewidth in Semiconductors
(b) Vol. 94, No. 2, p. 701–709 (1979).
See also 22, p. 403.

K. N. Kocharyan, A. A. Mirzakhanyan, and E. G. Sharoyan
*The Zero-Field Splitting of Mo^{3+} Ground State in Corundum
(b) Vol. 94, No. 2, p. K129–K131 (1979).
See also *13.4*, p. *174*; 22.6, p. 482.

S. Oseroff, R. Calvo, J. Stankiewicz, Z. Fisk, and D. C. Johnston
*Magnetic and Electric Properties of EuB_6
(b) Vol. 94, No. 2, p. K133–K135 (1979).
See also *14*, p. *197*; *18.2*, p. *277*; 21.7, p. 392.

R. Rai, R. Morehouse, and E. Dibaj
*Dynamic Shift of the ESR Field in the One-Dimensional Magnetic Copper Dipyridine Dichloride
(b) Vol. 94, No. 2, p. K159–K164 (1979).
See also 18.1, p. 268; 18.4, p. 287; 22.8, p. 495.

A. V. Dvurechenskii, L. P. Potapova, and V. V. Kalinin
Efficiency of Defect Build-up in Ion-Implanted Polycrystalline Silicon Films
(a) Vol. 54, No. 2, p. 457–462 (1979).
See also *1.5*, p. *35*; 10.2, p. 111; *11*, p. *123*; 22.1.2, p. 422.

A. V. Zalesskii, V. V. Vanchikov, V. G. Krivenko, and A. N. Ivashchenko
Spin Echo Spectra of ^{57}Fe and ^{19}F Nuclei in a Canted Antiferromagnet FeF_3
(a) Vol. 54, No. 2, p. 471–476 (1979).
See also *18.4*, p. *288*; 22.5, p. 462.

A. Ślęzak, J. Lech, and I. Bójko
*Electron Paramagnetic Resonance of Mn^{2+} Ions in Aragonite Single Crystals
(a) Vol. 54, No. 2, p. 755–759 (1979).
See also 22, p. 403.

G. Völkel, W. Brunner, W. Windsch, and D. Steudel
*Electron Spin-Echo (ESE) Investigations of Critical Paramagnetic Relaxation Anomalies in X-Irradiated Triglycine Sulphate (TGS:X)
(b) Vol. 95, No. 1, p. 99—105 (1979).
See also *14.4.1*, p. *246*; 22.9, p. 512.

S. Ueda and T. Shimizu
*NMR Studies of the Effect of Alloying Cu, Ag, Tl, I, and Ge into Amorphous As_2Se_3
(b) Vol. 95, No. 1, p. 279—282 (1979).
See also *2*, p. *47*; 22, p. 404.

M. Razeghi, J. P. Buisson, and B. Houlier
*EPR Investigation of Gd^{3+} and Eu^{2+} in the α- and β-Phase of Lead Phosphate
(b) Vol. 95, No. 1, p. 283—289 (1979).
See also *1.2*, p. *19*; 22.8.1, p. 501.

L. Kubler, A. Jaéglé, and J. J. Koulmann
*The Effect of Oxygen on the Properties of Evaporated Amorphous Silicon. ESR Studies
(b) Vol. 95, No. 1, p. 307—315 (1979).
See also 2, p. 47; *13.4*, p. *175*; 22.1.2, p. 423.

A. Lupei and V. Lupei
*A $5f^1$ Ion in Non-Octahedral Symmetry
(b) Vol. 95, No. 1, p. K69—K71 (1979).
See also 22.5.3, p. 474.

V. K. Jain
*EPR of Cr^{3+} in $(NH_4)_xRb_{1-x}Al(SO_4)_2 \cdot 12H_2O$ Single Crystals
(b) Vol. 95, No. 1, p. K79—K82 (1979).
See also 22.8.1, p. 501.

M. Dubiel, G. Berg, and F. Fröhlich
The Formation of Dimer and Trimer Aggregates in Sr-Doped NaCl Crystals
(a) Vol. 55, No. 1, p. 153—159 (1979).
See also *10.2*, p. *112*; 14.4, p. 242; 22.5.2, p. 470.

R. Erdmann and H. Alexander
Photo-EPR of Dislocations in Silicon
(a) Vol. 55, No. 1, p. 251—259 (1979).
See also *10.2*, p. *113*; 13.4, p. 176; 22.1.2, p. 423.

E. K. Henner and I. G. Shaposhnikov
Clustering of Magnetic Ions in Diluted Solid Paramagnets
(a) Vol. 55, No. 1, p. 315—323 (1979).
See also *13.4*, p. *176*; 18.1, p. 268.

J. Szuber and B. Długosz
*Dominant Interaction and Average Distances between Surface Paramagnetic Centers on Low Pressure Baked (111) Si Surface
(a) Vol. 55, No. 1, p. K35—K38 (1979).
See also *13.3*, p. *161*; 22.1.2, p. 424.

D. GOURIER, D. VIVIEN, and J. LIVAGE
*ESR of Paramagnetic Defects in the Conduction Planes of Na^+, K^+, and H_3O^+ β-Alumina
(a) Vol. **56**, No. 1, p. 247–257 (1979).
See also 22.6.1, p. 487.

J. PURĀNS, J. KLIAVA, and I. MILLERE
*Computer Simulation of Mn^{2+} EPR Spectra in Glasses with $g_{ef} = 4.29$
(a) Vol. **56**, No. 1, p. K25–K28 (1979).
See also *2*, p. *49*; 22.6, p. 485.

M. SUZUKI, K. MARUYAMA, M. KUMEDA, and T. SHIMIZU
Properties of a-$Si_{100-x}Al_x$ Prepared by High Pressure RF Sputtering
(a) Vol. **56**, No. 1, p. K49–K52 (1979).
See also *2*, p. *49*; *14.3.1*, p. *221*; 22.1.2, p. 426.

R. BLINC, M. BURGAR, J. SLAK, V. RUTAR, and F. MILIA
*^{13}C NMR Study of the Incommensurate Phase Transition in $[N(CH_3)_4]_2ZnCl_4$
(a) Vol. **56**, No. 1, p. K65–K69 (1979).
See also *1.2*, p. *21*; 14.4.1, p. 247; 22.9, p. 514.

M. MORENO
*The Relation between Bounding and g-Shift for Tl^{2+} and Pb^{3+} in Alkali Halides
(b) Vol. **96**, No. 2, p. 647–651 (1979).
See also 13.4, p. 179; 22.5.2, p. 472.

W. J. KULIŃSKI, S. K. HOFFMANN, and A. B. WIĘCKOWSKI
*EPR Studies of VO^{2+} Ions in Rochelle Salt Crystal
(b) Vol. **96**, No. 1, p. 745–755 (1979).
See also 14.4.1, p. 247.

S. K. MISRA and B. MIKOLAJCZAK
*EPR of Mn^{2+} Doped Single Crystals of $Ni(NH_4)_2(SO_4)_2 \cdot 6\,H_2O$ and $NiSO_4 \cdot 7\,H_2O$
(b) Vol. **96**, No. 2, p. 807–815 (1979).
See also 13.4, p. 179; 22.8.1, p. 503.

ZH. EGEMBERDIEV, Ü. HALDRE, T. LEHTO, K. REALO, S. REIFMAN, V. SEEMAN, and S. ZAZUBOVICH
Polarized Luminescence and EPR Study of Sn^{3+} Centres in $KCl:SnCl_2$ Crystals
(b) Vol. **96**, No. 2, p. 867–875 (1979).
See also *10.2*, p. *116*; 20.1, p. 323; *20.3*, p. *337*; 22.5.4, p. 476.

Y. OCHIAI and E. MATSUURA
*Orbach Relaxation Process in Heavily Doped n-Type Silicon
(a) Vol. **56**, No. 2, p. K109–K112 (1979).
See also 22.1.2, p. 427.

20. Optical Properties

Review Articles

Original Papers and Short Notes

L. N. OVANDER and N. S. TYU
*Theory of Raman Scattering on Polar Phonons
(b) Vol. 91, No. 2, p. 763–772 (1979).
See also 6, p. 65; *13.5*, p. *181*.

S. GUHA, P. K. SEN, and S. GHOSH
*Modulational Instability of a Finite Amplitude Alfvén Wave in Semimetals
(b) Vol. 91, No. 2, p. K135–K138 (1979).
See also *13*, p. *137*; 21.7, p. 391.

K. S. SRIVASTAVA, O. K. HARSH, and V. KUMAR
*Chemical Shifts of the X-Ray L-Absorption Edges of Pb and Bi
(b) Vol. 91, No. 2, p. K169–K172 (1979).
See also 21, p. 339; 21.7, p. 391; 22, p. 394.

TH. KLEINEFELD and H. H. MENDE
*Calculations for Optimizing the Magnification of the Magneto-Optic Kerr Effect by Interference Layers on Iron Whiskers
(a) Vol. 51, No. 2, p. 487–490 (1979).
See also *18.2*, p. *271*; 21.1.1, p. 371.

M. S. VIJAYARAGHAVAN, V. SIVARAMAKRISHNAN, and K. V. S. RAMA RAO
*Magneto-Optical Properties of Cu-Substituted Manganese–Antimony Films
(a) Vol. 51, No. 2, p. 543–547 (1979).
See also *18.2.1*, p. *281*; 22.7, p. 489.

R. ENDERLEIN, K. PEUKER, and F. BECHSTEDT
*General Theory of Light Scattering in Solids
(b) Vol. 92, No. 1, p. 149–158 (1979).
See also *6*, p. *65*; 22, p. 395.

P. SUORTTI
*Scattering of X-Rays near the K Absorption Edge (II): Dispersion Factors f' and f''
(b) Vol. 92, No. 1, p. 259–268 (1979).
See also 21.1, p. 356.

Y. L. RAO and C. MANDE
*K X-Ray Absorption Discontinuity in Crystalline and Amorphous Selenium
(b) Vol. 92, No. 1, p. K61–K64 (1979).
See also 2, p. 41; 22.1.3, p. 428.

NGUYEN VAN TRONG
*Theory of Inelastic Reflection of Particles by Rayleigh Waves
(b) Vol. 93, No. 1, p. 45–52 (1979).
See also *13.3*, p. *159*.

J. WENK and L. MERTEN
*Stimulated Raman Effect in Piezoelectric Cubic Crystals
(b) Vol. 93, No. 1, p. 175–181 (1979).
See also *6*, p. *68*; 13.5, p. 182.

V. V. Nemoshkalenko, Vl. N. Antonov, V. N. Antonov, and W. John
*Electronic Structure and Soft X-Ray Emission Spectrum of H.C.P. Hafnium
(b) Vol. 94, No. 1, p. K19–K22 (1979).
See also *13.1*, p. *150*; 21.4, p. 384.

A. Ślebarski, K. Ławniczak, and J. Auleytner
*Soft X-Ray Emission Spectra of Aluminium and Manganese from $Gd(Al_{1-x}Mn_x)_2$ Intermetallic Compounds
(a) Vol. 54, No. 1, p. 79–83 (1979).
See also 21, p. 345; 21.4, p. 384.

M. S. Sodha and R. R. Sharma
*Stimulated Brillouin Scattering of Helicon Waves in a Semiconductor
(b) Vol. 94, No. 2, p. 667–671 (1979).
See also 7, p. 79; *13*, p. *140*; 22.2.3, p. 440.

A. I. Mitsek, D. I. Sirota, and N. P. Kolmakova
Phenomenological Theory of Electromagnetic and Sound Energy Attenuation in Many-Domain Magnetics
(b) Vol. 94, No. 2, p. 739–747 (1979).
See also 7, p. 79; *18.2*, p. *277*; *18.4*, p. *287*; 21.1, p. 362; 22.5, p. 462; 22.6, p. 482.

G. Dräger, O. Brümmer, and J. Bonitz
Polarized K-Emission Valence Bands, Structure, and Bonding of Some Hexagonal Close-Packed Metals
(b) Vol. 94, No. 2, p. K111–K114 (1979).
See also *4*, p. *57*; *13.1*, p. *151*; 21.1, p. 362; 21.3, p. 381.

E. Sobczak
*X-Ray Bremsstrahlung Isochromat and X-Ray Photoelectron Investigations of Ion Implanted Germanium
(a) Vol. 54, No. 2, p. K151–K153 (1979).
See also 13.1, p. 151; *17*, p. *258*; 22.1.1, p. 412.

E. Donato, E. S. Giuliano, R. Ruggeri, and A. Stancanelli
*Soft X-Ray Spectra in Nb–Mo Alloy
(b) Vol. 95, No. 1, p. K37–K41 (1979).
See also 21, p. 348.

K. K. Pukhov and V. P. Sakun
Theory of Nonradiative Multiphonon Transitions in Impurity Centers with Extremely Weak Electron–Phonon Coupling
(b) Vol. 95, No. 2, p. 391–402 (1979).
See also 6, p. 74; *13.4*, p. *176*; 22, p. 404.

R. Podloucky and A. Neckel
Compton Profiles of FeAl
(b) Vol. 95, No. 2, p. 541–548 (1979).
See also *13.1*, p. *152*; 21.1.1, p. 375.

T. K. Hatwar, S. K. Malik, M. N. Ghatikar, and B. D. Padalia
*X-Ray Absorption Edge Shifts of Cobalt and Rare Earth Ions in Rare Earth–Cobalt Intermetallic Compounds
(b) Vol. 95, No. 2, p. 621–625 (1979).
See also 21.1, p. 365; 21.4, p. 384.

I. M. Dykman and L. A. Prokhnitskii
Electromagnetic Waves in an Inhomogeneous Semiconductor under Electric and Magnetic Fields
(b) Vol. 96, No. 1, p. 55—60 (1979).
See also *13*, p. *141*; *14.3*, p. *216*; 22, p. 405.

Sh. N. Gifeisman and Yu. G. Malkoch
The Theory of Magneto-Optical Effects in Shallow F-Centres
(b) Vol. 96, No. 1, p. 73—90 (1979).
See also 13.4, p. 177; 13.5.3, p. 193; *18*, p. *263*; 22, p. 405.

R. M. Nayak and B. D. Padalia
*Chemical Shifts of the K-Absorption Edge of Mn in Some Mixed Oxidic Spinels
(b) Vol. 96, No. 1, p. 259—262 (1979).
See also *4*, p. *57*; 22.8.2, p. 507.

V. L. Bonch-Bruevich, V. D. Karaivanov, and Ya. G. Proykova
Random Field Parameters and Pseudopotential
(b) Vol. 96, No. 1, p. 271—278 (1979).
See also *2*, p. *48*; *13*, p. *141*; 14, p. 198.

E. Donato, E. S. Giuliano, R. Ruggeri, and A. Stancanelli
Appearance Potential Spectroscopy in Metals and Alloys. Multiple Scattering Approach
(b) Vol. 96, No. 1, p. 313—320 (1979).
See also *13.1*, p. *153*; 21, p. 351.

A. P. Marchetti and M. Scozzafava
The Optically Detected Magnetic Resonance of Nickel-Doped Silver Chloride
(b) Vol. 96, No. 1, p. 441—445 (1979).
See also 13.4, p. 178; *19*, p. *296*; 22.5.1, p. 464.

M. Załużny
*On the Optical Orientation of Free Carriers in Size Quantized $Pb_{1-x}Sn_xTe$ Films
(b) Vol. 96, No. 1, p. K51—K55 (1979).
See also *13*, p. *141*; 22, p. 406.

A. R. Badzian and A. Kłokocki
Splitting of the Valence Band of Copper by X-Ray K-Emission Spectroscopy
(b) Vol. 96, No. 2, p. 529—535 (1979).
See also *13.1*, p. *154*; 21.1, p. 367.

R. Claus
*On the Lattice Dynamical Background of Optical Activity in Crystals
(b) Vol. 96, No. 2, p. 545—554 (1979).
See also 6, p. 76; *13.5*, p. *182*.

S. Guha and P. Namjoshi
*Parametric Decay of a Helicon in n-InSb
(b) Vol. 96, No. 2, p. 713—716 (1979).
See also *13*, p. *142*; 22.2.3, p. 441.

I. Egri and A. Stahl
Real Space Wave Equation for Exciton-Polaritons of Wannier Type
(b) Vol. **96**, No. 2, p. K83—K86 (1979).
See also *13.5.1*, p. *190*; 22, p. 407.

W. Jamroz and J. Karniewicz
*Three-Dimensional Model of Nonlinear Magneto-Electric Susceptibilities
(b) Vol. **96**, No. 2, p. K87—K90 (1979).
See also *14.4*, p. *243*; 22.2.1, p. 436; 22.4.3, p. 457.

S. Guha and P. K. Sen
*Effect of Transverse Magnetic Field on the Electromagnetic Wave Propagation in Semiconductor Plasmas at High Electric Field
(a) Vol. **56**, No. 2, p. K117—K121 (1979).
See also *13*, p. *142*; 14.3, p. 218; 22.2.3, p. 441.

20.1 Spectra. Optical Constants (X-Ray Spectra See 20)

Review Articles

O. A. Golikova
Boron and Boron-Based Semiconductors
(a) Vol. **51**, No. 1, p. 11–40 (1979).
See also *1*, p. *9*; 13.4, p. 163; *14.3*, p. *209*; 14.3.4, p. 231; 15, p. 249; 22, p. 393; 22.1.3, p. 428; 22.2, p. 430.

R. H. Enns and S. S. Rangnekar
The Three-Wave Interaction in Nonlinear Optics
(b) Vol. **94**, No. 1, p. 9–28 (1979).
See also 6, p. 63; *20.2*, p. *325*.

Original Papers and Short Notes

D. Laplaze
*Second-Order Raman Spectra of LiH
(b) Vol. **91**, No. 1, p. 59–69 (1979).
See also *6*, p. *63*; 22.5, p. 460.

E. Jahne, W. Pilz, M. Giehler, and L. Hildisch
*Non-Isodisplacement of P Atoms in Long-Wavelength Optical Phonons in $In_{1-x}Ga_xP$
(b) Vol. **91**, No. 1, p. 155–165 (1979).
See also *6*, p. *63*; 22.2.4, p. 442.

J. Voigt, F. Spiegelberg, and M. Senoner
*Band Parameters of CdS and CdSe Single Crystals Determined from Optical Exciton Spectra
(b) Vol. **91**, No. 1, p. 189–199 (1979).
See also *13.5.1*, p. *183*; 22.4.1, p. 450; 22.4.2, p. 453.

A. J. C. Sampaio and R. Luzzi
*Resonant Magneto-Raman Scattering in Polar Semiconductors
(b) Vol. **91**, No. 1, p. 305–310 (1979).
See also *6*, p. *63*; 22.2.3, p. 439.

W. Szuszkiewicz
*Interband Absorption and the Γ-Point Band Structure Parameters in HgSe
(b) Vol. **91**, No. 1, p. 361–369 (1979).
See also *13.1*, p. *144*; 22.4.2, p. 453.

R. Sebastian, G. A. Savari Raj, and T. M. Haridasan
*U Centres in Alkali Iodides
(b) Vol. **91**, No. 1, p. 371–376 (1979).
See also *13.4*, p. *163*; 22.5.2, p. 465.

N. A. Bakhyshov, N. M. Gasanly, B. M. Yavadov, V. I. Tagirov, and Sh. M. Efendiev
*Mixed One- and Two-Mode Behaviour of Optical Phonons in $TlGaS_{2x}Se_{2(1-x)}$ and $TlInS_{2x}Se_{2(1-x)}$ Layer Solid Solutions
(b) Vol. **91**, No. 1, p. K1–K3 (1979).
See also *6*, p. *63*; 22.8, p. 491.

M. SCROCCO
*A Raman Spectroscopic Study of K_2SO_4 in the D_{2h}^{16} and C_{2v}^{9} Phases
(b) Vol. **91**, No. 1, p. K21 – K25 (1979).
See also *6*, p. *64*; 22.8.1, p. 498.

A. E. BAKHYSHOV, M. F. AGAEVA, and A. M. DARVISH
*Electrical and Optical Properties of $TlInSe_2$ Single Crystals
(b) Vol. **91**, No. 1, p. K31 – K34 (1979).
See also *14.3*, p. *210*; 16, p. 252; 22.8, p. 491.

W. HÖRIG, H. NEUMANN, E. RECCIUS, H. WEINERT, G. KÜHN, and B. SCHUMANN
*Temperature Dependence of the Absorption Edge in $CuGaS_2$
(a) Vol. **51**, No. 1, p. 57 – 62 (1979).
See also 13.1, p. 144; 22.8, p. 491.

J. SCHOENES, E. KALDIS, W. THÖNI, and P. WACHTER
*Optical, Magnetic, and Magnetooptical Properties of the Europium Silicate Glass $Eu_{0.14}Si_{0.31}O_{0.55}$
(a) Vol. **51**, No. 1, p. 173 – 181 (1979).
See also 2, p. 40; *18*, p. *261*; 22.6, p. 477.

N. M. NESTERENKO and V. I. FOMIN
*Soft Modes and the Raman Spectrum in $KSc(MoO_4)_2$ at Ferroelastic Phase Transitions
(a) Vol. **51**, No. 1, p. K101 – K104 (1979).
See also 6, p. 64; *14.4.1*, p. *244*; 22.8.1, p. 498.

I SAVATINOVA and E. ANACHKOVA
*Effect of Ordering on the OH Raman Bands of $K_4Fe(CN)_6 \cdot 3\ H_2O$
(b) Vol. **91**, No. 2, p. 413 – 419 (1979).
See also *6*, p. *64*; 14.4.1, p. 244; 22.8, p. 492.

P. KLEINERT
*Cluster Approximation for the Spectral Density of Mixed Diatomic Systems
(b) Vol. **91**, No. 2, p. 455 – 465 (1979).
See also *13.4*, p. *164*; 22, p. 394.

J. M. DE SIEBENTHAL and H. BILL
Investigation of Sulfur Molecular Paramagnetic Centers in $SrCl_2$
(b) Vol. **91**, No. 2, p. 479 – 486 (1979).
See also *10.2*, p. *103*; 19, p. 290; 22.5.3, p. 473.

G. O. MÜLLER, H. H. WEBER, and I. HÖRICKE
*Linear Optical Response of Highly Excited Semiconductors. Electron–Hole Liquid in CdS
(b) Vol. **91**, No. 2, p. 531 – 539 (1979).
See also 13.5, p. 181; 22.4.1, p. 450.

E. CASTELLUCCI and V. SCHETTINO
*Angular Dispersion of the Internal Phonon Modes in Ferroelectric Sodium Nitrite
(b) Vol. **91**, No. 2, p. 641 – 645 (1979).
See also *6*, p. *64*; 14.4.1, p. 245.

A. D. BROTHERS and J. B. BRUNGARDT
*Excitons in WS_2 Films. Pressure and Temperature Effects
(b) Vol. **91**, No. 2, p. 675 – 679 (1979).
See also *13.5.1*, p. *183*; 22, p. 394.

A. VAIDYANATHAN, Y. F. TSAY, and S. S. MITRA
Electronic Structure and Optical Properties of Rhombohedral Arsenic and Their Variations with the Internal Displacement Parameter
(b) Vol. 92, No. 1, p. 73–81 (1979).
See also *13.1*, p. *145*; 14.4, p. 238; 22.1, p. 408.

H. WIEDER, M. CARDONA, and C. R. GUARNIERI
*Vibrational Spectrum of Hydrogenated Amorphous Si–C Films
(b) Vol. 92, No. 1, p. 99–112 (1979).
See also *2*, p. *41*; 4, p. 55; 22.3, p. 444.

A. STAHL
*Boundary Conditions for Polarization Waves in Spatially Dispersive Dielectrics
(b) Vol. 92, No. 1, p. 113–122 (1979).
See also *6*, p. *65*; 13.5.1, p. 183.

V. V. HIZHNYAKOV and A. V. SHERMAN
On the Theory of Resonant Secondary Radiation of Excitons Strongly Interacting with Phonons
(b) Vol. 92, No. 1, p. 177–183 (1979).
See also *6*, p. *65*; *13.5.1*, p. *184*; 20.3, p. 328; 22, p. 395.

M. BACCI
*Calculation of First-Order and Second-Order Jahn-Teller Coupling Constants in $MgO:V^{2+}$ and $KMgF_3:V^{2+}$
(b) Vol. 92, No. 1, p. 193–198 (1979).
See also *13.4*, p. *166*; 22.5.4, p. 475; 22.6, p. 478.

G. MARICHY, G. CHASSAGNE, and D. DURAND
*Optical Properties of Metal Aggregates in Li^+ and K^+ Irradiated MgO Crystals
(b) Vol. 92, No. 1, p. 221–231 (1979).
See also *13.4*, p. *166*; 22.6, p. 478.

R. WOJTAL, A. GOLNIK, and J. A. GAJ
*Acceptor Photoionization and Light-to-Heavy Hole Absorption in $Cd_{1-x}Mn_xTe$
(b) Vol. 92, No. 1, p. 241–247 (1979).
See also *13.4*, p. *166*; 22.8, p. 492.

M. JOUANNE, J. F. MORHANGE, and M. BALKANSKI
*First and Second Order Raman Scattering by Pair Modes in Silicon
(b) Vol. 92, No. 1, p. 255–258 (1979).
See also *6*, p. *66*; 22.1.2, p. 417.

B. KOILLER and H. S. BRANDI
*A Self-Consistent Model for the Electronic Structure of the U-Center in Alkali Halides
(b) Vol. 92, No. 1, p. 279–285 (1979).
See also *13.4*, p. *166*; 22.5.2, p. 466.

M. SIU LI and M. DE SOUZA
*Off-Center Cu^+ in Mixed Crystals
(b) Vol. 92, No. 1, p. 287–291 (1979).
See also *10.2*, p. *105*; 22.5.4, p. 475.

S. Radhakrishna and K. Hariharan
*Optical and Vibronic Spectra of MnO_4^{2-} in Cesium Bromide
(b) Vol. 92, No. 1, p. 293–296 (1979).
See also 6, p. 66; *13.4*, p. *166*; 22.5.2, p. 466.

Zh. Egemberdiev, V. Seeman, S. Zazubovich, and N. Jaanson
Polarized Luminescence and EPR Study of Sn^+ Centres in $KCl:SnCl_2$ Crystals
(b) Vol. 92, No. 1, p. 335–345 (1979).
See also *10.2*, p. *105*; 19, p. 291; *20.3*, p. *328*; 22.5.4, p. 475.

W. Hörig, W. Möller, H. Neumann, E. Reccius, and G. Kühn
*Optical Properties of $CuIn_{1-x}Ga_xSe_2$ Mixed Crystals
(b) Vol. 92, No. 1, p. K1–K4 (1979).
See also *13.1*, p. *145*; 22.8, p. 492.

G. M. Harutunian
*The Absorption of a Weak Electromagnetic Wave in the Presence of an Intensive One in a Three-Band Semiconductor
(b) Vol. 92, No. 1, p. K9–K13 (1979).
See also 13.1, p. 146; *14.3.4*, p. *232*; 22, p. 396.

C. Ance
*Variation des transitions optiques de $Ga_{1-x}Al_xAs$ et $Ga_{1-x}Al_xSb$
(Variation of the optical transitions of $Ga_{1-x}Al_xAs$ and $Ga_{1-x}Al_xSb$)
(b) Vol. 92, No. 1, p. K33–K38 (1979).
See also *13.1*, p. *146*; 22.2.4, p. 442.

R. Birsoy and L. E. Murr
*Coloring of Natural Fluorites
(a) Vol. 52, No. 1, p. 77–84 (1979).
See also *10.2*, p. *105*; 22.5.3, p. 473.

Y. Mlik, A. Daoud, and M. Couzi
Raman Spectra and Structural Phase Transitions in Chain Compounds $(CH_3)_4NMnCl_3$ and $(CH_3)_4NCdCl_3$
(a) Vol. 52, No. 1, p. 175–188 (1979).
See also *1.2*, p. *17*; 6, p. 66; 22.9, p. 510.

Hj. Bernhardt
*Defect Electron Centres in Lead Molybdate Crystals
(a) Vol. 52, No. 1, p. 223–229 (1979).
See also 10.2, p. 105; 13.5.3, p. 193; *20.3*, p. *328*; 22.8.1, p. 499.

G. Frank, H. Köstlin, and A. Rabenau
*X-Ray and Optical Measurements in the In_2O_3–SnO_2 System
(a) Vol. 52, No. 1, p. 231–238 (1979).
See also *1.3*, p. *22*; 22.6.1, p. 487.

L. A. Demberel, A. S. Popov, D. B. Kushev, and N. N. Zheleva
Deep Levels in Fe-Doped InP
(a) Vol. 52, No. 1, p. 341–345 (1979).
See also *13.4*, p. *167*; *14.3.4*, p. *232*; 20.3, p. 328; 22.2.2, p. 437.

A. J. Twarowski and A. C. Albrecht
*Thermoelectric Absorption Spectroscopy of Polycrystalline Tetracene Films
(a) Vol. 52, No. 1, p. K35 – K37 (1979).
See also 22.9, p. 510.

W. Rühle, K. Lösch, and J. U. Fischbach
A New Optical Method to Investigate Deep Traps in GaAs and InP
(a) Vol. 52, No. 1, p. K65 – K67 (1979).
See also *13.4*, p. *167*; 22.2.1, p. 433; 22.2.2, p. 437.

A. Stępniak and J. Skwarcz
*Optical and Magnetic-Optical Properties of $Sm_{0.55}Tb_{0.45}FeO_3$
(a) Vol. 52, No. 1, p. K77 – K79 (1979).
See also *18.3*, p. *283*; 22.8.2, p. 505.

A. Kurowska
*Intraband Absorption in Lead Chalcogenide Mixed Crystals
(b) Vol. 92, No. 2, p. 371 – 377 (1979).
See also *13.1*, p. *146*; 22.8, p. 492.

T. Sauder, M. Certier, and B. Stébé
*Excitons Bound to Neutral Donor Impurities in CuCl
(b) Vol. 92, No. 2, p. 411 – 416 (1979).
See also *13.5.1*, p. *184*; 22.5, p. 460.

N. Koumvakalis and W. von der Osten
First-Order Resonant Raman Scattering in Mixed Silver Halides
(b) Vol. 92, No. 2, p. 441 – 445 (1979).
See also *6*, p. *66*; 13.5.1, p. 184; *20.3*, p. *329*; 22.5.1, p. 463; 22.5.4, p. 475.

M. Ghomi, E. Rzepka, and L. Taurel
*Study of Initial Stages of F-Center Aggregation in NaCl by Raman Scattering
(b) Vol. 92, No. 2, p. 447 – 453 (1979).
See also *6*, p. *66*; 10.2, p. 106; 22.5.2, p. 466.

Nguyen van Hieu and Nguyen Ái Viet
*Electronic Raman Scattering on Donor Levels in Three-Band Semiconductors with Direct Band Gaps
(b) Vol. 92, No. 2, p. 537 – 544 (1979).
See also *13.1*, p. *146*; 22.2, p. 430.

M. Hoclet, P. Plumelle, M. Vandevyver, R. Triboulet, and Y. Marfaing
*Lattice Dynamics of Cd-Rich $Cd_{1-x}Hg_xTe$ Alloys
(b) Vol. 92, No. 2, p. 545 – 549 (1979).
See also *6*, p. *67*; 22.4.4, p. 458.

R. Murray and B. L. Evans
The Electro-Absorption Edge Spectra of $2H\text{-}MoS_2$
(b) Vol. 92, No. 2, p. 551 – 556 (1979).
See also *13.5.1*, p. *184*; 14.3.4, p. 233; 22, p. 396.

K. R. Allakhverdiev, E. Yu. Salaev, N. Yu. Safarov, L. K. Vodopyanov, and L. V. Golubev
*Light Scattering in $TlInSe_2$ Single Crystals
(b) Vol. 92, No. 2, p. K91 – K93 (1979).
See also *6*, p. *67*; 22.8, p. 492.

A. Subrahmanyam and K. V. Rao
*F-Band Absorption and Thermoluminescence of KCl Single Crystals under DC Field and X-Ray Irradiation
(a) Vol. **52**, No. 2, p. K147–K150 (1979).
See also *20.3*, p. *329*; 22.5.2, p. 466.

O. Litzman and V. Holý
*Some Special Models of the Spatial Theory of Dispersion
(b) Vol. **93**, No. 1, p. 135–146 (1979).
See also *14.4*, p. *239*.

W. Ekardt and H.-J. Brocksch
Phonon-Assisted Two-Photon Absorption of the Excitonic Molecule in CuCl (I): Bare-Exciton Approach
(b) Vol. **93**, No. 1, p. 203–211 (1979).
See also *13.5.1*, p. *185*; 22.5, p. 461.

P. N. Koul, T. K. Hatwar, M. N. Ghatikar, and B. D. Padalia
*Chemical Shifts of the L_3-Absorption Edge of Pt in the Rare Earth-Platinum Intermetallic Compounds
(b) Vol. **93**, No. 1, p. 223–228 (1979).
See also 1.3, p. 23; 21.4, p. 383.

J. Gallardo and D. Mattis
Green's Theorem Calculation of Energy of Exciton near Hard Walls
(b) Vol. **93**, No. 1, p. 229–235 (1979).
See also *13.5.1*, p. *185*; 14.3.2, p. 223.

F. Dochy
*Theoretical Calculation of the Absorption Energy of the F-Centre in Alkali Halides
(b) Vol. **93**, No. 1, p. 325–335 (1979).
See also *10.2*, p. *107*; 13.4, p. 169; 22.5.2, p. 467.

Y. Itoh, S. Adachi, and C. Hamaguchi
*Analysis of Resonant Brillouin Scattering in GaAs
(b) Vol. **93**, No. 1, p. 381–389 (1979).
See also 13.5.1, p. 185; 22.2.1, p. 433.

D. N. Popov, K. I. Terzijsky, and N. S. Backalova
*Spectral and Temperature Dependence of the Initial Yield of the Z_3-Center Formation in Additively Coloured KCl:Ca
(b) Vol. **93**, No. 1, p. K35–K37 (1979).
See also *13.4*, p. *169*; 22.5.2, p. 467.

L. Mărculescu and S. Frunză
*Optical Spectra of Lead Centres in NaBr Crystals
(b) Vol. **93**, No. 1, p. K59–K62 (1979).
See also *20.3*, p. *330*; 22.5.2, p. 467.

K. R. Allakhverdiev, S. S. Babaev, and M. M. Tagiev
Two-Mode Behaviour of the Optical Phonons in GaS_xSe_{1-x} Solid Solutions
(b) Vol. **93**, No. 1, p. K67–K70 (1979).
See also *6*, p. *68*; 22.3.1, p. 447.

A. V. LELYAKOV
Absorption Band Shape of Free Excitons in Indirect Band Gap Semiconductors at High Excitation Density
(b) Vol. **93**, No. 2, p. 523–530 (1979).
See also *13.5.1*, p. *185*; 22.5.1, p. 463.

I. S. GORBAN, V. A. GORYNYA, V. I. LUGOVOI, N. P. KRASNOLOB, G. I. SALIVON, and I. I. TYCHINA
*One- and Two-Phonon Raman Scattering in Ternary Phosphides $ZnSiP_2$, $ZnGeP_2$, $CdSiP_2$
(b) Vol. **93**, No. 2, p. 531–538 (1979).
See also *6*, p. *69*; 22.8, p. 494.

R. ENDERLEIN, P. RENNER, and M. SCHEELE
*Electro-Optic Effects Due to the Stark Splitting of Energy Bands
(b) Vol. **93**, No. 2, p. 539–547 (1979).
See also 13.1, p. 148; 13.5.1, p. 185; *14.3.4*, p. *234*; 22.6, p. 480.

A. M. BEREZHKOVSKII and A. A. OVCHINNIKOV
Certain Features of Semiconductor Behaviour in Strong Parallel Magnetic and Electric Fields (III): Interband Absorption
(b) Vol. **93**, No. 2, p. 549–551 (1979).
See also *13*, p. *139*; 22, p. 399.

J. M. HVAM
Excitonic Molecule Transitions in ZnO
(b) Vol. **93**, No. 2, p. 581–590 (1979).
See also *13.5.1*, p. *185*; 20.3, p. 331; 22.4, p. 448.

M. GIEHLER and W. PILZ
*Infrared Reflection and Raman Scattering Studies of the Vibrational Properties and Structure of $As_2S_3Ge_x$ Glasses
(b) Vol. **93**, No. 2, p. 641–651 (1979).
See also *6*, p. *69*; 22, p. 400.

B. DIETRICH and J. VOIGT
Spatial Dispersion Effects in the Transmission Spectra of Semiconductors in the Region of Exciton Resonances
(b) Vol. **93**, No. 2, p. 669–677 (1979).
See also *13.5.1*, p. *186*; 22.4.1, p. 451.

A. BIANCONI, S. STIZZA, R. BERNARDINI, and S. NANNARONE
*Electronic Structure of Ti_2O_3 from Single Crystal Thermoreflectance Spectroscopy
(b) Vol. **93**, No. 2, p. 767–774 (1979).
See also *13.1*, p. *149*; 22.6, p. 480.

M. ILIEV and H. PINK
*Thermoreflectance of $CdCr_2S_4$ near the Curie Point
(b) Vol. **93**, No. 2, p. 799–804 (1979).
See also *18.2*, p. *275*; 22.8, p. 494.

R. S. TURTELLI and A. R. B. DE CASTRO
*LO Phonon Raman Lineshape in Highly Excited GaAs
(b) Vol. **93**, No. 2, p. 811–815 (1979).
See also *6*, p. *69*; 20.3, p. 331; 22.2.1, p. 433.

M. J. Thwaites, R. D. Tomlinson, and M. J. Hampshire
*The Observation of Optical Transitions from Valence Band States in $CuInTe_2$
(b) Vol. **94**, No. 1, p. 211–214 (1979).
See also *13.1*, p. *150*; 22.8, p. 495.

J. Zarembowitch, J. Gouteron, and A. M. Lejus
*Raman Spectra of Lanthanide Sesquioxide Single Crystals with A-Type Structure
(b) Vol. **94**, No. 1, p. 249–256 (1979).
See also *6*, p. *70*; 22.6, p. 481.

Y. Hara and M. Nicol
*Raman Spectra and the Structure of Rutile at High Pressures
(b) Vol. **94**, No. 1, p. 317–322 (1979).
See also *6*, p. *70*; 22.6, p. 481.

K. R. Allakhverdiev, A. N. Abbasov, T. R. Mekhtiev, and R. Kh. Nani
*Far-Infrared Reflection Spectra of $CdInGaS_4$ Single Crystals
(b) Vol. **94**, No. 1, p. K31–K34 (1979).
See also *6*, p. *71*; 22.8, p. 495.

V. Nekvasil
*On Low-Lying Crystal Field Levels of Ho^{3+} in Aluminium and Gallium Garnets
(b) Vol. **94**, No. 1, p. K41–K43 (1979).
See also *13.4*, p. *172*; 22.8.2, p. 506.

Y. Kayanuma
*Anomalous Reduction of the Two Phonon Pinning Interaction in Magneto-Optics
(b) Vol. **94**, No. 1, p. K53–K56 (1979).
See also *13.5.3*, p. *193*; 22.5.1, p. 463.

P. Bussemer
*A Simple Model for the Explanation of Urbach's Rule
(b) Vol. **94**, No. 1, p. K77–K79 (1979).
See also 6, p. 71.

M. Giehler and W. Pilz
*Major Difference of the Vibrational Spectra and Structure of $(As_3S_3)_{1-x}(GeS_2)_x$ and $As_2S_3Ge_x$ Glasses
(b) Vol. **94**, No. 1, p. K81–K84 (1979).
See also 2, p. 45; *6*, p. *71*; 22, p. 402.

V. K. Tripathi and D. P. Singh
*DC Field Induced Enhancement in the Transparency of n-InSb
(b) Vol. **94**, No. 1, p. K85–K89 (1979).
See also *14.3.4*, p. *234*; 22.2.3, p. 440.

J. Weszka, I. Miotkowski, K. Kloc, and W. Żdanowicz
*Raman Scattering in CdP_2 Single Crystals
(b) Vol. **94**, No. 1, p. K97–K99 (1979).
See also *6*, p. *71*; 22.3, p. 445.

Č. Koňák
Changes of Optical Properties at an Incommensurate–Commensurate Phase Transition in $(NH_4)_2BeF_4$ Crystals
(a) Vol. **54**, No. 1, p. 99–104 (1979).
See also *1.2*, p. *18*; 14.4.1, p. 246; 22.8.1, p. 500.

V. B. Sharan, D. N. S. Srivastwa, and S. C. Sen
On the Origin of the 375 nm Emission Band in KCl:Tl
(a) Vol. **54**, No. 1, p. 129–137 (1979).
See also *13.4*, p. *172*; *20.3*, p. *332*; 22.5, p. 461.

M. Mutha Reddy, K. Somaiah, and V. Hari Babu
Thermoluminescence of BaFBr Crystals
(a) Vol. **54**, No. 1, p. 245–250 (1979).
See also 10.2, p. 110; *20.3*, p. *332*; 22.5.4, p. 476.

V. S. Sivasankar and P. W. Whippey
Thermoluminescence and Optical Absorption of X-Ray Irradiated $NaBrO_3$
(a) Vol. **54**, No. 1, p. 369–373 (1979).
See also *11*, p. *122*; *20.3*, p. *333*; 22.8.1, p. 501.

J. Hajtó and P. J. S. Ewen
*Natural Optical Activity and Related Phenomena in As_2S_3 Glasses
(a) Vol. **54**, No. 1, p. 385–390 (1979).
See also *2*, p. *45*; 22, p. 402.

A. H. Edwards
F-Center Photoionization in the Semicontinuum Model
(b) Vol. **94**, No. 2, p. 367–370 (1979).
See also *13.1*, p. *151*; 22.5.2, p. 469.

R. M. Luchytskii, V. S. Manzhara, and P. M. Staryk
*Investigation of n-$Pb_{1-x}Sn_xSe$ Absorption Spectra
(b) Vol. **94**, No. 2, p. 381–386 (1979).
See also 6, p. 71; 13.1, p. 151; 22, p. 402.

Ch. Uihlein and S. Feierabend
Magnetoreflectance of Exciton-Polaritons in Cubic II–VI Compounds (II): Semiclassical Treatment
(b) Vol. **94**, No. 2, p. 421–428 (1979).
See also *13.5.1*, p. *187*; 22.4, p. 449.

S. Nakashima, A. Aziza, M. le Postollec, and M. Balkanski
*Raman and Infrared Spectra in Organic Layer Crystals of Squaric Acid
(b) Vol. **94**, No. 2, p. 529–539 (1979).
See also 1.2, p. 19; 4, p. 57; 22.9, p. 512.

B. Esser and D. Sprenger
*On the Theory of Absorption and Luminescence Spectra in Amorphous Semiconductors
(b) Vol. **94**, No. 2, p. 583–593 (1979).
See also 2, p. 46; *20.3*, p. *333*; 22, p. 403.

V. May, K. Henneberger, and F. Henneberger
Renormalization of Polaritons Due to Virtual Formation of Biexcitons at High Densities of Excitation
(b) Vol. **94**, No. 2, p. 611–620 (1979).
See also *13.5.1*, p. *187*; 20.2, p. 325; 22.5, p. 461.

R. L. Sarkar and S. Chatterjee
*Electronic Structure and Optical Properties of AlP, AlAs, and AlSb
(b) Vol. **94**, No. 2, p. 641–647 (1979).
See also *13.1*, p. *151*; 22.2.1, p. 434; 22.2.2, p. 438; 22.2.3, p. 440.

W. Wojdowski, T. Łukasiewicz, W. Nazarewicz, and J. Żmija
*Infrared Studies of Lattice Vibrations in $Bi_{12}GeO_{20}$ and $Bi_{12}SiO_{20}$ Crystals
(b) Vol. **94**, No. 2, p. 649–658 (1979).
See also *6*, p. *72*; 22.8, p. 495.

P. N. Ram and M. D. Tiwari
*Impurity-Induced Infrared Absorption in AgBr
(b) Vol. **94**, No. 2, p. 721–729 (1979).
See also *6*, p. *72*; 22.5.1, p. 464.

D. M. G. Pinatti, S. C. Zílio, and M. de Souza
*Point Defects in NaCl Crystals Doped with Divalent Ions. The NaCl–NaF:Pb^{++} Crystal
(b) Vol. **94**, No. 2, p. 749–755 (1979).
See also *10.2*, p. *111*; 22.5.2, p. 469.

P. Kleinert
*The Influence of Contact Excitons on the Electroabsorption in Disordered Semiconductors
(b) Vol. **94**, No. 2, p. K145–K147 (1979).
See also 2, p. 46; *13.5.1*, p. *187*; 14.3.4, p. 235; 22, p. 403.

B. Koiller and H. S. Brandi
*A Self-Consistent Model for the Optical Excitations of the U_2 and U_1 Centers in Alkali-Halides
(b) Vol. **94**, No. 2, p. K179–K183 (1979).
See also *13.4*, p. *174*; 22.5.2, p. 469.

L. G. Suslina, A. G. Plyukhin, O. Goede, and D. Hennig
*Compositional Fluctuation-Induced Broadening of Bound-Exciton Lines in II–VI Semiconductor Mixed Crystals
(b) Vol. **94**, No. 2, p. K185–K189 (1979).
See also *13.5.1*, p. *187*; 22.4.4, p. 459.

Hj. Bernhardt
On the Polaron Character of the Coloration Centres in $LiNbO_3$ Crystals
(a) Vol. **54**, No. 2, p. 597–603 (1979).
See also *10.2*, p. *111*; *13.5.3*, p. *193*; 20.3, p. 334; 22.8.2, p. 506.

J. Prakash
Photo-Induced Polarization and Photo-Induced Depolarization of S^{2-}-Anion Vacancy Dipoles in Potassium Iodide
(a) Vol. **54**, No. 2, p. 681–688 (1979).
See also *10.2*, p. *111*; 14.4, p. 242; 22.5.2, p. 469.

A. S. RAO and R. J. KEARNEY
*Logarithmic Derivative Reflectance Spectra of BaO and SrO
(b) Vol. **95**, No. 1, p. 243–250 (1979).
See also 13.5.1, p. 188; 22.6, p. 483.

N. M. GASANLY, B. M. DZHAVADOV, V. I. TAGIROV, and E. A. VINOGRADOV
*Long-Wave Lattice Vibrations of $TlInS_{2x}Se_{2(1-x)}$ and $TlGaS_{2x}Se_{2(1-x)}$ Layer Solid Solutions
(b) Vol. **95**, No. 1, p. K27–K30 (1979).
See also *6*, p. *73*; 22.8, p. 496.

N. M. GASANLY, N. F. GAKHRAMANOV, B. M. DZHAVADOV, V. I. TAGIROV, and E. A. VINOGRADOV
*Infrared Spectra of the Layer Compound InS
(b) Vol. **95**, No. 1, p. K89–K92 (1979).
See also *6*, p. *73*; 22.3, p. 445.

I. V. MOCHALOV
Spectroscopic Characteristics of the Erbium Ions and Laser Action at the Transition $^4S_{3/2} \rightarrow {}^4I_{9/2}$ ($\lambda_e \approx 1.66$ μm) in the Crystals of Gadolinium, Yttrium, and Lutecium Orthoaluminates
(a) Vol. **55**, No. 1, p. 79–87 (1979).
See also 13.4, p. 175; *20.2*, p. *326*; 20.3, p. 335; 22.8.1, p. 502.

G. MARTENS, P. RABE, G. TOLKIEHN, and A. WERNER
EXAFS in Photoelectron Yield Spectra at K Edges of Cu, Ni, and Ge
(a) Vol. **55**, No. 1, p. 105–108 (1979).
See also *17*, p. *258*; 21.1, p. 364; 22.1.1, p. 412.

H.-G. UNRUH, W. ELLER, and G. KIRF
Spectroscopic and Dielectric Investigations of K_2SeO_4
(a) Vol. **55**, No. 1, p. 173–180 (1979).
See also 6, p. 73; *14.4 1*, p. *246*; 22.8.1, p. 502.

L. NIKOLOVA, T. TODOROV, D. POPOV, and K. TERSIISKI
*Laser Induced Fast Reversible Processes in Additively Coloured KCl Containing F_A Centres
(a) Vol. **55**, No. 1, p. 333–337 (1979).
See also *10.2*, p. *113*; 22.5.2, p. 470.

N. M. BEZDETNY, V. G. SILVESTROV, S. V. SAFONOVA, A. KH. ZEINALLY, and A. L. TIMOFEEV
*The Influence of γ-Irradiation on Photorefraction in Iron Doped Lithium Niobate
(a) Vol. **55**, No. 1, p. K45–K47 (1979).
See also *11*, p. *123*; 22.8.1, p. 502.

E. A. VINOGRADOV, G. N. ZHIZHIN, N. N. MELNIK, S. I. SUBBOTIN, V. V. PANFILOV, K. R. ALLAKHVERDIEV, E. YU. SALAEV, and R. KH. NANI
*Raman Scattering and Phase Transformations of $TlGaSe_2$ and $TlGaS_2$ under Pressure
(b) Vol. **95**, No. 2, p. 383–389 (1979).
See also 1.2, p. 20; *6*, p. *74*; 22.3.1, p. 447.

K. BÄRNER, R. STEARNS, and R. BRAUNSTEIN
*Evidence for Excitonic Absorption in Metallic $MnAs_{0.88}P_{0.12}$
(b) Vol. **95**, No. 2, p. 483–489 (1979).
See also *13.5.1*, p. *188*; 22.8, p. 496.

V. Z. Polinger and G. I. Bersuker
Multimode Jahn-Teller Effect for an E-Tem with Strong Vibronic Coupling (II): Band Shapes of the Infrared and Raman Spectra
(b) Vol. **96**, No. 1, p. 153–161 (1979).
See also *6*, p. *75*; *13.4*, p. *178*.

E. A. Popov and M. M. Kotlyarskii
*Optical Absorption of the Layer Antiferromagnet Rb_2MnCl_4
(b) Vol. **96**, No. 1, p. 163–167 (1979).
See also *18.4*, p. *288*; 22.8, p. 496.

K. R. Allakhverdiev, S. S. Babaev, E. Yu. Salaev, and M. M. Tagyev
*Angular Bihaviour of the Polar Optical Phonons in $A^{III}B^{VI}$ Layered Semiconductors
(b) Vol. **96**, No. 1, p. 177–182 (1979).
See also *6*, p. *75*; 22.3, p. 446.

H. J. Mattausch and Ch. Uihlein
Fine Structure of P Excitons in CuBr
(b) Vol. **96**, No. 1, p. 189–200 (1979).
See also *13.5.1*, p. *188*; 22.5, p. 462.

G. Reuter, F. Lohse, and J. M. Spaeth
*Generation of Dichroic H Centres in KCl by Polarized Optical Bleaching of Interstitial Atomic Hydrogen
(b) Vol. **96**, No. 1, p. 243–250 (1979).
See also *10.2*, p. *115*; 13.4, p. 178; 19, p. 296; 22.5.2, p. 471.

T. Tsuboi
*Identification of the Absorption Bands in KI:Pb^{2+} Crystals
(b) Vol. **96**, No. 1, p. 321–328 (1979).
See also 13.4, p. 178; 22.5.2, p. 472.

O. Ziep and D. Genzow
*Calculation of the Interband Absorption in Lead Chalcogenides Using a Multiband Model
(b) Vol. **96**, No. 1, p. 359–368 (1979).
See also *13.1*, p. *153*; 22, p. 405.

W. Stössel and H. J. Wagner
*The Influence of Spatial Dispersion and of the Angle of Incidence on Thermomodulation Spectra of the $A_{n=1}$-Exciton in CdS
(b) Vol. **96**, No. 1, p. 369–378 (1979).
See also *13.5.1*, p. *189*; 22.4.1, p. 451.

A. Kuroiwa, H. Saito, and S. Shionoya
*Exciton Reflection Spectra under High-Intensity Excitation in CdS and CdSe
(b) Vol. **96**, No. 1, p. 483–486 (1979).
See also *13.5.1*, p. *189*; 20.3, p. 336; 22.4.1, p. 451; 22.4.2, p. 454.

E. Jäger
*Faraday Effect in Inhomogeneously Magnetized Solids
(b) Vol. **96**, No. 1, p. K11–K16 (1979).
See also *18.2*, p. *279*; 21, p. 352.

O. Goede, D. Hennig, and L. John
*Disorder Effects on Free Excitons in $CdS_{1-x}Se_x$ Mixed Crystals
(b) Vol. **96**, No. 2, p. 671 – 681 (1979).
See also *13.5.1*, p. *189*; 22.4.4, p. 459.

J. Vindevoghel and M. Vindevoghel
Contribution of Interband Transitions to the Conductivity of p-Si at Submillimeter Wavelengths
(b) Vol. **96**, No. 2, p. 717 – 725 (1979).
See also *13.1*, p. *154*; *14.3*, p. *217*; 22.1.2, p. 426.

A. Zwick and M. A. Renucci
*Long-Wavelength Optical Phonons in $ZrSe_3$
(b) Vol. **96**, No. 2, p. 757 – 767 (1979).
See also *6*, p. *76*; 22, p. 407.

J. Szczyrbowski
*On the Shape of the Absorption Tail in Amorphous Solids
(b) Vol. **96**, No. 2, p. 769 – 774 (1979).
See also *2*, p. *49*; 13.1, p. 154.

V. D. Iskra
*Phonon-Assisted Interband Optical Absorption in Disordered Semiconductors
(b) Vol. **96**, No. 2, p. 843 – 856 (1979).
See also *13.5.1*, p. *189*; 22, p. 407.

Zh. Egemberdiev, Ü. Haldre, T. Lehto, K. Raelo, S. Reifman, V. Seeman, and S. Zazubovich
Polarized Luminescence and EPR Study of Sn^{3+} Centres in $KCl:SnCl_2$ Crystals
(b) Vol. **96**, No. 2, p. 867 – 875 (1979).
See also *10.2*, p. *116*; 19, p. 297; *20.3*, p. *337*; 22.5.4, p. 476.

Doan Nhat Quang
*On the Theory of Electroreflectance for Heavily Doped Degenerate Semiconductors
(b) Vol. **96**, No. 2, p. K73 – K76 (1979).
See also *14.3.4*, p. *237*; 22, p. 407.

I. Savatinova, E. Anachkova, J. Rohleder, and B. Jakubovski
*C≡N Vibrations in Potassium Ferrocyanide Trihydrate
(b) Vol. **96**, No. 2, p. K97 – K101 (1979).
See also *6*, p. *77*; 14.4.1, p. 247; 22.8, p. 497.

J. Y. Harbec, C. Deville Cavellin, and S. Jandl
*Raman Spectra and Crystal Symmetry of ZrS_3
(b) Vol. **96**, No. 2, p. K117 – K120 (1979).
See also *6*, p. *77*; 22, p. 407.

D. Wruck and P. Gaworzewski
Electrical and Infrared Spectroscopic Investigations of Oxygen-Related Donors in Silicon
(a) Vol. **56**, No. 2, p. 557 – 564 (1979).
See also *13.4*, p. *180*; 14.3, p. 218; 22.1.2, p. 426.

H.-J. Kos and R. Nink
*Correlation of Thermoluminescence and Optical Absorption of Z_3 Centers in LiF:Mg, Ti Crystals
(a) Vol. **56**, No. 2, p. 593–596 (1979).
See also *10.2*, p. *117*; 20.3, p. 337; 22.5.2, p. 472.

V. A. Krylov, W. Ulrici, and L. S. Sochava
*Optical Spectra of Off-Centre Ni^{2+} Ions in SrO
(a) Vol. **56**, No. 2, p. 615–621 (1979).
See also *10.2*, p. *117*; 22.6, p. 485.

S. G. Stoyanov and I. S. Gorban
Diffuse Reflectance and Thermoreflectance Spectra of Pure and Ga-Diluted YIG near the Curie Point
(a) Vol. **56**, No. 2, p. 655–661 (1979).
See also *13.1*, p. *155*; 18.3, p. 285; 22.8.2, p. 508.

K. Zellama, J. F. Morhange, P. Germain, and J. C. Bourgoin
Growth Rate of Crystallization in Amorphous Germanium Produced by Ion Implantation. A Raman Spectroscopy Study
(a) Vol. **56**, No. 2, p. 717–723 (1979).
See also 1.2, p. 21; *2*, p. *50*; 6, p. 77; *11*, p. *126*; 22.1.1, p. 413.

A. A. Kaminskii, S. E. Sarkisov, T. I. Butaeva, G. A. Denisenko, B. Hermoneit, J. Bohm, W. Grosskreutz, and D. Schultze
Growth, Spectroscopy, and Stimulated Emission of Cubic $Bi_4Ge_3O_{12}$ Crystals Doped with Dy^{3+}, Ho^{3+}, Er^{3+}, Tm^{3+}, or Yb^{3+} Ions
(a) Vol. **56**, No. 2, p. 725–736 (1979).
See also *20.2*, p. *326*; *20.3*, p. *337*; 22.8, p. 497.

K. Somaiah, P. Veeresham, K. L. N. Prasad, and V. Hari Babu
Thermoluminescence and Optical Absorption of BaFCl Crystals
(a) Vol. **56**, No. 2, p. 737–742 (1979).
See also 10.2, p. 117; *20.3*, p. *337*; 22.5.4, p. 476.

20.2 Lasers

Review Article

R. H. Enns and S. S. Rangnekar
*The Three-Wave Interaction in Nonlinear Optics
(b) Vol. 94, No. 1, p. 9–28 (1979).
See also 6, p. 63; 20.1, p. 304.

Original Papers and Short Notes

K. Schindler, K. Schmidt, and G. Staupendahl
*Stimulated Scattering of CO_2-Laser Radiation in Tellurium
(b) Vol. 91, No. 1, p. K17–K19 (1979).
See also 22.1.3, p. 428.

V. Kremenitskii, S. Odoulov, and M. Soskin
*Dynamic Gratings in Cadmium Telluride
(a) Vol. 51, No. 1, p. K63–K66 (1979).
See also 22.4.3, p. 455.

A. A. Kaminskii, S. E. Sarkisov, I. V. Mochalov, L. K. Aminov, and A. O. Ivanov
*Anisotropy of Spectroscopic Characteristics in the Biaxial $YAlO_3$–Nd^{3+} Laser Crystals
(a) Vol. 51, No. 2, p. 509–520 (1979).
See also 20.3, p. 327; 22.8, p. 492.

H. J. Eichler, J. Eichler, K. Knof, and Ch. Noack
Lifetimes of Laser-Induced Population Density Gratings in Ruby
(a) Vol. 52, No. 2, p. 481–486 (1979).
See also *13.4*, p. *168*; *20.3*, p. *329*; 22.6, p. 479.

A. A. Kaminskii, V. A. Fedorov, S. E. Sarkisov, J. Bohm, P. Reiche, and D. Schultze
*Stimulated Emission of Ho^{3+} and Er^{3+} Ions in $Gd_3Ga_5O_{12}$ Crystals and Cascade Laser Action of Ho^{3+} Ions over the $^5S_2 \rightarrow {}^5I_5 \rightarrow {}^5I_6 \rightarrow {}^5I_8$ Scheme
(a) Vol. 53, No. 2, p. K219–K222 (1979).
See also 22.8, p. 495.

V. May, K. Henneberger, and F. Henneberger
Renormalization of Polaritons Due to Virtual Formation of Biexcitons at High Densities of Excitation
(b) Vol. 94, No. 2, p. 611–620 (1979).
See also *13.5.1*, p. *187*; 20.1, p. 317; 22.5, p. 461.

S. P. Tandon, P. P. Vaishnava, and K. Tandon
Optical Properties of Laser Materials: Solid State Vibrational Spectra of Rare Earth: p–Nitrobenzoic Acid in the Region 800 to 200 cm^{-1}
(a) Vol. 54, No. 2, p. K117–K118 (1979).
See also *6*, p. *72*; *20.1*, p. *318*; 21.4, p. 384.

20.3 Luminescence
(X-Ray Spectra See 20)

Original Papers and Short Notes

V. A. Vilkotskii, D. S. Domanevskii, R. D. Kakanakov, V. V. Krasovskii, and V. D. Tkachev
*Burstein-Moss Effect and Near-Band-Edge Luminescence Spectrum of Highly Doped Indium Arsenide
(b) Vol. **91**, No. 1, p. 71–81 (1979).
See also *13.1*, p. *143*; 22.2.1, p. 432.

L. Bosi, A. Longoni, and M. Nimis
*Addendum to: Lifetime of Color Centers in NaBr
(b) Vol. **91**, No. 1, p. 377 (1979).
See also *10.2*, p. *103*; 13.4, p. 164; 22.5.2, p. 465.

B. B. Laud and V. W. Kulkarni
*Photo- and Electro-Luminescence of Polycrystalline BaS:Cu Phosphors
(a) Vol. **51**, No. 1, p. 269–274 (1979).
See also 22.4.1, p. 450.

A. S. Popov and R. T. Yakimova
*Photoluminescence Spectra of n-Type GaAs:Sn Crystals
(a) Vol. **51**, No. 1, p. K17–K21 (1979).
See also 22.2.1, p. 432.

A. J. Harris and H. Thomas
*On the Electroluminescent Efficiency of SiC Diode Types
(a) Vol. **51**, No. 1, p. K49–K51 (1979).
See also *14.3.3*, p. *225*; 22.3, p. 444.

O. Breitenstein and K. Unger
*Calculation of the Broadening of Donor–Acceptor Pair Lines Due to the Potentials of Ionized Impurities
(b) Vol. **91**, No. 2, p. 557–562 (1979).
See also *13.4*, p. *164*; 22, p. 394.

V. M. Grekhov, V. V. Luchinin, N. S. Peev, Yu. M. Suleimanov, Yu. M. Tairov, and E. V. Vronskaya
*Influence of Synthesis Conditions on the Energy Distribution in Photoluminescence Spectra of SiC Epitaxial Layers
(a) Vol. **51**, No. 2, p. 429–433 (1979).
See also 22.3, p. 444.

A. A. Kaminskii, S. E. Sarkisov, I. V. Mochalov, L. K. Aminov, and A. O. Ivanov
Anisotropy of Spectroscopic Characteristics in the Biaxial $YAlO_3$–Nd^{3+} Laser Crystals
(a) Vol. **51**, No. 2, p. 509–520 (1979).
See also *20.2*, p. *325*; 22.8, p. 492.

K. D. Glinchuk, A. V. Prokhorovich, and V. I. Vovnenko
*Study of Non-Linear Extrinsic Luminescence in GaAs (II): The Role of Auger Recombination
(a) Vol. **51**, No. 2, p. 645–655 (1979).
See also *13.4*, p. *165*; 22.2.1, p. 432.

Yu. E. Perlin, A. A. Kaminskii, V. N. Enakii, and D. N. Vylegzhanin
*Nonradiative Multiphonon Relaxation in $Y_3Al_5O_{12}:Nd^{3+}$
(b) Vol. 92, No. 2, p. 403–410 (1979).
See also 6, p. 66; *13.4*, p. *167*; 22.8.2, p. 505.

G. Weber and W. Rühle
Effect of Uniaxial Stress on Free and Bismuth-Bound Excitons in InP
(b) Vol. 92, No. 2, p. 425–431 (1979).
See also 12.2, p. 134; *13.4*, p. *167*; *13.5.1*, p. *184*; 22.2.2, p. 437.

N. Koumvakalis and W. von der Osten
*First-Order Resonant Raman Scattering in Mixed Silver Halides
(b) Vol. 92, No. 2, p. 441–445 (1979).
See also *6*, p. *66*; 13.5.1, p. 184; 20.1, p. 309; 22.5.1, p. 463; 22.5.4, p. 475.

V. D. Kulakovskii and A. V. Malyavkin
*The Electronic Structure of the Bound Exciton and Bound Multiexciton Complexes in Al-Doped Si
(b) Vol. 92, No. 2, p. 455–465 (1979).
See also *13.5.1*, p. *184*; 22.1.2, p. 418.

T. Tashiro, S. Takeuchi, M. Saidoh, and N. Itoh
*Luminescence Induced by Tunnelling Recombination between Neutral Silver Atoms and V_k Centres in Alkali Halides
(b) Vol. 92, No. 2, p. 611–618 (1979).
See also 13.4, p. 168; 22.5.2, p. 466.

H. J. Eichler, J. Eichler, J. Knof, and Ch. Noack
*Lifetimes of Laser-Induced Population Density Gratings in Ruby
(a) Vol. 52, No. 2, p. 481–486 (1979).
See also *13.4*, p. *168*; 20.2, p. 325; 22.6, p. 479.

L. A. Demberel, A. S. Popov, and D. B. Kushev
Deep Levels in Fe-Doped GaP
(a) Vol. 52, No. 2, p. 653–656 (1979).
See also *13.4*, p. *168*; *14.3.4*, p. *233*; 20.1, p. 310; 22.2.2, p. 437.

H. B. Tripathi, H. C. Kandpal, and A. K. Agarwal
*Non-Radiative Energy Transfer from $Dy^{3+} \rightarrow Ho^{3+}$ in Calibo Glass
(a) Vol. 52, No. 2, p. 697–701 (1979).
See also *13.4*, p. *168*; 22.8, p. 493.

A. Subrahmanyam and K. V. Rao
*F-Band Absorption and Thermoluminescence of KCl Single Crystals under DC Field and X-Ray Irradiation
(a) Vol. 52, No. 2, p. K147–K150 (1979).
See also *20.1*, p. *311*; 22.5.2, p. 466.

R. Bindemann, E. Hempel, and K. Kreher
*Determination of the Dielectric Constant of GaP from S–C Donor–Acceptor Pair Spectra
(a) Vol. 52, No. 2, p. K201–K203 (1979).
See also *13.4*, p. *169*; 14.4, p. 239; 22.2.2, p. 437.

A. G. Makhanek and G. A. Skripko
*Application of Two-Photon Spectroscopy in the Study of Trivalent Rare-Earth Ions in Crystals
(a) Vol. **53**, No. 1, p. 243–252 (1979).
See also 13.4, p. 170; *20.1*, p. *312*; 22.5.3, p. 473.

B. S. Acharya and M. L. Mukherjee
*Isothermal Decay of Luminescence and Photostimulated Decay of Luminescence in KCl and KCl:Sn Single Crystals
(a) Vol. **53**, No. 1, p. 373–378 (1979).
See also 22.5.2, p. 467.

L. Huldt, N. G. Nilsson, B. O. Sundström, and W. Żdanowicz
*Photoluminescence of Zinc Phosphide — Zn_3P_2
(a) Vol. **53**, No. 1, p. K15–K18 (1979).
See also 22.3, p. 445.

S. Bhushan, A. N. Pandey, and B. Rao Kaza
*Low Temperature Studies of Photo- and Electroluminescence of Rare Earth (Yb, Nd, Pr, or Sm) Doped ZnO Electroluminophors
(a) Vol. **53**, No. 1, p. K57–K60 (1979).
See also 22.4, p. 448.

J. M. Hvam
Excitonic Molecule Transitions in ZnO
(b) Vol. **93**, No. 2, p. 581–590 (1979).
See also *13.5.1*, p. *185*; 20.1, p. 313; 22.4, p. 448.

W. Busse, H. E. Gumlich, A. Geoffroy, and R. Parrot
*Spectral Distribution and Decay Times of the Luminescence of Mn^{2+} on Different Lattice Sites in ZnS
(b) Vol. **93**, No. 2, p. 591–596 (1979).
See also *13.4*, p. *170*; 22.4.1, p. 450.

H. Schrey and C. Klingshirn
*Ground and Excited States of the Biexcition and the M-Band Luminescence in Highly Excited ZnO
(b) Vol. **93**, No. 2, p. 679–689 (1979).
See also *13.5.1*, p. *186*; 22.4, p. 448.

M. Casalboni, F. Crisanti, U. Grassano, C. Manfroncelli, A. Scacco, and A. Tanga
*Luminescence of $KBr:In^+$
(b) Vol. **93**, No. 2, p. 755–759 (1979).
See also *13.4*, p. *170*; 22.5.2, p. 468.

G. Boulon, J. C. Gacon, D. Trottier, and A. Vedrine
*Eu^{2+} Luminescence in KY_3F_{10}
(b) Vol. **93**, No. 2, p. 775–779 (1979).
See also 22.5, p. 461.

R. S. Turtelli and A. R. B. de Castro
LO Phonon Raman Lineshape in Highly Excited GaAs
(b) Vol. **93**, No. 2, p. 811–815 (1979).
See also *6*, p. *69*; *20.1*, p. *313*; 22.2.1, p. 433.

J. A. Groenink, C. Hakfoort, and G. Blasse
*The Luminescence of Calcium Molybdate
(a) Vol. 54, No. 1, p. 329–336 (1979).
See also *13.4*, p. *173*; 22.8.1, p. 500.

V. S. Sivasankar and P. W. Whippey
*Thermoluminescence and Optical Absorption of X-Ray Irradiated $NaBrO_3$
(a) Vol. 54, No. 1, p. 369–373 (1979).
See also *11*, p. *122*; 20.1, p. 316; 22.8.1, p. 501.

M. A. S. Sweet and D. Urquhart
Shallow Traps in ZnS Single Crystals
(a) Vol. 54, No. 1, p. K81–K84 (1979).
See also *13.4*, p. *173*; 22.4.1, p. 451.

W. Schmid
*Determination of Thermodynamical Parameters of the EHD in Silicon from Luminescence Data
(b) Vol. 94, No. 2, p. 413–419 (1979).
See also 6, p. 71; *13.5.1*, p. *187*; 22.1.2, p. 422.

S. Asano et N. Yamashita
*Luminescence du centre dimère des ions Sn^{2+} dans le phosphore CaSe
(Luminescence of the Dimer Centre of Sn^{2+} Ions in the CaSe Phosphor)
(b) Vol. 94, No. 2, p. 577–582 (1979).
See also *13.4*, p. *174*; 22.4.2, p. 454.

B. Esser and D. Sprenger
*On the Theory of Absorption and Luminescence Spectra in Amorphous Semiconductors
(b) Vol. 94, No. 2, p. 583–593 (1979).
See also 2, p. 46; *20.1*, p. *316*; 22, p. 403.

W. Rehm and R. Fischer
*Fast Radiationless Recombination in Amorphous Silicon
(b) Vol. 94, No. 2, p. 595–602 (1979).
See also 2, p. 46; 22.1.2, p. 422.

N. Magnea, D. Bensahel, J. L. Pautrat, and J. C. Pfister
*Optical Identification of Substitutional Acceptors in Refined ZnTe
(b) Vol. 94, No. 2, p. 627–639 (1979).
See also *13.4*, p. *174*; 13.5.1, p. 187; 22.4.3, p. 455.

G. S. Zavt and A. F. Ellervee
*Pb^{2+} and Bi^{3+} Impurity Centres in Alkali-Earth Oxides. Vibronic Spectra, Lattice Dynamics, and Electron–Phonon Interaction
(b) Vol. 94, No. 2, p. 757–768 (1979).
See also 6, p. 72; *13.4*, p. *174*; 22.6, p. 482.

S. B. S. Sastry and S. Sapru
*Thermoluminescence Glow and Emission Studies on Suprapure Rubidium Bromide Irradiated at Room Temperature
(b) Vol. 94, No, 2, p. K149–K154 (1979).
See also *10.2*, p. *111*; 22.5.2, p. 469.

I. E. Lacis and J. A. Valbis
*Vacuum-Ultraviolett Luminescence of Be-Doped MgO Epitaxial Layers
(b) Vol. 95, No. 1, p. K21—K26 (1979).
See also 22.6, p. 483.

S. V. Moharil, D. G. Wakde, and B. T. Deshmukh
*Thermoluminescence in Mixed KCl:KBr Crystals
(b) Vol. 95, No. 1, p. K73—K78 (1979).
See also *10.2*, p. *112*; 22.5.2, p. 470.

I. V. Mochalov
Spectroscopic Characteristics of the Erbium Ions and Laser Action at the Transition $^4S_{3/2} \rightarrow {}^4I_{9/2}$ ($\lambda_e \approx 1.66$ μm) in the Crystals of Gadolinium, Yttrium, and Lutecium Orthoaluminates
(a) Vol. 55, No. 1, p. 79—84 (1979).
See also 13.4, p. 175; 20.1, p. 319; *20.2*, p. *326*; 22.8.1, p. 502

B. V. R. Chowdari and S. Selvasekarapandian
*Thermoluminescence of Cu^+ Doped CsCl and CsBr Crystals
(a) Vol. 55, No. 1, p. K103—K106 (1979).
See also 22.5.2, p. 470.

R. K. Sinha, R. K. Gartia, and B. S. Acharya
*Thermoluminescence of Z_1 Centers in Ca-Doped $KCl_{0.5}Br_{0.5}$ Mixed Crystals
(a) Vol. 55, No. 1, p. K109—K113 (1979).
See also *10.2*, p. *113*; 22.5.4, p. 476.

R. N. Tripathi and S. Sivaraman
*Fluorescence Spectra of CaS:(Tb, Eu) Phosphors under X-Ray Excitation
(a) Vol. 55, No. 1, p. K123—K126 (1979).
See also 11, p. 124; 22.4.1, p. 451.

G. Pauli, K. F. Renk, G. Klimke, and H. J. Kreuzer
Resonance Interaction of Electronic Two-Level States and Short-Wave Phonon Radiation in Ruby
(b) Vol. 95, No. 2, p. 503—508 (1979).
See also *6*, p. *74*; *13.4*, p. *177*; 22.6, p. 484.

R. Schwabe, F. Thuselt, H. Weinert, and R. Bindemann
*Electron–Hole Liquid in GaP. The Influence of the Isoelectronic Impurity Nitrogen
(b) Vol. 95, No. 2, p. 571—583 (1979).
See also *13.5.1*, p. *188*; 22.2.2, p. 438.

L. Bosi and M. Nimis
*To What Extent Are the Observed Lifetimes of F-Centers Intrinsic?
(b) Vol. 95, No. 2, p. 615—619 (1979).
See also *13.4*, p. *177*; 22.5.2, p. 470.

V. D. Negryi and Yu. A. Osipyan
*Dislocation Emission in CdS
(a) Vol. 55, No. 2, p. 583—588 (1979).
See also *10.2*, p. *114*; 22.4.1, p. 451.

ZH. EGEMBERDIEV, Ü. HALDRE, T. LEHTO, K. REALO, S. REIFMAN, V. SEEMAN, and S. ZAZUBOVICH
*Polarized Luminescence and EPR Study of Sn^{3+} Centres in KCl:$SnCl_2$ Crystals
(b) Vol. **96**, No. 2, p. 867–875 (1979).
See also *10.2*, p. *116*; 19, p. 297; 20.1, p. 323; 22.5.4, p. 476.

J. MAZZASCHI, J. BARRAU, M. BROUSSEAU, J. COLLET, and H. MAAREF
*Luminescence Line Shape of Free Excitons in Natural Diamond
(b) Vol. **96**, No. 2, p. K61–K64 (1979).
See also *13.5.1*, p. *189*; 22.1, p. 408.

H.-J. KOS and R. NINK
Correlation of Thermoluminescence and Optical Absorption of Z_3 Centers in LiF:Mg, Ti Crystals
(a) Vol. **56**, No. 2, p. 593–596 (1979).
See also *10.2*, p. *117*; *20.1*, p. *324*; 22.5.2, p. 472.

A. A. KAMINSKII, S. E. SARKISOV, T. I. BUTAEVA, G. A. DENISENKO, B. HERMONEIT, J. BOHM, W. GROSSKREUTZ, and D. SCHULTZE
*Growth, Spectroscopy, and Stimulated Emission of Cubic $Bi_4Ge_3O_{12}$ Crystals Doped with Dy^{3+}, Ho^{3+}, Er^{3+}, Tm^{3+}, or Yb^{3+} Ions
(a) Vol. **56**, No. 2, p. 725–736 (1979).
See also 20.1, p. 324; *20.2*, p. *326*; 22.8, p. 497.

K. SOMAIAH, P. VEERESHAM, K. L. N. PRASAD, and V. HARI BABU
*Thermoluminescence and Optical Absorption of BaFCl Crystals
(a) Vol. **56**, No. 2, p. 737–742 (1979).
See also 10.2, p. 117; 20.1, p. 324; 22.5.4, p. 476.

R. P. KHARE
*Electroluminescence of CaO:Ag and CaO:Au
(a) Vol. **56**, No. 2, p. K97–K99 (1979).
See also 22.6, p. 486.

B. P. CHANDRA
*Molecular Triboluminescence in Benzil Crystals
(a) Vol. **56**, No. 2, p. K105–K108 (1979).
See also 22.9, p. 514.

21. Metals. Alloys

Review Article

Original Papers and Short Notes

K. ELK, J. RICHTER, and V. CHRISTOPH
Influence of Electron Correlation on the Electrical Resistivity of Disordered Transition Metal Alloys
(b) Vol. **91**, No. 1, p. K39–K42 (1979).
See also 13, p. 136; *14.1*, p. *199*.

W. BADEN, P. C. SCHMIDT, and ALARICH WEISS
The Intermetallic System $LiCd_{1-x}Tl_x$. X-Ray Investigations and Measurements of the Knight Shift of ^{205}Tl and ^{113}Cd
(a) Vol. **51**, No. 1, p. 183–190 (1979).
See also 1.2, p. 15; *1.3*, p. *22*; 19, p. 290.

S. TANIGAWA, K. HINODE, R. NAGAI, K. KANBE, M. DOYAMA, and N. SHIOTANI
Positron Annihilation in Amorphous Alloys
(a) Vol. **51**, No. 1, p. 249–255 (1979).
See also *2*, p. *40*.

S. A. MAHMOUD, K. H. GEORGY, F. M. MANSY, and R. KAMEL
Effect of Cold-Work on the Mechanism Controlling the High-Temperature Creep in Al–2 wt% Mg
(a) Vol. **51**, No. 1, p. 257–260 (1979).
See also *10.1*, p. *92*; *12.1*, p. *129*.

B. B. YU and R. F. DAVIS
Near-Surface Effect in Self-Diffusion in NbC_x
(a) Vol. **51**, No. 1, p. 261–267 (1979).
See also *9*, p. *85*.

G. SCHANZ, V. MÜLLER, and E. J. UNTERHORST
Nuclear-Acoustic-Resonance Determination of the Core Polarization Contribution to the Knight Shift and Hyperfine Fields in Single-Crystal Tantalum
(b) Vol. **91**, No. 2, p. 421–425 (1979).
See also *18.1*, p. *265*; *19*, p. *290*.

V. CHRISTOPH
Temperature Dependence of the Electrical Resistivity of Disordered Metallic Alloys
(b) Vol. **91**, No. 2, p. 593–603 (1979).
See also *14.1*, p. *199*.

W. M. WEBER
Anisotropic K Absorption in Gallium
(b) Vol. **91**, No. 2, p. 667–673 (1979).
See also *20*, p. *298*.

H. P. AUBAUER
Definition of a Local Concentration, Short-Range Order, Clustering, and Microstructure of Alloys for an Arbitrary Duration and Temperature of Experiment
(b) Vol. **91**, No. 2, p. K139–K142 (1979).
See also *1.3*, p. *22*; *1.4*, p. *26*.

K. S. SRIVASTAVA, O. K. HARSH, and V. KUMAR
Chemical Shifts of the X-Ray L-Absorption Edges of Pb and Bi
(b) Vol. **91**, No. 2, p. K169–K172 (1979).
See also *20*, p. *299*; 21.7, p. 391; 22, p. 394.

W. F. W. M. van Heugten, L. M. Caspers, and J. Th. M. de Hosson
Atomic Configuration of a 1/2 ⟨111⟩ Screw Dislocation in Pure Mo and in Mo Containing He Interstitials
(b) Vol. 92, No. 1, p. 199–209 (1979).
See also *10.1*, p. *93*.

C. N. Tomé, A. M. Monti, and E. J. Savino
Vacancy and Interstitial Configuration in Hexagonal Close Packed Metals
(b) Vol. 92, No. 1, p. 323–334 (1979).
See also *10.1*, p. *93*.

O. A. Troitskii, V. I. Spitsyn, N. V. Sokolov, and V. G. Ryzhkov
Application of High-Density Current in Plastic Working of Metals
(a) Vol. 52, No. 1, p. 85–93 (1979).
See also *14.1*, p. *199*.

J. Gastaldi and C. Jourdan
Observation of the Annealing Twin Growth by Synchrotron Radiation X-Ray Topography
(a) Vol. 52, No. 1, p. 139–148 (1979).
See also *1.4*, p. *27*; *10.1*, p. *93*.

G. Dlubek, O. Brümmer, and Ashraful Alam
A Study of Recovery and Recrystallization of Plastically Deformed $\alpha + \beta$ Brass by Positron Annihilation
(a) Vol. 52, No. 1, p. 217–221 (1979).
See also *10.1*, p. *93*.

F. Vávra
Influence of Metallic Impurities on Defect Production and Recovery in Niobium Neutron-Irradiated at 4.6 K
(a) Vol. 52, No. 1, p. 269–276 (1979).
See also *10.1*, p. *93*; 11, p. 120; 14.1, p. 200.

R. A. Varin
Spreading of Extrinsic Grain Boundary Dislocations in Austenitic Steel
(a) Vol. 52, No. 1, p. 347–356 (1979).
See also *1.4*, p. *27*; 10.1, p. 94.

F. Fouquet, P. Merle, and J. Merlin
Characterization of Stages during the Precipitation of the Θ'-Phase in an Al–4 wt% Cu Alloy
(a) Vol. 52, No. 1, p. K1–K5 (1979).
See also *1.3*, p. *22*; 10.1, p. 94; 12.1, p. 129; 21.1, p. 357.

F. Dettmann and P. Pertsch
Nb–NbO_x–PbIn Tunnel Junctions with Ultra-Thin Niobium Electrodes
(a) Vol. 52, No. 1, p. K17–K20 (1979).
See also *14.2*, p. *205*.

M. Ohta, M. Yamada, T. Kanadani, and A. Sakakibara
Effects of the Fluctuation of Solute Concentration in Al–10 wt% Zn Alloy on Aging at Low Temperature
(a) Vol. 52, No. 1, p. K21–K24 (1979).
See also *1.3*, p. *23*; *14.1*, p. *200*; 21.1, p. 357.

A. J. Drehman and W. L. Johnson
Effects of Thermal Relaxation on an Amorphous Superconducting Zr–Rh Alloy
(a) Vol. 52, No. 2, p. 499–507 (1979).
See also 2, p. 42; *14.2*, p. *205*.

A. Bose, G. Frohberg, and H. Wever
Self Diffusion of ^{60}Co and ^{72}Ga in the Ordered Electronic Compound CoGa
(a) Vol. 52, No. 2, p. 509–520 (1979).
See also *9*, p. *86*; 21.1, p. 357.

R. H. J. Fastenau, C. M. van Baal, P. Penning, and A. van Veen
*On the Sink Concentration Dependence of Reaction Constants for Point Defect Trapping
(a) Vol. 52, No. 2, p. 577–586 (1979).
See also *10.1*, p. *95*.

M. Bogdan, A. Nicula, and D. Lupu
Spin–Lattice Relaxation Time and Quadrupole Coupling of Deuterium in $PdD_{0.70}$
(a) Vol. 52, No. 2, p. 615–619 (1979).
See also *9*, p. *86*; 19, p. 291.

O. V. Presnyakova, V. I. Zaitsev, and N. A. Doroshenko
Aluminium Polygonization under High Pressure
(a) Vol. 52, No. 2, p. 627–633 (1979).
See also *1.4*, p. *28*; *10.1*, p. *95*.

F. Dettmann
Enhancement of Magnetic Field in "Window"-Type Tunnel Junctions
(a) Vol. 52, No. 2, p. K161–K164 (1979).
See also *14.2*, p. *206*.

C. Boiziau, V. Dose, and H. Scheidt
The Density of Unoccupied Electronic States in Tantal
(b) Vol. 93, No. 1, p. 197–201 (1979).
See also *13.1*, p. *147*.

V. A. Gasparov and M. H. Harutunian
Precision Investigations of the Fermi Surface of Molybdenum
(b) Vol. 93, No. 1, p. 403–414 (1979).
See also *13.2*, p. *156*; 19, p. 292.

V. Ya. Chumanov, Sh. Z. Izmailov, G. P. Pokhil, E. I. Sirotinin, and A. F. Tulinov
On the Determination of Energy Losses by Charged Particles from the Backscattered Energy Spectra
(a) Vol. 53, No. 1, p. 51–62 (1979).
See also *11*, p. *121*.

D. van Dyck, D. Colaïtis, P. Delavignette, and S. Amelinckx
On the Diffraction Features of Two-Layer Systems. Application to $Ni_{3\pm x}Te_2$
(a) Vol. 53, No. 1, p. 105–112 (1979).
See also *1.3*, p. *23*.

R. Ciach, H. Löffler, J. Salawa, and G. Wendrock
Modulated Structures and Position of the Coherent Spinodal Curve in Al–Zn(22 at%)–Cu(x) Alloys
(a) Vol. **53**, No. 2, p. 441–446 (1979).
See also *1.2*, p. *18*; 10.1, p. 96; 21.1, p. 360.

A. A. Aly, G. N. Podus, and A. F. Sirenko
Thermoactivated Creep and Stress Relaxation in Cu and Al
(a) Vol. **53**, No. 2, p. 461–467 (1979).
See also *10.1*, p. *96*; 21.1, p. 360.

J. Sosnowski and M. Drys
Superconductivity of the A15 Phase in $Nb_{78}Ga_{22-x}Fe_x$ and $Nb_{80}Ga_{20-x}Fe_x$ Alloys
(a) Vol. **53**, No. 2, p. 481–485 (1979).
See also *14.2*, p. *206*.

E. Bonetti, E. Evangelista, and P. Gondi
Internal Friction and Dynamic Modulus of Polycrystalline Al during Deformation in the Temperature Range of the Kê Peak
(a) Vol. **53**, No. 2, p. 653–658 (1979).
See also *10.1*, p. *96*; 12.1, p. 130.

P. Lukáč and Z. Trojanová
Solid Solution Hardening of Cadmium Single Crystals
(a) Vol. **53**, No. 2, p. K143–K145 (1979).
See also *10.1*, p. *97*.

G. Dlubek, O. Brümmer, and N. Meyendorf
Vacancy–Impurity Binding Enthalpy in $\underline{Ni}Sb$ by Means of Positron Annihilation
(a) Vol. **53**, No. 2, p. K157–K160 (1979).
See also *10.1*, p. *97*; 21.1, p. 360.

I. I. Eru, S. A. Peskovatskii, and A. V. Poladich
Current–Voltage Characteristics of Long Superconducting Microbridges of Variable Thickness and with a Normal Bank
(a) Vol. **53**, No. 2, p. K185–K187 (1979).
See also *14.2*, p. *206*.

H. Sato
Sondheimer Oscillation in Aluminium Single Crystals
(b) Vol. **94**, No. 1, p. 309–315 (1979).
See also 13.2, p. 156; *14.1*, p. *201*.

P. Otschik
Problems of the Calculation of Superconducting Parameters of Nb by Tunnelling Measurements
(b) Vol. **94**, No. 1, p. K61–K66 (1979).
See also *14.2*, p. *206*.

A. Ślebarski, K. Ławniczak, and J. Auleytner
Soft X-Ray Emission Spectra of Aluminium and Manganese from $Gd(Al_{1-x}Mn_x)_2$ Intermetallic Compounds
(a) Vol. **54**, No. 1, p. 79–83 (1979).
See also *20*, p. *301*; 21.4, p. 384.

C. SALUSTRI
Plasmons in Three-Component Systems
(b) Vol. 94, No. 2, p. 517–524 (1979).
See also *13.5.2*, p. *192*; 21.1, p. 362.

K. BERNDT, U. MARX, and O. BRÜMMER
Photoelectron Spectra of Core Electrons from Palladium–Silver Alloys
(b) Vol. 94, No. 2, p. 541–548 (1979).
See also *13*, p. *139*; 17, p. 257; 21.6, p. 388.

K. JOHN and W. KELLER
Density of States of Expanded Liquid Mercury. An Effective Medium Approach
(b) Vol. 94, No. 1, p. 603–609 (1979).
See also *2*, p. *46*; *13*, p. *140*.

R. RAMJI and A. RAJPUT
Third-Order Elastic Constants, Thermal Expansion, and Bulk Modulus of Body-Centered Cubic Crystals
(b) Vol. 94, No. 2, p. 691–699 (1979).
See also *12.1*, p. *131*; 21.2, p. 379.

SEUNG-AM CHO
Atomic Vibration and Fusion Behaviour of A_3 (H.C.P.) Metals
(b) Vol. 94, No. 1, p. K123–K127 (1979).
See also *6*, p. *72*; *8*, p. *82*; 21.1, p. 363; 21.3, p. 381.

J. ZIELIŃSKI
Electrical Resistivity in the Anderson Model
(b) Vol. 94, No. 2, p. K141–K143 (1979).
See also *14.1*, p. *201*.

R. RANGANATHAN and S. DATTAGUPTA
Theory of Ultrasonic Scattering by Grains in Polycrystals
(a) Vol. 54, No. 2, p. 537–546 (1979).
See also 1.4, p. 29; *7*, p. *79*.

E. SCHACHINGER, H. W. WEBER, and H. ZLÖBL
Dynamic Destruction of Superconductivity in Pb–n% Sb Alloys
(a) Vol. 54, No. 2, p. 557–563 (1979).
See also *14.2*, p. *207*; 18, p. 263.

O. A. TROITSKII, V. I. SPITSYN, M. M. MOISEENKO, and A. D. TSIGANOV
Fractionation of Stable Isotopes and Impurities during Zone Recrystallization of Metals and Preparation of Single Crystals Rich in Individual Isotopes
(a) Vol. 54, No. 2, p. 651–654 (1979).
See also *3*, p. *53*; 9, p. 87; 21.1, p. 363.

L. D. LIVSHITS and A. G. FOMICHEV
Micro-Indentation of Aluminium under Hydrostatic Pressures up to 8500 kp/cm^2
(a) Vol. 54, No. 2, p. K123–K127 (1979).
See also *12.1*, p. *131*.

V. M. Chernov and M. M. Savin
Elastic Interaction of Edge and Screw Dislocations with Point Defects in Cubic Crystals
(a) Vol. 55, No. 1, p. 113–122 (1979).
See also *10.1*, p. *98*.

H. C. Kim and T. Kishi
The Grain Size and Flow Stress Dependence of Acoustic Emission Energy Release in Polycrystalline Aluminium
(a) Vol. 55, No. 1, p. 189–195 (1979).
See also 7, p. 79; *10.1*, p. *98*.

O. Buck, L. A. Ahlberg, L. J. Graham, G. A. Alers, C. A. Wert, and K. C. Hsieh
Harmonic and Anharmonic Elastic Properties of the α-Nb–H System
(a) Vol. 55, No. 1, p. 223–230 (1979).
See also *12.1*, p. *131*.

B. K. Samantaray and G. D. Nigam
A Simple Relation for the Additivity of the Debye Characteristic Temperature of Binary Alloys
(a) Vol. 55, No. 1, p. K23–K26 (1979).
See also *1.3*, p. *24*; *6*, p. *73*; 8, p. 83.

M. A. Chubisov and A. F. Akkerman
Computer Simulation of Low Energy Hydrogen and Helium Ion Reflection from Monocrystalline Surfaces
(a) Vol. 55, No. 1, p. K53–K58 (1979).
See also *11*, p. *123*; 22.1.2, p. 424.

P. Varotsos
Comments on the Migration Volume of Vacancies in Solids
(a) Vol. 55, No. 1, p. K63–K66 (1979).
See also *10*, p. *90*; 22, p. 404.

V. L. Laska, V. P. Ledovskoi, N. A. Potsar, E. Seidowski, and A. Kh. Chizhov
Quasi-Closed Volume as the Method of Preparing the Complex Superconducting Films
(a) Vol. 55, No. 1, p. K107–K108 (1979).
See also *3*, p. *53*; *14.2*, p. *207*.

N. Igata, N. Kuramochi, N. Masamura, A. Kohyama, K. Miyahara, S. Sato, and H. Yoshida
The α′-Peak of Deformed Niobium
(a) Vol. 55, No. 1, p. K127–K129 (1979).
See also *10.1*, p. *98*; *12.1*, p. *131*.

G. Borstel, W. Braun, M. Neumann, and G. Seitz
Band Structure of Rhodium and Photoemission from Its Low Index Surfaces
(b) Vol. 95, No. 2, p. 453–460 (1979).
See also *13.1*, p. *152*; 17, p. 258.

R. Richter, M. Wolf, and F. Goedsche
On the Spin-Disorder Resistivity of Amorphous Ferromagnets
(b) Vol. 95, No. 2, p. 473–482 (1979).
See also *2*, p. *48*; *14.1*, p. *202*; 18.2, p. 278.

M. Kiwi, E. Pestana, and R. Ramírez
Magnitude of Localized Magnetic Moments in Metals
(b) Vol. 95, No. 2, p. 497–502 (1979).
See also *13.4*, p. *176*; 18, p. 263.

A. K. Tewari, S. K. Roy, and D. L. Bhattacharya
Resonance Energy Shift for Mössbauer Impurity Atoms Trapped by Vacancies
(b) Vol. 95, No. 2, p. 607–614 (1979).
See also *5*, p. *61*; 10.1, p. 98.

O. K. Alekseeva, V. A. Onishchuk, V. P. Shantarovich, I. Ya. Dekhtyar, and V. I. Shevchenko
Influence of Absorbed Hydrogen on Positron Lifetime Spectra in Pd, Pd–Ag, Pd–Cu
(b) Vol. 95, No. 2, p. K135–K139 (1979).
See also *4*, p. *57*; *10.1*, p. *99*.

Á. Barna, P. B. Barna, G. Radnóczi, F. M. Reicha, and L. Tóth
Formation of Aluminium Thin Films in the Presence of Oxygen and Nickel
(a) Vol. 55, No. 2, p. 427–435 (1979).
See also *1.5*, p. *36*; *3*, p. *54*.

A. Seeger
A Theory of the Snoek-Köster Relaxation (Cold-Work Peak) in Metals
(a) Vol. 55, No. 2, p. 457–468 (1979).
See also *10.1*, p. *99*; 21.1.1, p. 375.

G. Dlubek, O. Kabisch, O. Brümmer, and H. Löffler
Precipitation and Dissolution Processes in Age-Hardenable Al Alloys. A Comparison of Positron Annihilation and X-Ray Small-Angle Scattering Investigations (I): Al–Zn(x) (x = 3, 4.5, 6, 10, and 18 at%)
(a) Vol. 55, No. 2, p. 509–518 (1979).
See also *10.1*, p. *99*; 21.1, p. 365.

B. Jouffrey
Methodical Aspects of HVEM in In-Situ Experiments in the Field of Condensed Matter
(a) Vol. 55, No. 2, p. 549–562 (1979).
See also *1.4*, p. *31*; 22, p. 405.

G. Thomas and K. H. Westmacott
The Role of Environmental Conditions in In-Situ Experiments in the High-Voltage Electron Microscope
(a) Vol. 55, No. 2, p. 563–568 (1979).
See also *1.4*, p. *31*; 11, p. 124; 21.1.1, p. 376.

J. Mucha and J. Rafałowicz
Influence of the Lattice Component on the Thermal Conductivity Minimum of Aluminium
(a) Vol. 55, No. 2, p. 631–634 (1979).
See also *8*, p. *83*.

P. Bartuška, R. Kroggel, H. Löffler, M. Simerská, V. Syneček, and G. Wendrock
On the Room Temperature Stability of Metastable Phases Formed at 200 °C in Al–Zn Alloys
(a) Vol. 55, No. 2, p. 673–679 (1979).
See also *1.3*, p. *25*; *10.1*, p. *99*; 21.1, p. 365.

V. A. Goltsov, N. I. Timofeev, and I. Yu. Machikina
The Phenomenon of Controllable Hydrogen Phase "Naklep" of Metals and Alloys
(a) Vol. **55**, No. 2, p. 691–698 (1979).
See also *1.2*, p. *20*; *1.3*, p. *25*; 10.1, p. 99.

F. Okuyama
A Novel Technique for Preparing Tungsten Field Cathodes with [001] Orientation
(a) Vol. **55**, No. 2, p. 793–800 (1979).
See also *17.1*, p. *260*.

P. Varotsos
Entropy and Volume of Solution of Substitutional Impurities in Solids
(a) Vol. **55**, No. 2, p. K139–K142 (1979).
See also *10*, p. *90*; 22, p. 405.

P. D. Pathak and N. P. Shah
Debye Temperatures of Silver and Aluminium at High Temperatures. Some New Correlations
(a) Vol. **55**, No. 2, p. K159–K162 (1979).
See also *6*, p. *75*; *8*, p. *83*; 21.6, p. 389.

A. V. Kobelev, R. M. Kobeleva, and V. F. Ukhov
On the Adhesion Theory for Two Metallic Surfaces
(b) Vol. **96**, No. 1, p. 169–176 (1979).
See also *13.3*, p. *161*; 21.1, p. 366; 21.2, p. 379.

G. Lautenschläger and E. Mrosan
Pre-Asymptotic Charge Oscillations around 3d Transition Metal Impurities in Aluminium
(b) Vol. **96**, No. 1, p. 183–187 (1979).
See also *13.4*, p. *178*.

E. Donato, E. S. Giuliano, R. Ruggeri, and A. Stancanelli
Appearance Potential Spectroscopy in Metals and Alloys. Multiple Scattering Approach
(b) Vol. **96**, No. 7, p. 313–320 (1979).
See also *13.1*, p. *153*; 20, p. 302.

M. Fähnle, B. Barbara, and H. Kronmüller
Thermally Activated Domain Wall Motion and Coercive Field in Materials with Extremely Narrow Domain Walls
(b) Vol. **96**, No. 1, p. 343–350 (1979).
See also *1.4*, p. *32*; 14.4.1, p. 247; 18, p. 264.

E. Borchi, S. de Gennaro, G. Pelosi, and A. Rettori
The Lattice Thermal Conductivity of Aluminium at Low Temperatures
(b) Vol. **96**, No. 1, p. 425–428 (1979).
See also *6*, p. *75*; *8*, p. *83*.

K. Stachulec and A. Stachulec
Wave-Phase Dependence of the Electrical Resistivity of Metallic Thin Films
(b) Vol. **96**, No. 1, p. K1–K5 (1979).
See also *14.1*, p. *202*.

H. Rauh
Investigation of the Electrical Resistance of Dislocations in Metals (I): General Theory
(b) Vol. **96**, No. 2, p. 627–640 (1979).
See also *10.1*, p. *101*; 13.2, p. 157; *14.1*, p. *203*.

A. Swaroop and D. M. Esterling
Interatomic Potentials from Experimental Phonon Spectra (II): General Results
(b) Vol. **96**, No. 2, p. 691–696 (1979).
See also 2, p. 49; *4*, p. *58*; 6, p. 76; 12.1, p. 131; 21.1, p. 368; 21.2, p. 380; 21.6, p. 390.

L. Ya. Vinnikov and O. V. Zharikov
Investigation of Pinning at Grain Boundaries in Pb–In Alloys
(a) Vol. **56**, No. 2, p. 433–442 (1979).
See also *10.1*, p. *101*; *14.2*, p. *208*.

B. Tirbonod and B. Vittoz
Introduction of a Dislocation–Defect Elastic Force in the Dragging Model of Internal Friction Peaks
(a) Vol. **56**, No. 2, p. 501–504 (1979).
See also *10.1*, p. *101*; *12.1*, p. *132*.

S. M. Ohr
Black–White Contrast of Dislocation Loops in Anisotropic Cubic Crystals
(a) Vol. **56**, No. 2, p. 527–530 (1979).
See also *1.4*, p. *32*; *10.1*, p. *101*.

E. Schachinger and H. Zlöbl
Dynamic Destruction of Superconductivity in Pb–n% In Alloys
(a) Vol. **56**, No. 2, p. 565–568 (1979).
See also *14.2*, p. *208*; 18, p. 264.

J. D. Livingston
Stresses and Magnetic Domains in Amorphous Metal Ribbons
(a) Vol. **56**, No. 2, p. 637–645 (1979).
See also 2, p. 50; *12.1*, p. *132*; *18.2*, p. *280*.

21.1 Transition Metals (Group Sc to Zn) and Their Alloys

Original Papers and Short Notes

F. Goedsche, R. Richter, and G. Vojta
Force–Force Correlation Function Approach to the Electrical Resistivity of Disordered Transition Metal Alloys
(b) Vol. 91, No. 2, p. 581–591 (1979).
See also 13, p. 136; *14.1*, p. *199*.

Ch. Müller, G. Seifert, G. Lautenschläger, H. Wonn, P. Ziesche, and E. Mrosan
Band Structure and Cluster Calculations of FeAl Systems
(b) Vol. 91, No. 2, p 605–613 (1979).
See also *13.1*, p. *144*; 20, p. 298; 21.1.1, p. 370.

J. Zieliński
The Effects of Magnetic Field and Direct Interaction between Impurities in the Anderson Model. Resistivity
(b) Vol. 91, No. 2, p. 615–625 (1979).
See also *14.1*, p. *199*; *18.1*, p. *265*; 21.6, p. 387.

P. Suortti
Scattering of X-Rays near the K Absorption Edge (I): Fluorecence and Resonant Raman Scattering in Transition Metals
(b) Vol. 91, No. 2, p. 657–666 (1979).
See also *20*, p. *298*.

U. Mizutani, T. Kondow, and M. Uda
Experimental Observation of the Fermi Surface–Brillouin Zone Interaction in Pure Zinc by Means of X-Ray Photoelectron Spectroscopy
(b) Vol. 91, No. 2, p. 693–696 (1979).
See also *13.1*, p. *144*; 17, p. 256.

N. I. Kulikov
Band Structure and Electronic Properties of Transition Metal Hydrides
(b) Vol. 91, No. 2, p. 753–762 (1979).
See also *13.1*, p. *145*; 14.2, p. 204; 18.2, p. 271; 20.1, p. 306.

E. Kolley and W. Kolley
Renormalization of the Spin Wave Stiffness in Ni-Based Alloys
(b) Vol. 91, No. 2, p. K173–K177 (1977).
See also *13.5.4*, p. *195*; 18.2, p. 271.

L. E. Murr and O. T. Inal
The Role of Substrate Dislocations and Grain Boundaries in the Nucleation and Growth of Thin Electrochemical Overgrowths. TEM and FIM Studies
(a) Vol. 51, No. 2, p. 345–358 (1979).
See also *1.4*, p. *26*; 1.5, p. 34; 1.6, p. 38.

M. Ignat and F. Durand
Creep of the Lamellar Al–$CuAl_2$ Composite (II): Stress and Temperature Dependence
(a) Vol. 51, No. 2, p. 567–572 (1979).
See also *10.1*, p. *92*; 21, p. 340.

H. S. Chen and Y. Waseda
Structure of Glassy Zr–Cu and Nb–Ni Alloys
(a) Vol. 51, No. 2, p. 593–599 (1979).
See also *2*, p. *41*.

R. C. Ecob, J. V. Bee, and B. Ralph
The Structure of the β-Phase in Dilute Copper–Titanium Alloys
(a) Vol. 52, No. 1, p. 201–210 (1979).
See also *1.2*, p. *17*; *1.4*, p. *27*.

S. Mager, E. Wieser, T. Zemčík, O. Schneeweiss, P. N. Stetsenko, and V. V. Surikov
Investigations of the Magnetic Moments in $(Fe_{1-x}Mn_x)_3Al$ Alloys
(a) Vol. 52, No. 1, p. 249–258 (1979).
See also *18*, p. *261*; 21.1.1, p. 371.

A. Caplain and W. Chambron
Study of the Interaction between Vacancies and Boron in a Ni–20 at% Fe Alloy by the Induced Magnetic Anisotropy
(a) Vol. 52, No. 1, p. 299–310 (1979).
See also *10.1*, p. *94*; 18.2, p. 272; 21.1.1, p. 371.

F. Fouquet, P. Merle, and J. Merlin
Characterization of Stages during the Precipitation of the Θ'-Phase in an Al–4 wt% Cu Alloy
(a) Vol. 52, No. 1, p. K1–K5 (1979).
See also *1.3*, p. *22*; 10.1, p. 94; 12.1, p. 129; 21, p. 341.

M. Ohta, M. Yamada, T. Kanadani, and A. Sakakibara
Effects of the Fluctuation of Solute Concentration in Al–10 wt% Zn Alloy on Aging at Low Temperature
(a) Vol. 52, No. 1, p. K21–K24 (1979).
See also *1.3*, p. *23*; *14.1*, p. *200*; 21, p. 341.

A. Vehanen, J. Yli-Kauppila, P. Hautojärvi, and V. S. Mikhalenko
A Positron Study of Deformed NiGe Alloy
(a) Vol. 52, No. 1, p. K73–K75 (1979).
See also *10.1*, p. *94*; 22.1.1, p. 411.

J. Morkowski and A. M. Oleś
On the Origin of Localized Magnetic States in Transition Metal Alloys (I): Wolff Model
(b) Vol. 92, No. 2, p. 525–536 (1979).
See also *13.4*, p. *167*; *18*, p. *261*.

A. Oleś and A. Bombik
Diluted Cr-Alloys and Two-Band Model
(b) Vol. 92, No. 2, p. K81–K83 (1979).
See also *18.1*, p. *266*; *18.4*, p. *286*.

A. Bose, G. Frohberg, and H. Wever
Self-Diffusion of ^{60}Co and ^{72}Ga in the Ordered Electronic Compound CoGa
(a) Vol. 52, No. 2, p. 509–520 (1979).
See also *9*, p. *86*; 21, p. 343.

D. Hafner and H. Hoffmann
Microscopic and Macroscopic Inhomogeneity of Magnetization and Anisotropy in Amorphous Rare Earth/Transition Metal Films
(a) Vol. 52, No. 2, p. 549–558 (1979).
See also 2, p. 43; *18.2.1*, p. *281*; *20.1*, p. *310*; 21.4, p. 383.

N. Rusović and H. Warlimont
Young's Modulus of β_2-NiAl Alloys
(a) Vol. **53**, No. 1, p. 283–288 (1979).
See also *12.1*, p. *130*.

M. I. Darby, M. N. Read, and K. N. R. Taylor
The Energy of Hydrogen in Zirconium–Rhodium
(a) Vol. **53**, No. 1, p. 391–395 (1979).
See also *10.1*, p. *95*; *13.4*, p. *170*.

J. L. Besson and J. Petit
Accurate Determination of Activation Parameters of the Bordoni Peaks in Nickel
(a) Vol. **53**, No. 1, p. K61–K66 (1979).
See also *10.1*, p. *96*.

V. I. Zaitsev, E. I. Lyafer, and V. A. Fomchenko
Influence of Compression Anisotropy Effects on Zn Creep
(a) Vol. **53**, No. 1, p. K99–K101 (1979).
See also *10.1*, p. *96*; *12.1*, p. *130*.

H. Schmidt and E. Mann
Theory of Anisotropic Electron–Phonon Interaction in Copper (I): Electron–Phonon Matrix Elements
(b) Vol. **93**, No. 2, p. 561–573 (1979).
See also *6*, p. *69*; *13.1*, p. *148*; 14.1, p. 200.

D. Brown, S. Fairbairn and G. J. Morgan
A Quasi-Polycrystalline Approximation for Electronic States in Liquid Transition Metals and Tetrahedrally Bonded Semiconductors
(b) Vol. **93** No. 2, p. 617–627 (1979).
See also *2*, p. *44*; *13.1*, p. *148*; 22.1.1, p. 411; 22.1.2, p. 419.

K. M. Kesharwani
Localized Phonons Due to Substitutional Impurities in Dilute Copper Alloys
(b) Vol. **93**, No. 2, p. 691–701 (1979).
See also *6*, p. *69*; 10.1, p. 96.

W. Schüle and R. Scholz
Properties of Vacancies and Divacancies in F.C.C. Metals
(b) Vol. **93**, No. 2, p. K119–K123 (1979).
See also 9, p. 87; *10.1*, p. *96*; 14.1, p. 201; 21, p. 344; 21.6, p. 388.

W. John and D. Hamann
Relativistic Electron–Phonon Interaction in Transition Metals
(b) Vol. **93**, No. 2, p. K143–K148 (1979).
See also *6*, p. *70*; *13.1*, p. *149*.

D. Colaitis, D. van Dyck, P. Delavignette, and S. Amelinckx
Direct Structure Imaging of Long Period $Ni_{3\pm x}Te_2$ Superstructures
(a) Vol. **53**, No. 2, p. 423–431 (1979).
See also 1.2, p. 18; *1.4*, p. *28*.

R. Ciach, H. Löffler, J. Salawa, and G. Wendrock
Modulated Structures and Position of the Coherent Spinodal Curve in Al–Zn(22 at%)–Cu(x) Alloys
(a) Vol. **53**, No. 2, p. 441–446 (1979).
See also *1.2*, p. *18*; 10.1, p. 96; 21, p. 345.

J. Białoń and B. Wysłocki
Investigation of Domain Walls in Thin Permalloy Films by the Low-Angle Electron Deflection Method
(a) Vol. **53**, No. 2, p. 455–459 (1979).
See also *18.2.1*, p. *281*; 21.1.1, p. 373.

A. A. Aly, G. N. Podus, and A. F. Sirenko
Thermoactivated Creep and Stress Relaxation in Cu and Al
(a) Vol. **53**, No. 2, p. 461–467 (1979).
See also *10.1*, p. *96*; 21, p. 345.

K. Twarowski and H. K. Lachowicz
Magnetostriction of Amorphous $Gd_{1-x}Co_x$ RF Sputtered Thin Films
(a) Vol. **53**, No. 2, p. 599–603 (1979).
See also *2*, p. *44*; *18.2.1*, p. *281*; 18.3, p. 284; 21.4, p. 383.

M. Fumeron, J. P. Guillot, A. Dauger, and J. Caisso
X-Ray Scattering Study of the Shape of Gunier-Preston Zones and Their Size Effect in an Al–6.8 at% Zn Alloy
(a) Vol. **53**, No. 2, p. 605–610 (1979).
See also *1.3*, p. *23*; 10.1, p. 96; 12.1, p. 130.

R. Malmhäll, S. M. Bhagat, K. V. Rao, and G. Bäckström
Transport Properties of Amorphous Ferromagnets — Hall and Ohmic Resistivities
(a) Vol. **53**, No. 2, p. 641–651 (1979).
See also *2*, p. *44*; *14.1*, p. *201*; 18.2, p. 276; 21.1.1, p. 374.

O. Bostanjoglo and H.-P. Gemünd
Fast Easy-Axis Switching of Oxidized Ni–Fe Films
(a) Vol. **53**, No. 2, p. K125–K128 (1979).
See also *18.2.1*, p. *282*; 21.1.1, p. 374.

I. C. Baianu, K. Rubinson, and J. Patterson
The Observation of Structural Relaxation in FeNiPB Glass by Ferromagnetic Resonance and X-Ray Scattering
(a) Vol. **53**, No. 2, p. K133–K138 (1979).
See also 1.3, p. 24; *2*, p. *44*; *18.2*, p. *276*; 19, p. 293; 21.1.1, p. 374.

E. Török and G. Hausch
Bulk Modulus Anomaly of Fe–Ni and Fe–Pt Invar Alloys
(a) Vol. **53**, No. 2, p. K147–K151 (1979).
See also *12.1*, p. *130*; 18.1, p. 267; *18.2*, p. *276*; 21.1.1, p. 374; 21.6, p. 388.

G. Dlubek, O. Brümmer, and N. Meyendorf
Vacancy–Impurity Binding Enthalpy in NiSb by Means of Positron Annihilation
(a) Vol. **53**, No. 2, p. K157–K160 (1979).
See also *10.1*, p. *97*; 21, p. 345.

W. Kleinert and R. Franke
Change of Electrical Resistivity and Stress Amplitude during Push–Pull Fatigue of Polycrystalline Nickel
(a) Vol. **53**, No. 2, p. K177 – K179 (1979).
See also *10.1*, p. *97*; *14.1*, p. *201*.

V. T. Cherepin, M. A. Vasilev, and O. A. Yakubtsov
Appearance Potential Spectroscopy Study of the Oxidation of CoNi Alloy Surfaces
(a) Vol. **53**, No. 2, p. K207 – K210 (1979).
See also *13.3*, p. *159*; 17, p. 257.

O. V. Farberovich, S. I. Kurganskii, and E. P. Domashevskaya
Problems of the OPW Method (I): Transition Metals
(b) Vol. **94**, No. 1, p. 51 – 62 (1979).
See also *13.1*, p. *149*.

H. Schmidt and E. Mann
Anisotropy of the Electron–Phonon Interaction in Copper (II): Scattering and Transport Properties
(b) Vol. **94**, No. 1, p. 95 – 105 (1979).
See also *6*, p. *70*; *13.1*, p. *149*; 14.1, p. 201.

W. Schiller, V. Christoph, and R. Sellger
Force-Correlation Function Method Applied to the Hall Coefficient of Paramagnetic Transition Metals
(b) Vol. **94**, No. 1, p. 215 – 220 (1979).
See also *14.1*, p. *201*.

I. Mertig and E. Mrosan
Calculation of Cyclotron Masses of Copper Using a Tetrahedron Method
(b) Vol. **94**, No. 1, p. K23 – K25 (1979).
See also *13.2*, p. *157*.

H. R. Khan, U. Köbler, K. Lüders, Ch. J. Raub, and Z. Scücz
Magnetic Susceptibility of V_xMo_{1-x} Solid Solutions
(b) Vol. **94**, No. 1, p. K27 – K29 (1979).
See also *18.1*, p. *267*.

M. Maskiewicz, B. Mrygoń, and K. Wentowska
Rounding of Specific Heat of Nickel near the Critical Temperature
(a) Vol. **54**, No. 1, p. 111 – 115 (1979).
See also *8*, p. *82*.

F. F. Lavrentev, O. P. Salita, and S. V. Sokolski
Jump-Like Plastic Deformation and Dislocation Structure Instability in Zinc Crystals with Forest Dislocations
(a) Vol. **54**, No. 1, p. 145 – 152 (1979).
See also *10.1*, p. *97*.

E. Nembach
Analysis of the Extraordinary Viscous Magnetic Domain Wall Motion in NiFe Films
(a) Vol. **54**, No. 1, p. 167 – 170 (1979).
See also *18.2.1*, p. *282*; 21.1.1, p. 374.

SEUNG-AM CHO
Atomic Vibration and Fusion Behaviour of A_3 (H.C.P.) Metals
(b) Vol. **94**, No. 2, p. K123–K127 (1979).
See also *6*, p. *72*; *8*, p. *82*; 21, p. 347; 21.3, p. 381.

Y. D. YAO
Temperature-Dependent Resistivity in Ni-Rich Alloys
(b) Vol. **94**, No. 2, p. K137–K140 (1979).
See also *14.1*, p. *201*; 18.1, p. 268; 18.2, p. 277.

V. S. SUNDARAM, C. L. BARRETO, J. D. ROGERS, and G. G. KLEIMAN
ESCA Studies of Pt–Cu Alloys
(b) Vol. **94**, No. 2, p. K191–K193 (1979).
See also *13*, p. *140*; 21.6, p. 388.

C. PĂPUŞOI and P. APOSTOL
Independence Properties of the Magnetic Hysteretic and Anhysteretic Remanence in Electrolytic Cd–Ni–P Layers
(a) Vol. **54**, No. 2, p. 477–486 (1979).
See also *18.2.1*, p. *282*.

M. GUYMONT, D. GRATIAS, and A. BISSON
Pre-Transition Fluctuations in AuCu
(a) Vol. **54**, No. 2, p. 573–576 (1979).
See also *1.2*, p. *19*; 1.4, p. 29; 21.6, p. 388.

C. RADOMSKY and H. LÖFFLER
On the Extra-Resistivity of Coherent Precipitates
(a) Vol. **54**, No. 2, p. 615–620 (1979).
See also 10.1, p. 98; *14.1*, p. *201*.

T. LEFFERS and B. N. SINGH
Loop Growth and Point-Defect Profiles during HVEM Irradiation
(a) Vol. **54**, No. 2, p. 621–631 (1979).
See also *10.1*, p. *98*; 11, p. 123.

O. A. TROITSKII, V. I. SPITSYN, M. M. MOISEENKO, and A. D. TSIGANOV
Fractionation of Stable Isotopes and Impurities during Zone Recrystallization of Metals and Preparation of Single Crystals Rich in Individual Isotopes
(a) Vol. **54**, No. 2, p. 651–654 (1979).
See also *3*, p. *53*; 9, p. 87; 21.4, p. 347.

A. K. SINGH, C. SURYANARAYANA, and O. N. SRIVASTAVA
Electron Microscopic Studies of Phase Transformations in NiSe Thin Films
(a) Vol. **54**, No. 2, p. K103–K105 (1979).
See also *1.2*, p. *19*; *1.5*, p. *35*.

A. S. ERMOLENKO, A. Z. MENSHIKOV, and YU. A. DOROFEEV
Magnetic Heterogeneities in $Y(Co_{1-x}Ni_x)_5$ Alloys and Their Relation to Coercive Force
(a) Vol. **54**, No. 2, p. K113–K116 (1979).
See also *18.2*, p. *277*; 21.4, p. 384.

B. NOWAK and R. LAGNIER
Low-Temperature Heat Capacity Studies of the Dihydride Phase of Titanium–Niobium–Hydrogen Alloys
(a) Vol. **54**, No. 2, p. K167–K169 (1979).
See also *8*, p. *82*; 21, p. 348.

O. Bostanjoglo, W. Giese, and R. Liedtke
Effect of Codeposited Gases on the Crystallization of Amorphous FeNi Films
(a) Vol. 55, No. 1, p. K7–K9 (1979).
See also 1.2, p. 19; *1.5*, p. *36*; *2*, p. *47*; 21.1.1, p. 375.

Y. Obi, H. Morita, and H. Fujimori
Exchange Anisotropy in Ni–Mn–Si–B Amorphous Alloys
(a) Vol. 55, No. 1, p. K67–K70 (1979).
See also *2*, p. *48*; *18*, p. *263*.

V. T. Cherepin, M. A. Vasilev, and O. A. Yakubtsov
Auger Electron and Appearance Potential Spectroscopic Study of Fe–Cr Alloy Surfaces
(a) Vol. 55, No. 1, p. K83–K86 (1979).
See also *13.3*, p. *161*; 17, p. 258; 21.1.1, p. 375.

Kin-ichi Masuda
Chemisorption of Ordered Atomic Layers on a Model Transition Metal
(b) Vol. 95, No. 2, p. 413–425 (1979).
See also *13.3*, p. *161*.

T. K. Hatwar, S. K. Malik, M. N. Ghatikar, and B. D. Padalia
X-Ray Absorption Edge Shifts of Cobalt and Rare Earth Ions in Rare Earth–Cobalt Intermetallic Compounds
(b) Vol. 95, No. 2, p. 621–625 (1979).
See also *20*, p. *301*; 21.4, p. 384.

A. Laupheimer and M. Wilkens
TEM Studies of Displacement Cascades in Ion-Bombarded Cu_3Pd and Fe_3Al
(a) Vol. 55, No. 2, p. 497–507 (1979).
See also *1.3*, p. *24*; *11*, p. *124*; 21.1.1, p. 376.

G. Dlubek, O. Kabisch, O. Brümmer, and H. Löffler
Precipitation and Dissolution Processes in Age-Hardenable Al Alloys. A Comparison of Positron Annihilation and X-Ray Small-Angle Scattering Investigations (I): Al–Zn(x) ($x = 3$, 4.5, 6, 10, and 18 at%)
(a) Vol. 55, No. 2, p. 509–518 (1979).
See also *10.1*, p. *99*; 21, p. 350.

J. F. Dinhut, P. Lepinay, and J. C. Desoyer
Dynamic Aging of Fe–Co–2% V
(a) Vol. 55, No. 2, p. 621–629 (1979).
See also *10.1*, p. *99*; 21.1.1, p. 376.

P. Lukáč and V. Jesenák
Distribution of Ion-Bombardment Implanted ^{85}Kr in Si, Cu, and SiO_2
(a) Vol. 55, No. 2, p. 647–652 (1979).
See also *11*, p. *124*; 22.1.2, p. 424; 22.6, p. 484.

P. Bartuška, R. Kroggel, H. Löffler, M. Simerská, V. Syneček, and G. Wendrock
On the Room Temperature Stability of Metastable Phases Formed at 200 °C in Al–Zn Alloys
(a) Vol. 55, No. 2, p. 673–679 (1979).
See also *1.3*, p. *25*; *10.1*, p. *99*; 21, p. 350.

A. van den Beukel and G. J. den Otter
Modulus Effect and Internal Friction of Copper and Some Copper Alloys Cold-Worked at 78 K
(a) Vol. 56, No. 1, p. 117–127 (1979).
See also *10.1*, p. *100*; *12.1*, p. *131*.

H.-G. Fabian, R. Kroggel, H. Löffler, and O. Simmich
On the Description of the Growth and Decomposition Kinetics of Supersaturated Alloys
(a) Vol. 56, No. 1, p. 143–148 (1979).
See also *10.1*, p. *100*; 21, p. 352.

M. B. Dutt, S. K. Sen, and A. K. Barua
The Diffusion of Nickel into Copper and Copper–Nickel Alloys
(a) Vol. 56, No. 1, p. 149–155 (1979).
See also *9*, p. *88*.

N. H. Quyen, P. Paufler, K.-H. Berthel, M. Bertram, U. Krämer, D. M. Nghiep, A. San Martin, A. Gladun, and K. Kleinstück
Influence of Plastic Deformation on Superconducting Properties of V_3Si Single Crystals
(a) Vol. 56, No. 1, p. 231–236 (1979).
See also *10.1*, p. *100*; *14.2*, p. *208*.

M. Bernardin, A. Dupasquier, A. Gallone, and P. Pizzi
Positron Annihilation Study of Ageing and Creep of Inconel X750
(a) Vol. 56, No. 1, p. 277–284 (1979).
See also *10.1*, p. *101*.

L. Takács
Electronic Structure of Transition Metal–Metalloid Crystalline and Glassy Alloys
(a) Vol. 56, No. 1, p. 371–377 (1979).
See also *2*, p. *48*; 5, p. 62; *13.1*, p. *154*; 18.2, p. 279; 21.1.1, p. 377.

V. E. Rode
Role of the Exchange Interaction between Ferro- and Antiferromagnetic Components in Invar Alloys
(a) Vol. 56, No. 1, p. 407–417 (1979).
See also *18.2*, p. *279*; *18.4*, p. *288*; 21.1.1, p. 377.

M. Wendt
Detection of Thin Surface Films by Electron Probe Microanalysis in Dependence on the Substrate
(a) Vol. 56, No. 1, p. K21–K23 (1979).
See also *1.5*, p. *37*.

A. R. Badzian and A. Kłokocki
Splitting of the Valance Band of Copper by X-Ray K-Emission Spectroscopy
(b) Vol. 96, No. 2, p. 529–535 (1979).
See also *13.1*, p. *154*; 20, p. 302.

E. Ya. Mikhlin
Suppression of Diffusion Mobility of Small Gas Bubbles in Solids
(a) Vol. 56, No. 2, p. 763–768 (1979).
See also *9*, p. *88*; 10.1, p. 102; 21.5, p. 386; 21.6, p. 390.

21.1.1 Fe and Fe Alloys

Review Article

Original Papers and Short Notes

C. Blaauw, C. G. White, and W. Leiper
Hyperfine Interactions of ^{57}Fe in $Fe_{1+x}Sb$
(b) Vol. **91**, No. 2, p. K129–K132 (1979).
See also *5*, p. *59*.

A. Dunlop, N. Lorenzelli, and J. C. Jousset
Investigation of Iron at 20 K by Fast Neutrons and Fission Fragments of Uranium 235
(a) Vol. **51**, No. 2, p. 479–485 (1979).
See also *11*, p. *119*.

Th. Kleinefeld and H. H. Mende
Calculations for Optimizing the Magnification of the Magneto-Optic Kerr Effect by Interference Layers on Iron Whiskers
(a) Vol. **51**, No. 2, p. 487–490 (1979).
See also *18.2*, p. *271*; *20*, p. *299*.

H. Kirihata and J. Akutsu
Formation of Exoelectron Emission Centers by Slow Electron Impact
(a) Vol. **51**, No. 2, p. K109–K112 (1979).
See also 13.3, p. 158; *17*, p. *256*.

Yu. P. Lobastov and V. I. Pushkarskii
The Influence of Internal Field Non-Uniformity in Ferromagnetic Samples on the Magnetization Process in the High Magnetic Field Region
(a) Vol. **51**, No. 2, p. K155–K159 (1979).
See also *18.2*, p. *271*.

L. Cser and L. J. Naszódi
Relationship between Electron Density Distribution and Atomic Order in Iron–Aluminium and Iron–Silicon Alloys
(b) Vol. **92**, No. 1, p. K55–K56 (1979).
See also *1.3*, p. *22*; 5, p. 59.

W. Decker, J. Diehl, A. Dunlop, W. Frank, H. Kronmüller, W. Mensch, H. E. Schaefer, B. Schwendemann, A. Seeger, H. P. Stark, F. Walz, and M. Weller
Interstitial Migration in Iron at 220 K
(a) Vol. **52**, No. 1, p. 239–248 (1979).
See also *10.1*, p. *93*; 11, p. 120; 14.1, p 200.

S. Mager, E. Wieser, T. Zemčík, O. Schneeweiss, P. N. Stetsenko, and V. V. Surikov
Investigations of the Magnetic Moments in $(Fe_{1-x}Mn_x)_3Al$ Alloys
(a) Vol. **52**, No. 1, p. 249–258 (1979).
See also *18*, p. *261*; 21.1, p. 357.

A. Caplain and W. Chambron
Study of the Interaction between Vacancies and Boron in a Ni–20 at% Fe Alloy by the Induced Magnetic Anisotropy
(a) Vol. **52**, No. 1, p. 299–310 (1979).
See also *10.1*, p. *94*; 18.2, p. 272; 21.1, p. 357.

I. G. Ritchie, J. F. Dufresne, and P. Moser
Internal Friction in Deformed Hydrogen-Doped Iron
(a) Vol. **52**, No. 1, p. 331–340 (1979).
See also *10.1*, p. *94*; *12.1*, p. *129*.

A. V. ANDREEV, A. V. DERYAGIN, R. Z. LEVITIN, A. S. MARKOSYAN, and M. ZELENÝ
Magnetic Anisotropy of the Intermetallic Compound UFe_2
(a) Vol. 52, No. 1, p. K13 – K15 (1979).
See also *18.2*, p. *272*; 21.5, p. 386.

J. SCHNEIDER, B. SPRINGMANN, K. ZAVETA, G. WENZEL, and E. KÖHLER
Magnetic After-Effect in Amorphous Fe–Co Based Alloys
(a) Vol. 52, No. 1, p. K25 – K29 (1979).
See also *2*, p. *42*; *18.2*, p. *272*.

L. M. CASPERS, A. VAN VEEN, M. R. YPMA, and R. FASTENAU
Interaction of Helium with Small Self-Interstitial Platelets in α-Fe
(a) Vol. 52, No. 1, p. K61 – K64 (1979).
See also *10.1*, p. *94*.

H. KIRIHATA
Effects of Gas Pressure and Sample Temperature during Slow Electron Excitation on Exoelectron Emission
(a) Vol. 52, No. 1, p. K97 – K99 (1979).
See also 13.3, p. 159; *17*, p. *256*.

J. M. ALBERDI, J. M. BARANDIARÁN, and J. ILARRAZ
Temperature Dependence of Coercivity in Some Rare-Earth–Fe_2 Intermetallic Compounds
(a) Vol. 52, No. 1, p. K101 – K103 (1979).
See also *18.2*, p. *272*; *18.3*, p. *283*; 21.4, p. 383.

D. K. BOWEN and V. VITEK
The Construction of Central-Force Interatomic Potentials from Whisker Deformation Data
(b) Vol. 92, No. 2, p. 645 – 652 (1979).
See also 4, p. 56; *10.1*, p. *94*; *12.1*, p. *129*.

V. E. ANTONOV, I. T. BELASH, B. K. PONOMAREV, E. G. PONYATOVSKII, and V. G. THIESSEN
Magnetic Properties of Hydrogen Solid Solutions in Fe–Ni–Mn Alloys
(a) Vol. 52, No. 2, p. 703 – 710 (1979).
See also *18.2*, p. *273*; *18.4*, p. *287*; 21.1, p. 358.

A. A. BAHGAT
A New Method for Quantitative Analysis of the Mössbauer Effect
(a) Vol. 52, No. 2, p. K217 – K220 (1979).
See also *5*, p. *60*.

YA. M. KSENDZOV
The Electron Structure of 3d Metals
(b) Vol. 93, No. 1, p. 415 – 425 (1979).
See also *13.1*, p. *147*; 13.5.2, p. 191; 21.1, p. 358.

A. URBANIAK-KUCHARCZYK
Spin Correlation Function in Iron
(b) Vol. 93, No. 1, p. K25 – K27 (1979).
See also *18.2*, p. *274*.

V. G. Gavrilyuk, V. G. Prokopenko, and O. N. Razumov
Mössbauer Study of the Effect of Alloying Elements on the Carbide Phase in Cold-Worked Steel
(a) Vol. **53**, No. 1, p. 147–154 (1979).
See also *5*, p. *60*; 10.1, p. 95.

J. D. Stephenson, T. Tuomi, V. Kelhä, and M. Tilli
Quasi-Lifetime Topography of Magnetic Domain Movements in Grain Oriented (100) [001] Fe–3 wt% Si Polycrystals Using White Synchrotron Radiation (II): The Influence of Small Tensile Stress
(a) Vol. **53**, No. 1, p. 271–276 (1979).
Erratum
(a) Vol. **55**, No. 1, p. 339 (1979).
See also *1.4*, p. *28*; 12.1, p. 130; *18.2*, p. *274*.

M. J. Besnus, J. M. Bouton, and R. Clad
Crystal Structure and Magnetic Properties of $Y_6(Fe_{1-x}Al_x)_{23}$ Compounds
(a) Vol. **53**, No. 1, p. 351–354 (1979).
See also *1.3*, p. *23*; *18.2*, p. *274*; 21.4, p. 383.

F. P. Livi, L. Amaral, and S. G. Fries
The Magnetic Hyperfine Field at Hf Sites in the (Zr, Hf)Fe_2 Laves Pseudo-Binary Compound
(a) Vol. **53**, No. 1, p. 379–382 (1979).
See also *18.2*, p. *275*; 21, p. 344.

H. O. K. Kirchner, H. Mughrabi, F. Prinz, and H. Schmid
Erratum to: Dislocation Loops in α-Iron
(a) Vol. **53**, No. 1, p. 411 (1979).
See also *10.1*, p. *95*; 12.1, p. 130.

D. H. Sastry and C. V. Shastry
Cottrell-Stokes Law during High Temperature Deformation of α-Iron
(a) Vol. **53**, No. 1, p. K5–K7 (1979).
See also *10.1*, p. *96*.

K. Hoselitz
Crystallisation of Magnetic Fe–Si–B Metallic Glasses
(a) Vol. **53**, No. 1, p. K23–K26 (1979).
See also *2*, p. *44*; *18.2*, p. *275*.

Kamna Aggarwal and R. G. Mendiratta
Mössbauer Investigations on Thin Films of $FeSi_2$
(a) Vol. **53**, No. 1, p. K95–K98 (1979).
See also *1.5*, p. *35*; *5*, p. *60*.

J. Białoń and B. Wysłocki
Investigation of Domain Walls in Thin Permalloy Films by the Low-Angle Electron Deflection Method
(a) Vol. **53**, No. 2, p. 455–459 (1979).
See also *18.2.1*, p. *281*; 21.1, p. 360.

V. V. Golubkov, V. I. Petrov, and G. V. Spivak
Dynamics of Néel Wall Transitions into Cross-Tie Walls
(a) Vol. 54, No. 1, p. K1 – K3 (1979).
See also *18.2.1*, p. *282*.

R. Krishnan, M. Tessier, and J. Sztern
Magnetic Properties of Amorphous $Fe_{38}Ni_{38}B_{24}$ Alloy
(a) Vol. 54, No. 1, p. K77 – K79 (1979).
See also *2*, p. *46*; *18.2*, p. *277*.

A. Amamou
Mössbauer Effect and Short-Range Order in Fe–P–B and Fe–P–C Amorphous Alloys
(a) Vol. 54, No. 2, p. 565 – 572 (1979).
See also *2*, p. *46*; *5*, p. *61*; 18.2, p. 277.

A. Möbius, F. Goedsche, D. Elefant, K.-H. Berthel, and G. Vojta
Deviations from Matthiessen's Rule in the Low-Temperature Resistivity of Iron
(b) Vol. 95, No. 1, p. 203 – 214 (1979).
See also *14.1*, p. *201*.

L. M. Sandler, V. P. Popov, and Ya. V. Naglyuk
Magnetic Anisotropy Induced by Rolling Single Crystals of Ferromagnetic Alloys (I): Face-Centered Cubic Single Crystals
(a) Vol. 55, No. 1, p. 271 – 277 (1979).
See also *10.1*, p. *98*; *18.2*, p. *278*; 21.1, p. 364.

J. Echigoya and S. Hayashi
Anomaly of Dislocation Motion in Fe–35% Ni Invar Alloys
(a) Vol. 55, No. 1, p. 279 – 284 (1979).
See also *10.1*, p. *98*; 18.2, p. 278; 21.1, p. 364.

O. Bostanjoglo, W. Giese, and R. Liedtke
Effect of Codeposited Gases on the Crystallization of Amorphous FeNi Films
(a) Vol. 55, No. 1, p. K7 – K9 (1979).
See also 1.2, p. 19; *1.5*, p. *36*; *2*, p. *47*; 21.1, p. 365.

V. T. Cherepin, M. A. Vasilev, and O. A. Yakubtsov
Auger Electron and Appearance Potential Spectroscopic Study of Fe–Cr Alloy Surfaces
(a) Vol. 55, No. 1, p. K83 – K86 (1979).
See also *13.3*, p. *161*; 17, p. 258; 21.1, p. 365.

R. Podloucky and A. Neckel
Compton Profiles of FeAl
(b) Vol. 95, No. 2, p. 541 – 548 (1979).
See also *13.1*, p. *152*; 20, p. 301.

W. Arabczyk, H.-J. Müssig, and F. Storbeck
Sulphur Segregation on the (111) Iron Surface Studied by LEED, AES, and WFC
(a) Vol. 55, No. 2, p. 437 – 446 (1979).
See also *1.6*, p. *39*.

A. Seeger
A Theory of the Snoek-Köster Relaxation (Cold-Work Peak) in Metals
(a) Vol. 55, No. 2, p. 457 – 468 (1979).
See also *10.1*, p. *99*; 21, p. 350.

E. A. Bakulin, V. L. Hilarov, and V. V. Shcherbinina
Characteristic Energy Losses and Optical Properties of Intermetallic Compounds in the Fe–Ge System
(a) Vol. 56, No. 1, p. 259–265 (1979).
See also *14.4*, p. *242*; 22.1.1, p. 413.

L. Takács
Electronic Structure of Transition Metal–Metalloid Crystalline and Glassy Alloys
(a) Vol. 56, No. 1, p. 371–377 (1979).
See also *2*, p. *48*; 5, p. 62; *13.1*, p. *154*; 18.2, p. 279; 21.1, p. 367.

V. E. Rode
Role of the Exchange Interaction between Ferro- and Antiferromagnetic Components in Invar Alloys
(a) Vol. 56, No. 1, p. 407–417 (1979).
See also *18.2*, p. *279*; *18.4*, p. *288*; 21.1, p. 367.

G. Marchal, Ph. Mangin, M. Piecuch, and Chr. Janot
Some Comments about Crystallization of Fe–Si Amorphous Films
(a) Vol. 56, No. 1, p. K29–K31 (1979).
See also *1.5*, p. *37*; 2, p. 49; *5*, p. *62*.

E. Japa and K. Krop
Hyperfine Fields in Fe–Si Alloys
(b) Vol. 96, No. 2, p. K65–K68 (1979).
See also 5, p. 62; *18.2*, p. *280*.

T. Ohtani, K. Kosuge, and S. Kachi
Anomalous Thermal Hysteresis in Metal–Semiconductor Transition of $(Ni_{1-x}Fe_x)_{1-y}S$ $(0 \leqq x \leqq 0.30, 0 \leqq y \leqq 0.035)$
(b) Vol. 96, No. 2, p. K69–K72 (1979).
See also *14.1*, p. *203*; *14.3*, p. *217*; 21.1, p. 368.

W. O. Meyer and G. Inden
Quantitative Determination of Ordered Configurations in B.C.C. Fe–Co–Si Solid Solutions by Mössbauer Spectroscopy
(a) Vol. 56, No. 2, p. 481–486 (1979).
See also 1.3, p. 25; *5*, p. *62*.

K. H. J. Buschow and N. M. Beekmans
On the Thermal Stability and the Electrical Resistivity of Amorphous Th–Fe Alloys
(a) Vol. 56, No. 2, p. 505–511 (1979).
See also *2*, p. *50*; 14.1, p. 203; 21.4, p. 384.

Y. Tanji, Y. Nakagawa, Y. Saito, K. Nishimura, and K. Nakatsuka
Anomalous Thermodynamic Properties of Iron–Nickel (F.C.C.) Alloys
(a) Vol. 56, No. 2, p. 513–519 (1979).
See also 1.2, p. 21; *8*, p. *84*; 21.1, p. 368.

O. Bostanjoglo, U. Heinecke, and R. Liedtke
Possible Stabilization of Amorphous FeNi Films by Conduction Electrons
(a) Vol. 56, No. 2, p. 569–572 (1979).
See also 1.5, p. 37; *2*, p. *50*; 21.1, p. 368.

21.2 Alkali Metals

Original Papers and Short Notes

K. N. KHANNA and P. K. SHARMA
Binding Energy, Compressibility, Interionic Potential, and Transition Temperature of Simple Metals Obtained by the Pseudopotential Method
(b) Vol. 91, No. 1, p. 251–254 (1979).
See also *4*, p. *55*; *12.1*, p. *129*; 14.2, p. 204.

K. N. KHANNA and P. K. SHARMA
Erratum to: Susceptibility of Alkali Metals
(b) Vol. 91, No. 2, p. K185 (1979).
See also *18.1*, p. *266*.

J. THARUK, P. L. SRIVASTAVA, and N. R. MITRA
Harrison's Method. Electrical Transport Properties of Alkali Metals
(b) Vol. 93, No. 1, p. 391–395 (1979).
See also *14.1*, p. *200*; *15*, p. *250*.

G. KNUYT and L. M. STALS
The Na Vacancy Formation Energy Value from Phonon Data
(b) Vol. 93, No. 1, p. K9–K11 (1979).
See also 4, p. 56; 6, p. 68; *10.1*, p. *95*.

V. RAMAMURTHY and S. B. RAJENDRAPRASAD
On the Homology of Lattice Vibrations of Rare-Gas Solids
(b) Vol. 93, No. 2, p. 793–797 (1979).
See also *6*, p. *69*; 23, p. 515.

R. RAMJI RAO and A. RAJPUT
Third-Order Elastic Constants, Thermal Expansion, and Bulk Modulus of Body-Centered Cubic Crystals
(b) Vol. 94, No. 2, p. 691–699 (1979).
See also *12.1*, p. *131*; 21, p. 347.

K. N. KHANNA and P. L. SRIVASTAVA
Fermi Surface Distortion in Na and K
(b) Vol. 94, No. 2, p. K169–K172 (1979).
See also *13.2*, p. *157*.

A. V. KOBELEV, R. M. KOBELEVA, and V. F. UKHOV
On the Adhesion Theory for Two Metallic Surfaces
(b) Vol. 96, No. 1, p. 169–176 (1979).
See also *13.3*, p. *161*; 21, p. 351; 21.1, p. 366.

D. M. ESTERLING and A. SWAROOP
Interatomic Potentials from Experimental Phonon Spectra (I): Prototypes
(b) Vol. 96, No. 1, p. 401–411 (1979).
See also 2, p. 48; *4*, p. *57*; 6, p. 75; 12.1, p. 131; 21.1, p. 366; 21.1.1, p. 376.

21.3 Alkali-Earth Metals

Original Papers and Short Notes

G. M. Mladenov and M. Braun
Ion Bombardment Induced Line Radiation from Sputtered Beryllium
(a) Vol. **53**, No. 2, p. 631 – 640 (1979).
See also *17*, p. *257*.

P. Lukáč
The Critical Resolved Shear Stress of Mg–Cd Single Crystals
(a) Vol. **54**, No. 1. p. K33 – K35 (1979).
See also *10.1*, p. *97*; 21, p. 346.

G. Dräger, O. Brümmer, and J. Bonitz
Polarized K-Emission Valence Bands, Structure, and Bonding of Some Hexagonal Close-Packed Metals
(b) Vol. **94**, No. 2, p. K111 – K114 (1979).
See also *4*, p. *57*; *13.1*, p. *151*; 20, p. 301; 21.1, p. 362.

Seung-Am Cho
Atomic Vibration and Fusion Behaviour of A_3 (H.C.P.) Metals
(b) Vol. **94**, No. 2, p. K123 – K127 (1979).
See also *6*, p. *72*; *8*, p. *82*; 21, p. 347; 21.1, p. 363.

M. van Sande, G. van Tendeloo, S. Amelinckx, and P. Airo
High Resolution Electron Microscopic and Electron Diffraction Study of the Au–Mg System (I): The Monoclinic $Au_{15}Mg_4$ Phase
(a) Vol. **54**, No. 2, p. 499 – 512 (1979).
See also *1.3*, p. *24*; *1.4*, p. *29*; 21.6, p. 388.

21.4 Rare-Earth Metals

Original Papers and Short Notes

J. ILARRAZ and A. DEL MORAL
Magnetic Ordering Temperatures in Rare Earth–Fe_2 Laves Phases Determined by DTA
(a) Vol. **51**, No. 1, p. K41 – K43 (1979).
See also *18.1*, p. *265*; *18.3*, p. *283*; 21.1.1, p. 370.

L. G. KHVOSTANTSEV and N. A. NIKOLAEV
Behaviour of the Thermopower of Cerium under Pressure up to 7500 MPa
(a) Vol. **51**, No. 1, p. K57 – K61 (1979).
See also *15*, p. *249*.

I. O. BASHKIN, E. G. PONYATOVSKII, and M. E. KOST
On the Possibility of Hydrogen Ordering in Rare-Earth Hydrides
(b) Vol. **91**, No. 2, p. 401 – 405 (1979).
See also *1.2*, p. *15*; 14.1, p. 199; 14.3, p. 210.

R. RAMJI RAO and A. RAMANAND
Specific Heat of Holmium and Terbium
(a) Vol. **51**, No. 2, p. 579 – 582 (1979).
See also *8*, p. *80*.

A. V. DERYAGIN, N. V. KUDREVATYKH, R. Z. LEVITIN, and YU. F. POPOV
Magnetostriction of Single Crystals of Intermetallic R_2Co_{17} (R = Ho, Er, Tm, Lu) Compounds
(a) Vol. **51**, No. 2, p. K125 – K127 (1979).
See also *18.2*, p. *271*; 21.1, p. 356.

F. DETTMANN, W. RICHTER, G. ALBRECHT, and W. ZAHN
A Monolithic Thin Film DC-SQUID
(a) Vol. **51**, No. 2, p. K185 – K188 (1979).
See also *14.2*, p. *204*; 21, p. 340.

K. SUGAWARA
Effect of Crystal Field on the Inelastic Neutron Scattering Cross-Section (II)
(b) Vol. **92**, No. 1, p. 317 – 321 (1979).
See also *18.1*, p. *266*.

D. M. EAGLES
Erratum to: Widths of 5d Energy Bands in the Heavy Rare-Earth Metals
(b) Vol. **92**, No. 1, p. 353 (1979).
See also *13.1*, p. *145*.

H. HOFFMANN, A. J. OWEN, and F. SCHRÖPF
Electron Microscopy of Evaporated and Sputtered Gd/Co and Ho/Co Films
(a) Vol. **52**, No. 1, p. 161 – 174 (1979).
See also *1.5*, p. *34*; 2, p. 42; 18.2.1, p. 281; 21.1, p. 356.

H. Bartholin and O. Vogt
Magnetic Phase Diagrams of $Ce_x(La_{0.76}Y_{0.24})_{1-x}Bi$ Compounds
(a) Vol. **52**, No. 1, p. 315–320 (1979).
See also *18.4*, p. *286*.

J. M. Alberdi, J. M. Barandiarán, and J. Ilarraz
Temperature Dependence of Coercivity in Some Rare-Earth–Fe_2 Intermetallic Compounds
(a) Vol. **52**, No. 1, p. K101–K103 (1979).
See also *18.2*, p. *272*; *18.3*, p. *283*; 21.1.1, p. 372.

D. Hafner and H. Hoffmann
Microscopic and Macroscopic Inhomogeneity of Magnetization and Anisotropy in Amorphous Rare Earth/Transition Metal Films
(a) Vol. **52**, No. 2, p. 549–558 (1979).
See also 2, p. 43; *18.2.1*, p. *281*; *20.1*, p. *310*; 21.1, p. 357.

P. N. Koul, T. K. Hatwar, M. N. Ghatikar, and B. D. Padalia
Chemical Shifts of the L_3-Absorption Edge of Pt in the Rare Earth–Platinum Intermetallic Compounds
(b) Vol. **93**, No. 1, p. 223–228 (1979).
See also 1.3, p. 23; *20.1*, p. *311*.

M. Drulis and Z. Biegański
Low-Temperature Specific Heats of Cerium and Praseodymium Trihydrides. Second-Order Phase Transitions and Crystal Field Splitting Effects
(a) Vol. **53**, No. 1, p. 277–281 (1979).
See also *8*, p. *81*.

M. J. Besnus, J. M. Bouton, and R. Clad
Crystal Structure and Magnetic Properties of $Y_6(Fe_{1-x}Al_x)_{23}$ Compounds
(a) Vol. **53**, No. 1, p. 351–354 (1979).
See also *1.3*, p. *23*; *18.2*, p. *274*; 21.1.1, p. 373.

V. V. Nemoshkalenko, Vl. N. Antonov, V. N. Antonov, and W. John
Calculation of Soft X-Ray Emission Spectra by the Relativistic APW Method: Lutetium
(b) Vol. **93**, No. 2, p. 575–580 (1979).
See also *13.1*, p. *148*; *20*, p. *300*.

A. F. Barabanov and A. M. Tsvelik
On the Renormalization of Collective Excitations in Intermediate Valence Metals
(b) Vol. **93**, No. 2, p. 817–822 (1979).
See also 7, p. 78; *13.5.2*, p. *192*.

K. Twarowski and H. K. Lachowicz
Magnetostriction of Amorphous $Gd_{1-x}Co_x$ RF Sputtered Thin Films
(a) Vol. **53**, No. 2, p. 599–603 (1979).
See also *2*, p. *44*; *18.2.1*, p. *281*; 18.3, p. 284; 21.1, p. 360.

K. H. J. Buschow and A. M. van der Kraan
Magnetic Properties of Amorphous $Th_{1-x}Fe_x$ Alloys
(a) Vol. **53**, No. 2, p. 665–669 (1979).
See also 2, p. 44; 5, p. 61; *18.1*, p. *267*; *18.2*, p. *276*; 21.1.1, p. 374.

J. Fidler and H. Kronmüller
Nucleation and Pinning of Magnetic Domains at Co_7Sm_2 Precipitates in Co_5Sm Crystals
(a) Vol. **56**, No. 2, p. 545–556 (1979)
See also *18.2*, p. *280*; 21.1, p. 368.

21.5 Actinides

Original Papers and Short Notes

A. V. Andreev, A. V. Deryagin, R. Z. Levitin, A. S. Markosyan, and M. Zelený
Magnetic Anisotropy of the Intermetallic Compound UFe_2
(a) Vol. 52, No. 1, p. K13–K15 (1979).
See also *18.2*, p. *272*; 21.1.1, p. 372.

N. Shamir, M. Melamud, and H. Shaked
Crystalline Electric Field Studies of Uranium Compounds Using Neutron Inelastic Scattering at the NRCN
(a) Vol. 55, No. 1, p. K49–K52 (1979).
See also *13*, p. *140*; 18, p. 263; 22.1.2, p. 424; 22.5, p. 462; 22.6, p. 483.

H. Drulis and N. M. Suleimanov
Electron Paramagnetic Resonance of Er Localized Electron Magnetic States in Thorium Hydrides
(b) Vol. 95, No. 2, p. 377–382 (1979).
See also *19*, p. *296*.

E. Ya. Mikhlin
Suppression of Diffusion Mobility of Small Gas Bubbles in Solids
(a) Vol. 56, No. 2, p. 763–768 (1979).
See also *9*, p. *88*; 10.1, p. 102; 21.1, p. 369; 21.6, p. 390.

21.6 Noble Metals

Original Papers and Short Notes

M. Fukui, C. V. Y. So, and R. Normandin
Lifetimes of Surface Plasmons in Thin Silver Films
(b) Vol. **91**, No. 1, p. K61–K64 (1979).
See also *13.5.2*, p. *191*.

K. Breuer
On Size Effects in Small Gold Particles by Electron Energy Loss Analysis
(b) Vol. **91**, No. 1, p. K69–K72 (1979).
See also *13.3*, p. *158*; 14.4, p. 238.

G. van Tendeloo and S. Amelinckx
High Resolution Electron Microscopic Study of Ordering in Monoclinic Au_5Mn_2
(a) Vol. **51**, No. 1, p. 141–150 (1979).
See also *1.3*, p. *22*; *1.4*, p. *26*; 21.1, p. 354.

S. S. Sheinin and B. K. Jap
On the Validity of Assuming Symmetrical Laue Diffraction Conditions in Calculations of Diffraction Contrast
(b) Vol. **91**, No. 2, p. 407–412 (1979).
See also *1.1*, p. *11*; 10.1, p. 92.

J. Zieliński
The Effects of Magnetic Field and Direct Interaction between Impurities in the Anderson Model. Resistivity
(b) Vol. **91**, No. 2, p. 615–625 (1979).
See also *14.1*, p. *199*; *18.1*, p. *265*; 21.1, p. 355.

A. Korner, H. Schmid, and F. Prinz
Asymmetrical Equilibrium Positions and Line Tensions of Composite Dislocations
(a) Vol. **51**, No. 2, p. 613–621 (1979).
See also *10.1*, p. *92*; *12.1*, p. *129*.

S. Hashimoto and H. Iwasaki
A Microdomain Model Analysis for Diffuse X-Ray Scattering from Quenched CuPt Alloy
(a) Vol. **51**, No. 2, p. 673–682 (1979).
See also *1.4*, p. *27*; 21.1, p. 356.

S. C. Goel and V. S. Nanda
Application of a Low Temperature Equation of State for Monovalent Metals
(b) Vol. **92**, No. 1, p. K65–K67 (1979).
See also *8*, p. *81*; 21.1, p. 356.

J. G. Erlings and F. W. Schapink
Electron Diffraction from Periodic ⟨111⟩ Twist Grain Boundary Structures
(a) Vol. **52**, No. 2, p. 529–539 (1979).
See also *1.4*, p. *27*; 10.1, p. 95.

C. Bansal
Local Environment Effects in Ni–Cu and Ni–Au Alloys
(a) Vol. **52**, No. 2, p. K119–K123 (1979).
See also *18.1*, p. *267*; *18.2*, p. *273*; 21.1, p. 358.

E. N. Mgbenu
Activation Energy for Diffusion in Au–Ge Thin Film Couples
(a) Vol. **53**, No. 1, p. 397–401 (1979).
See also *9*, p. *86*; *14.3.1*, p. *220*; 22.1.1, p. 411.

W. Schüle and R. Scholz
Properties of Vacancies and Divacancies in F.C.C. Metals
(b) Vol. **93**, No. 2, p. K119–K123 (1979).
See also 9, p. 87; *10.1*, p. *96*; 14.1, p. 201; 21, p. 344; 21.1, p. 359.

R. Kutner, I. Sosnowska, and W. Kley
Erratum to: Next-Neighbour Jumps of Hydrogen in Quasi-Elastic Neutron Scattering from α-PdH_x
(b) Vol. **93**, No. 2, p. K187 (1979).
See also *9*, p. *87*.

E. Török and G. Hausch
Bulk Modulus Anomaly of Fe–Ni and Fe–Pt Invar Alloys
(a) Vol. **53**, No. 2, p. K147–K151 (1979).
See also *12.1*, p. *130*; 18.1, p. 267; *18.2*, p. *276*; 21.1, p. 360; 21.1.1, p. 374.

S. P. S. Badwal and H. J. de Bruin
Electrode Kinetics at the Pt/Yttria-Stabilized Zirconia Interface by Complex Impedance Dispersion Analysis
(a) Vol. **54**, No. 1, p. 261–270 (1979).
See also *13.3*, p. *160*; *14.3.4*, p. *235*; 22.6, p. 481.

K. Berndt, U. Marx, and O. Brümmer
Photoelectron Spectra of Core Electrons from Palladium–Silver Alloys
(b) Vol. **94**, No. 2, p. 541–548 (1979).
See also *13*, p. *139*; 17, p. 257; 21, p. 347.

V. S. Sundaram, C. L. Barreto, J. D. Rogers, and G. G. Kleiman
ESCA Studies of Pt–Cu Alloys
(b) Vol **94**, No. 2, p. K191–K193 (1979).
See also *13*, p. *140*; 21.1, p. 363.

M. van Sande, G. van Tendeloo, S. Amelinckx, and P. Airo
High Resolution Electron Microscopic and Electron Diffraction Study of the Au–Mg System (I): The Monoclinic $Au_{15}Mg_4$ Phase
(a) Vol. **54**, No. 2, p. 499–512 (1979).
See also *1.3*, p. *24*; *1.4*, p. *29*; 21.3, p. 381.

M. Guymont, D. Gratias, and A. Bisson
Pre-Transition Fluctuations in AuCu
(a) Vol. **54**, No. 2, p. 573–576 (1979).
See also *1.2*, p. *19*; 1.4, p. 29; 21.1, p. 363.

K. Heinemann, M. J. Yacamán, and H. Poppa
A New Approach for Simultaneous Orientation and Three-Dimensional Shape Determination of Small Metal Particles
(a) Vol. **54**, No. 2, p. 675–680 (1979).
See also *1.4*, p. *29*; *1.5*, p. *35*.

K. I. Wysokiński and M. Pilat
Effect of Temperature on the Off-Diagonal Alloy Properties
(b) Vol. **95**, No. 2, p. 369–375 (1979).
See also *13*, p. *140*; *14.1*, p. *202*.

M. Kumar and M. P. Hemkar
Temperature Variation of Phonon Limited Electrical Resistivity in Noble Metals
(b) Vol. **95**, No. 2, p. K107–K111 (1979).
See also 6, p. 74; *14.1*, p. *202*.

G. Honjo, K. Takayanagi, K. Kobayashi, and K. Yagi
On Cluster Mobilities in Nucleation and Growth Processes of Epitaxial Thin Films
(a) Vol. **55**, No. 2, p. 353–367 (1979).
See also *1.5*, p. *36*; *3*, p. *53*; 22, p. 404.

D. Kashchiev
On the Saturation Cluster Density in Deposition of Thin Films from Vapours
(a) Vol. **55**, No. 2, p. 369–374 (1979).
See also *1.5*, p. *36*; *3*, p. *53*; 10.2, p. 114; 22.5.2, p. 471.

G. van Tendeloo, M. van Sande, and S. Amelinckx
Imaging Hexagonal Based Superstructures. A New Stable Orthorhombic Structure in the Au–Mn System
(a) Vol. **55**, No. 2, p. 681–689 (1979).
See also *1.3*, p. *25*; 1.4, p. 31.

J. J. Jackson, E. A. Ryan, and Y. N. Lwin
A Study of Thermal Conversion of Interstitials in Platinum by Means of Electrical Resistivity Measurements
(a) Vol. **55**, No. 2, p. 735–746 (1979).
See also *10.1*, p. *99*; 11, p. 124; 14.1, p. 202.

P. D. Pathak and N. P. Shah
Debye Temperatures of Silver and Aluminium at High Temperatures. Some New Correlations
(a) Vol. **55**, No. 2, p. K159–K162 (1979).
See also *6*, p. *75*; *8*, p. *83*; 21, p. 351.

K. Urban
Radiation-Induced Processes in Experiments Carried out In-situ in the High-Voltage Electron Microscope
(a) Vol. **56**, No. 1, p. 157–168 (1979).
See also *10.1*, p. *100*; 11, p. 125; 21, p. 352; 21.1.1, p. 376.

K. Narayandas, M. Radhakrishnan, and C. Balasubramanian
Dependence of Defect Density and Activation Energy on Deposition Rates in Silver Films
(a) Vol. **56**, No. 1, p. 195–198 (1979).
See also 1.5, p. 37; *10.1*, p. *100*; *14.1*, p. *202*.

F. F. Komarov
Energy Loss of Relativistic Particles in Solids
(b) Vol. **96**, No 2, p. 555–568 (1979).
See also *11*, p. *125*; 13, p. 142; 22.1, p. 408.

21.7 Semi-Metals

(Semiconducting Alloys See 22.7)

Original Papers and Short Notes

H.-J. Gollnest, G. Kuka, W. Braune, and R. Herrmann
Temperature Dependence of the Relaxation Time of the Holes in Bismuth
(b) Vol. 91, No. 1, p. 177–183 (1979).
See also *19*, p. *289*.

T. Jung and W. Braune
On the Deformation Model of the Carrier Spectrum for Bismuth-Type Crystals (I): Electrons
(b) Vol. 91, No. 2, p. 563–569 (1979).
See also *13.1*, p. *144*; 13.2, p. 156.

S. Guha, P. K. Sen, and S. Ghosh
Modulational Instability of a Finite Amplitude Alfvén Wave in Semimetals
(b) Vol. 91, No. 2, p. K135–K138 (1979).
See also *13*, p. *137*; *20*, p. *299*.

K. S. Srivastava, O. K. Harsh, and V. Kumar
Chemical Shifts of the X-Ray L-Absorption Edges of Pb and Bi
(b) Vol. 91, No. 2, p. K169–K172 (1979).
See also *20*, p. *299*; 21, p. 339; 22, p. 394.

R. Balzer and H. Sigvaldason
Equilibrium Vacancy Concentration Measurements on Tin Single Crystals
(b) Vol. 92, No. 1, p. 143–147 (1979).
See also *8*, p. *80*; 10.1, p. 93.

T. Jung and W. Braune
On the Deformation Model of the Carrier Spectrum for Bismuth-Type Crystals (II): Holes
(b) Vol. 92, No. 1, p. 171–176 (1979).
See also *13.1*, p. *145*; 13.2, p. 156.

K. Bartkowski, D. Włosewicz, and J. Rafałowicz
Lorenz Function and Horizontal and Vertical Component of Thermal Resistivity of Tin Monocrystals
(a) Vol. 52, No. 2, p. 397–400 (1979).
See also *8*, p. *81*; *14.1*, p. *200*.

R. Tidecks and G. von Minnigerode
The Influence of High-Frequency Radiation on the *U–I* Characteristics of Superconducting Tin Whiskers
(a) Vol. 52, No. 2, p. 421–426 (1979).
See also *14.2*, p. *205*.

W. Braune, W. Schneider, J. Lebech, and K. Saermark
Magnetoplasma Waves in Bismuth and Semimetallic BiSb Alloys at Microwave Frequencies 9 to 100 GHz
(b) Vol. 93, No. 1, p. 269–278 (1979).
See also *13.5.2*, p. *191*; 13.5.4, p. 195; 19, p. 291.

22. Semiconductors and Ionic Crystals

Review Articles

O. A. Golikova
Boron and Boron-Based Semiconductors
(a) Vol. 51, No. 1, p. 11–40 (1979).
See also *1*, p. *9*; 13.4, p. 163; *14.3*, p. *209*; 14.3.4, p. 231; 15, p. 249; 20.1, p. 304; 22.1.3, p. 428; 22.2, p. 430.

W. Nolting
Theory of Ferromagnetic Semiconductors
(b) Vol. 96, No. 1, p. 11–54 (1979).
See also *13.5.4*, p. *195*; 18.2, p. 270.

Original Papers and Short Notes

K.-R. Schulze
On the Dielectric Response of Solids (I): New Approximations for the Treatment of Local Field Effects
(b) Vol. 91, No. 1, p. 211–221 (1979).
See also *13.1*, p. *143*; 14.4, p. 238; 21, p. 338.

O. Breitenstein and K. Unger
Distribution of the Perturbation Potential and the Density of States in High-Doped Semiconductors
(b) Vol. 91, No. 1, p. 243–250 (1979).
See also *13.1*, p. *143*; *14.3.4*, p. *231*.

T. Satow, O. Uemura, and M. Sanzyō
The Electrical Conductivity of Bismuth Chalcogenides in the Liquid State
(b) Vol. 91, No. 1, p. 325–330 (1979).
See also *2*, p. *40*; *14.3*, p. *209*.

H.-J. Wünsche and K. Henneberger
Bound Excitons in Local Density Approximation
(b) Vol. 91, No. 1, p. 331–337 (1979).
See also *13.4*, p. *163*; *13.5.1*, p. *183*.

C. K. Sarkar and D. Chattopadhyay
Effect of Band Non-Parabolicity on the Microwave Magnetoconductivity in a Large Quantizing Magnetic Field
(b) Vol. 91, No. 1, p. K9–K11 (1979).
See also *13.1*, p. *144*; *14.3*, p. *209*.

D. Ihle
On the Thermoelectric Power of Superionic Conductors
(b) Vol. 91, No. 1, p. K49–K52 (1979).
See also *15*, p. *249*.

F. L. Edelman, B. N. Zaitsev, V. Z. Latuta, and A. A. Khoromenko
The Structure of Silicon Nitride Films (II): Non-Stoichiometric Silicon Nitride
(a) Vol. 51, No. 1, p. 49–56 (1979).
See also *1.2*, p. *15*; *1.5*, p. *34*.

T. O. GEGECHKORI and Z. S. KACHLISHVILI
Influence of the Capture of Charge Carriers upon the *I–U* Curves of Hot Electron Semiconductors with Account of Phonon Heating
(a) Vol. 51, No. 2, p. 333–336 (1979).
See also 6, p. 65; *13*, p. *137*; *14.3.4*, p. *232*.

F. L. EDELMAN
The Structure of Silicon Nitride Films (III): Oxidation of Silicon Nitride – Defects
(a) Vol. 51, No. 2, p. 375–381 (1979).
See also *1.2*, p. *16*; *1.5*, p. *34*.

G. LO VECCHIO and G. MORANDI
On the Interpretation of TEM Images of p–n Junctions
(a) Vol. 51, No. 2, p. 383–390 (1979).
See also *1.4*, p. *26*; *14.3.3*, p. *226*.

X. AYMERICH-HUMET and F. SERRA-MESTRES
Resonant Tunneling Current for General Junction Potential Barrier
(a) Vol. 51, No. 2, p. 583–592 (1979).
See also 13.4, p. 165; *14.3.3*, p. *226*.

A. PINCEVIČIUS, A. SAKALAS, and J. VIŠČAKAS
Double Injection into Structures with a Low-Conductivity Region (II): Current–Voltage Characteristics of a Single-Level Semiconductor
(a) Vol. 51, No. 2, p. 623–628 (1979).
See also *14.3.4*, p. *232*.

E. A. LEBEDEV, E. NEBAUER, and P. SÜPTITZ
Broadening of Electrodiffusion Profiles at High Electric Fields in Glassy As_2Se_3:Cu
(a) Vol. 51, No. 2, p. K207–K211 (1979).
See also *2*, p. *41*; *9*, p. *85*; 14.3.4, p. 232.

R. PICCOLO
A Note on the Photomagnetoelectric Effect
(a) Vol. 51, No. 2, p. K217–K219 (1979).
See also 14.3.2, p. 223; *16*, p. *252*.

R. ENDERLEIN, K. PEUKER, and F. BECHSTEDT
General Theory of Light Scattering in Solids
(b) Vol. 92, No. 1, p. 149–158 (1979).
See also *6*, p. *65*; *20*, p. *299*.

V. V. HIZHNYAKOV and A. V. SHERMAN
On the Theory of Resonant Secondary Radiation of Excitons Strongly Interacting with Phonons
(b) Vol. 92, No. 1, p. 177–183 (1979).
See also *6*, p. *65*; *13.5.1*, p. *184*; 20.1, p. 307; 20.3, p. 328.

C. E. PATTON
Spin-Wave Instability Theory in Cubic Single Crystal Magnetic Insulators (I): General Theory
(b) Vol. 92, No. 1, p. 211–220 (1979).
See also *13.5.4*, p. *195*; 18.2, p. 272.

YU. D. ARBUZOV and V. M. EVDOKIMOV
On the Static Conductivity of a Heavily Doped Semiconductor
(b) Vol. 92, No. 2, p. 579–583 (1979).
See also *14.3*, p. *211*.

H.-R. TREBIN
Exciton–Phonon Interaction in the Degenerate-Band Case
(b) Vol. 92, No. 2, p. 601–610 (1979).
See also *13.5.1*, p. *184*.

S. GUHA and S. GHOSH
Longitudinal Phonon–Plasmon Interaction in Inhomogeneous Semiconductors
(b) Vol. 92, No. 2, p. K95–K97 (1979).
See also *6*, p. *67*; *13.5.2*, p. *191*.

Ľ. HRIVNÁK and B. ĎUROVOVÁ
High-Field Capture of Electrons by a Screened Coulomb Attractive Center
(b) Vol. 92, No. 2, p. K99–K101 (1979).
See also *13.4*, p. *168*; *14.3.4*, p. *233*.

V. K. BASHENOV, I. K. DOICHO, and A. G. PETUKHOV
Localized Electron States Associated with a Transition-Metal Impurity in Semiconductors
(b) Vol. 92, No. 2, p. K147–K150 (1979).
See also *13.4*, p. *168*.

R. GEMPERLE, P. NOVOTNÝ, and A. MEŇOVSKÝ
Bitter Figure Observations on U_3As_4 at Low Temperatures
(a) Vol. 52, No. 2, p. 587–596 (1979).
See also *18.2*, p. *273*.

A. MONTANER, M. GALTIER, C. BENOIT, and H. BILL
Optical Constants of Sodium Sulphide
(a) Vol. 52, No. 2, p. 597–601 (1979).
See also *6*, p. *67*; *20.1*, p. *310*.

V. G. ZHDANOV, B. T. KOLOMIETS, V. M. LYUBIN, and V. K. MALINOVSKII
Photoinduced Optical Anisotropy in Chalcogenide Vitreous Semiconducting Films
(a) Vol. 52, No. 2, p. 621–626 (1979).
See also *2*, p. *43*; *20.1*, p. *310*.

J. SHANKER and V. C. JAIN
Analysis of the Crystal Binding of Alkali Hydrides
(a) Vol. 52, No. 2, p. 675–680 (1979).
See also *4*, p. *56*; 12.2, p. 134.

A. A. URUSOVSKAYA and G. G. KNAB
Thermal Activation Analysis of Plastic Deformation of PbS Single Crystals
(a) Vol. 52, No. 2, p. K133–K136 (1979).
See also *10.2*, p. *106*.

N. M. RAVINDRA, SUSHIL AULUCK, and V. K. SRIVASTAVA
Temperature Dependence of the Energy Gap in PbS, PbSe, and PbTe
(a) Vol. 52, No. 2, p. K151–K155 (1979).
See also *13.1*, p. *147*.

R. Kutner and I. Sosnowska
Quasi-Elastic Neutron Scattering Laws at Finite Concentrations of Elements Diffusing in a Crystal
(b) Vol. **93**, No. 1, p. K39–K43 (1979).
See also *9*, p. *86*.

G. Kaden and H. Reimer
MOS Structures with Reserve-Biased External p–n Junction for Linear Gate Voltage Selection (I): Impulse-Current Behaviour
(a) Vol. **53**, No. 1, p. 195–206 (1979).
See also *13.3*, p. *159*; *14.3.3*, p. *227*; 14.3.4, p. 234.

M. Stock and W. Dultz
Raman Spectroscopic Studies of the Crystalline Phases of NaCN and KCN at Pressures up to 55 kbar
(a) Vol. **53**, No. 1, p. 237–241 (1979).
See also *1.2*, p. *17*; 20.1, p. 312.

C. Manolikas, J. van Landuyt, and S. Amelinckx
Electron Microscopy and Electron Diffraction Study of the Domain Structures, the Dislocation Fine Structure, and the Phase Transformations in β-$MoTe_2$
(a) Vol. **53**, No. 1, p. 327–338 (1979).
See also *1.1*, p. *13*; 1.2, p. 17.

Kh. I. Amirkhanov, Ya. B. Magomedov, Sh. M. Ismailov, and N. L. Kramynina
The Thermal Conductivity of Semiconductors of the System $As_2(Se_{1-x}Te_x)_3$ Depending on Temperature and Pressure
(a) Vol. **53**, No. 1, p. 361–366 (1979).
See also 2, p. 43; 6, p. 68; *8*, p. *81*.

L. Borisova and S. Dimitrova
Kinetic Coefficients in Solid Solutions of the $(Ag_{x/2}Pb_{1-x}Sb_{x/2})Te$ Type
(a) Vol. **53**, No. 1, p. 403–407 (1979).
See also *14.3*, p. *213*; 15, p. 250.

B. Rauschenbach and G. Blasek
Determination of Very Low Diffusion Coefficients in Glass with Means of Secondary-Ion Mass Spectrometry
(a) Vol. **53**, No. 1, p. K11–K13 (1979).
See also *2*, p. *44*; *9*, p. *87*; 17, p. 257.

L. D. Borisova
Thermoelectric Properties of Impurity Doped PbTe
(a) Vol. **53**, No. 1, p. K19–K22 (1979).
See also 14.3, p. 213; *15*, p. *250*.

A. M. Berezhkovskii and A. A. Ovchinnikov
Certain Features of Semiconductor Behaviour in Strong Parallel Magnetic and Electric Fields (III): Interband Absorption
(b) Vol. **93**, No. 2, p. 549–551 (1979).
See also *13*, p. *139*; 20.1, p. 313.

F. Mokross and H. Büttner
The Exciton–Neutral Donor and Biexciton Complexes in Polar Semiconductors
(b) Vol. 94, No. 1, p. 107–114 (1979).
See also 13.4, p. 171; *13.5.1*, p. *186*.

Doan Nhat Quang, B. Esser, and R. Keiper
On the Theory of the Complex Dielectric Function of a Heavily Doped Degenerate Semiconductor
(b) Vol. 94, No. 1, p. 115–123 (1979).
See also *13.4*, p. *171*; *20.1*, p. *314*.

H. Stolz and R. Zimmermann
Correlated Pairs and a Mass Action Law in Two-Component Fermi Systems. Excitons in an Electron–Hole Plasma
(b) Vol. 94, No. 1, p. 135–146 (1979).
See also *13.5.1*, p. *186*; 13.5.2, p. 192.

P. Kuivalainen, J. Sinkkonen, K. Kaski, and T. Stubb
Bound Magnetic Polaron in Magnetic Semiconductors
(b) Vol. 94, No. 1, p. 181–190 (1979).
See also *13.5.3*, p. *193*; 14.3, p. 213; *18.2*, p. *276*.

A. Stahl
Polariton Structure of Interband Transitions in Semiconductors
(b) Vol. 94, No. 1, p. 221–230 (1979).
See also *13.5*, p. *182*; 20, p. 300.

C. P. Mallett and P. Weinberger
A Preliminary Study of the Electronic Structure of Uranium Chalcogenides and Pnictides Using a Relativistic Single-Site Scattering Model
(b) Vol. 94, No. 1, p. 257–261 (1979).
See also 4, p. 57; *13.1*, p. *150*.

A. Conan, D. Delaunay, A. Bonnet, A. G. Moustafa, and M. Spiesser
Temperature Dependence of the Electrical Conductivity and Thermoelectric Power in $MoTe_2$ Single Crystals
(b) Vol. 94, No. 1, p. 279–286 (1979).
See also *14.3*, p. *214*; *15*, p. *250*.

H. van Cong
Erratum to: Influence of Band-Tail Parameters on Electronic Specific Heat and Transport Coefficients in Heavily Doped Semiconductors
(b) Vol. 94, No. 1, p. 337 (1979).
See also *8*, p. *82*; 13.4, p. 172; *14.3*, p. *214*.

M. Grünewald, P. Thomas, and D. Würtz
Simple Calculation of the Hopping Conductivity for an Exponential Band Tail
(b) Vol. 94, No. 1, p. K1–K4 (1979).
See also *2*, p. *45*; *14.3*, p. *214*.

S. Noor Mohammad
On the Screening of Impurity Ions in Semiconductors with Spatially Variable Dielectric Constants
(b) Vol. 94, No. 1, p. K5–K10 (1979).
See also 13.4, p. 172; *14.4*, p. *241*.

B. Esser and D. Sprenger
On the Theory of Absorption and Luminescence Spectra in Amorphous Semiconductors
(b) Vol. **94**, No. 2, p. 583–593 (1979).
See also **2**, p. **46** ; *20.1*, p. *316*; *20.3*, p. *333*.

V. K. Arora and H. N. Spector
Quantum-Limit Cyclotron Resonance Linewidth in Semiconductors
(b) Vol. **94**, No. 2, p. 701–709 (1979).
See also *19*, p. *294*.

J. Spałek and Z. Tarnawski
Static Magnetic Properties of Doped Magnetic Semiconductors. Mean Field Approach
(b) Vol. **94**, No. 2, p. 711–720 (1979).
See also *18.2*, p. *277*; 18.3, p. 284; 18.4, p. 287.

P. Kleinert
The Influence of Contact Excitons on the Electroabsorption in Disordered Semiconductors
(b) Vol. **94**, No. 2, p. K145–K147 (1979).
See also 2, p. 46; *13.5.1*, p. *187*; 14.3.4, p. 235; *20.1*, p. *317*.

V. F. Synorov, V. D. Linnik, and B. I. Sysoev
The Energy Spectrum of Discrete Centres in the High Inhomogeneous Electric Field of the Space Charge Region of a Semiconductor
(a) Vol. **54**, No. 2, p. 517–521 (1979).
See also *13.4*, p. *174*; *14.3.4*, p. *235*.

B. T. Kolomiets, V. Ch. Shpunt, and E. A. Smorgonskaya
Characteristic Features of the Thermal Switching Process in Chalcogenide Vitreous Semiconductors
(a) Vol. **54**, No. 2, p. 585–591 (1979).
See also *2*, p. *46*; *14.3.4*, p. *235*.

L. M. Lambert
Theory of ISC Measurements for p-Type Schottky Diodes with Deep Acceptor Impurities
(a) Vol. **54**, No. 2, p. 665–673 (1979).
See also *13.4*, p. *175*; *14.3.3*, p. *229*.

A. Ślęzak, J. Lech, and I. Bójko
Electron Paramagnetic Resonance of Mn^{2+} Ions in Aragonite Single Crystals
(a) Vol. **54**, No. 2, p. 755–759 (1979).
See also *19*, p. *294*.

M. Balarin
Temperature Dependence of the Pre-Exponential Factor in the Glow Curve Theory
(a) Vol. **54**, No. 2, p. K137–K140 (1979).
Erratum
(a) Vol. **56**, No. 2, p. K149 (1979).
See also *13.4*, p. *175*.

L. Tóth
Electron and Hole Drift Mobilities in $GeSe_2$ Glass
(a) Vol. **54**, No. 2, p. K159–K161 (1979).
See also *2*, p. *46*; *14.3*, p. *214*.

B. JOUFFREY
Methodical Aspects of HVEM in In-Situ Experiments in the Field of Condensed Matter
(a) Vol. 55, No. 2, p. 549–562 (1979).
See also *1.4*, p. *31*; 21, p. 350.

A. YU. LEIDERMAN
On the Theory of Double Injection in Semiconductors with Continuous Distribution of Trapping Levels within the Gap
(a) Vol. 55, No. 2, p. 661–671 (1979).
See also *13.4*, p. *177*; *14.3.4*, p. *236*.

P. VAROTSOS
Entropy and Volume of Solution of Substitutional Impurities in Solids
(a) Vol. 55, No. 2, p. K139–K142 (1979).
See also *10*, p. *90*; 21, p. 351.

I. M. DYKMAN and L. A. PROKHNITSKII
Electromagnetic Waves in an Inhomogeneous Semiconductor under Electric and Magnetic Fields
(b) Vol. 96, No. 1, p. 55–60 (1979).
See also *13*, p. *141*; *14.3*, p. *216*; 20, p. 302.

SH. N. GIFEISMAN and YU. G. MALKOCH
The Theory of Magneto-Optical Effects in Shallow F-Centres
(b) Vol. 96, No. 1, p. 73–90 (1979).
See also 13.4, p. 177; 13.5.3, p. 193; *18*, p. *263*; *20*, p. *302*.

H. BÖTTGER and V. V. BRYKSIN
Effective Medium Theory for the Hopping Conductivity in High Electrical Fields
(b) Vol. 96, No. 1, p. 219–224 (1979).
See also *14.3*, p. *216*.

O. ZIEP and D. GENZOW
Calculation of the Interband Absorption in Lead Chalcogenides Using a Multiband Model
(b) Vol. 96, No. 1, p. 359–368 (1979).
See also *13.1*, p. *153*; *20.1*, p. *321*.

H.-J. WÜNSCHE
Mean Component Approximation for the Correlation Energy of Multicomponent Electron–Hole Plasmas
(b) Vol. 96, No. 1, p. 379–387 (1979).
See also *13*, p. *141*.

V. M. FOMIN
Non-Linear Optical Properties of Band Charge Carriers Due to Scattering by Lattice Defects
(b) Vol. 96, No. 1, p. 389–399 (1979).
See also 6, p. 75; 13.4, p. 178; *20.2*, p. *326*.

O. ZIEP and D. GENZOW
On the Concentration Dependence of the Susceptibility Effective Mass in $Pb_{0.8}Sn_{0.2}Te$
(b) Vol. 96, No. 1, p. K25–K27 (1979).
See also *13.1*, p. *153*.

A. Zwick and M. A. Renucci
Long-Wavelength Optical Phonons in $ZrSe_3$
(b) Vol. 96, No. 2, p. 757–767 (1979).
See also *6*, p. *76*; *20.1*, p. *323*.

Y. Lépine and D. Matz
Mean Field Theory of a Single Fröhlich Polaron. Possible Existence of Phase Transitions
(b) Vol. 96, No. 2, p. 797–806 (1979).
See also *13.5.3*, p. *193*.

V. D. Iskra
Phonon-Assisted Interband Optical Absorption in Disordered Semiconductors
(b) Vol. 96, No. 2, p. 843–856 (1979).
See also *13.5.1*, p. *189*; *20.1*, p. *323*.

Doan Nhat Quang
On the Theory of Electroreflectance for Heavily Doped Degenerate Semiconductors
(b) Vol. 96, No. 2, p. K73–K76 (1979).
See also *14.3.4*, p. *237*; *20.1*, p. *323*.

I. Egri and A. Stahl
Real Space Wave Equation for Exciton-Polaritons of Wannier Type
(b) Vol. 96, No. 2, p. K83–K86 (1979).
See also *13.5.1*, p. *190*; 20, p. 303.

J. Y. Harbec, C. Deville Cavellin, and S. Jandl
Raman Spectra and Crystal Symmetry of ZrS_3
(b) Vol. 96, No. 2, p. K117–K120 (1979).
See also *6*, p. *77*; *20.1*, p. *323*.

E. M. Trukhanov, S. I. Stenin, and A. G. Noskov
Dislocation Interactions in Germanium Covered with Dielectric Films
(a) Vol. 56, No. 2, p. 443–450 (1979).
See also *1.5*, p. *37*; 2, p. 50; *10.2*, p. *117*; 22.1.1, p. 413; 22.6, p. 485.

A. Toneva and R. Kamburova
Electrical Properties of p-$Pb_{0.41}Sn_{0.59}Te$
(a) Vol. 56, No. 2, p. K113–K116 (1979).
See also *14.3*, p. *218*.

22.1 Elements

Original Papers and Short Notes

M. MÜLLNER, J. MAETZ, and H. JEX
Thermal Diffuse Scattering from Pyrolytic Graphite Investigated with Mössbauer γ-Ray Diffraction
(a) Vol. **56**, No. 2, p. 541–544 (1979).
See also *5*, p. *62*.

22.1.1 Germanium

Original Papers and Short Notes

N. Saito, H. Fujiyasu, and S. Yamada
Doping Effects of Sb and Ga on Evaporated Amorphous Ge
(a) Vol. **51**, No. 1, p. 235–242 (1979).
See also *2*, p. *40*; *14.3.1*, p. *219*; 15, p. 249.

P. U. Arifov, N. Yu. Arutyunov, and A. S. Baltenkov
The Annihilation of Positrons with Electrons Bound in Edge Dislocations
(b) Vol. **91**, No. 2, p. K91–K94 (1979).
See also *10.2*, p. *104*; *13.4*, p. *165*.

K. S. Dubey and H. A. Kassim
Lattice Thermal Resistivity Due to the Bound State Electrons
(b) Vol. **91**, No. 2, p. K149–K154 (1979).
See also *6*, p. *65*; *8*, p. *80*.

P. A. Tove, M. P. Ali, M. M. Ibrahim, and H. Norde
Electrical Properties of Junctions between Ge Films and Monocrystalline Silicon
(a) Vol. **51**, No. 2, p. 491–496 (1979).
See also *14.3.2*, p. *223*; *14.3.3*, p. *226*; 22.1.2, p. 416.

Y. Kamiura
New Shallow Acceptors Produced in Germanium by Quenching
(a) Vol. **51**, No. 2, p. 601–607 (1979).
See also *13.4*, p. *165*; *14.3*, p. *210*.

V. V. Mitin, Yu. A. Tkhorik, and Yu. M. Shvarts
Anisotropy of the Transverse Magnetoresistance in p-Ge Films on GaAs
(a) Vol. **51**, No. 2, p. K133–K135 (1979).
See also *14.3.1*, p. *219*.

R. M. Mehra, Radhey Shyam, and P. C. Mathur
Negative Magnetoresistance of an Amorphous Se–Te–Ge Sample
(a) Vol. **51**, No. 2, p. K165–K169 (1979).
See also *14.3*, p. *211*; 22.1.3, p. 428.

A. Manoogian and A. Leclerc
Determination of the Dilation and Vibrational Contributions to the Energy Band Gaps in Germanium and Silicon
(b) Vol. **92**, No. 1, p. K23–K27 (1979).
Erratum
(b) Vol. **94**, No. 2, p. K195 (1979).
See also 6, p. 66; *13.1*, p. *146*; 22.1.2, p. 417.

T. Soma
Equation of State under High Pressure for Metallic Si and Ge
(b) Vol. **92**, No. 1, p. K51–K53 (1979).
See also *1.2*, p. *16*; 4, p. 55; 12.1, p. 129; 22.1.2, p. 417.

P. L. Li and N. C. Halder
Frequency Response Characteristic of Amorphous Germanium Thin Films
(a) Vol. **52**, No, 1, p. 365–370 (1979).
See also 2, p. 42; *14.3.1*, p. *219*.

A. Vehanen, J. Yli-Kauppila, P. Hautojärvi, and V. S. Mikhalenko
A Positron Study of Deformed NiGe Alloy
(a) Vol. **52**, No. 1, p. K73–K75 (1979).
See also *10.1*, p. *94*; 21.1, p. 357.

S. Singh and G. S. Verma
Resonance Scattering of Phonons in Inhomogeneously Strained n-Ge
(b) Vol. **92**, No. 2, p. 627–631 (1979).
See also *6*, p. *67*; *8*, p. *81*.

E. N. Mgbenu
Activation Energy for Diffusion in Au–Ge Thin Film Couples
(a) Vol. **53**, No. 1, p. 397–401 (1979).
See also *9*, p. *86*; *14.3.1*, p. *220*; 21.6, p. 388.

F. Hashimoto, Y. Kamiura, and H. Takahashi
Effect of Quenching Temperature on the Annealing Rate of Quenched-In Acceptors in Germanium
(a) Vol. **53**, No. 1, p. K31–K33 (1979).
See also 9, p. 87; *10.2*, p. *108*.

D. Brown, S. Fairbairn, and G. J. Morgan
A Quasi-Polycrystalline Approximation for Electronic States in Liquid Transition Metals and Tetrahedrally Bonded Semiconductors
(b) Vol. **93**, No. 2, p. 617–627 (1979).
See also *2*, p. *44*; *13.1*, p. *148*; 21.1, p. 359; 22.1.2, p. 419.

G. Langouche, I.Dézsi, M. van Rossum, J. de Bruyn, and R. Coussement
Observation of a Strong Dose Dependence in ^{57}Co Implantations in Si and Ge
(b) Vol. **93**, No. 2, p. K107–K110 (1979).
See also *5*, p. *60*; *11*, p. *122*; 22.1.2, p. 420.

E. M. Trukhanov, S. I. Stenin, and A. G. Noskov
Dislocations and Stresses in a Crystal with an Island Film
(a) Vol. **53**, No. 2, p. 433–440 (1979).
See also *1.5*, p. *35*; *10.2*, p. *109*.

I. V. Gridneva, Yu. V. Milman, V. I. Trefilov, and S. I. Chugunova
Analysis of Dislocation Mobility under Concentrated Loads at Indentations of Single Crystals
(a) Vol. **54**, No. 1, p. 195–206 (1979).
See also *10.2*, p. *109*; 12.2, p. 135; 22.1.2, p. 421; 22.2.1, p. 434; 22.2.3, p. 440.

J. von Wienskowski and H. Froitzheim
Oxidation of Clean Cleaved Ge (111) Surfaces
(b) Vol. **94**, No. 2, p. 429–435 (1979).
See also *13.3*, p. *160*.

G. DUGGAN, G. J. MORGAN, and A. LETTINGTON
The Effective Screening Function in Si, Ge, and GaAs and the Effects of Exchange
(b) Vol. **94**, No. 2, p. 659—665 (1979).
See also *13*, p. *140*; *14.4*, p. *241*; 22.1.2, p. 422; 22.2.1, p. 434.

Y. KAMIURA and F. HASHIMOTO
Degree of Cooper Contamination in Germanium Heat-Treated at High Temperatures
(a) Vol. **54**, No. 2, p. 697—700 (1979).
See also *10.2*, p. *111*; *13.4*, p. *175*.

E. SOBCZAK
X-Ray Bremsstrahlung Isochromat and X-Ray Photoelectron Investigations of Ion Implanted Germanium
(a) Vol. **54**, No. 2, p. K151—K153 (1979).
See also 13.1, p. 151; *17*, p. *258*; *20*, p. *301*.

S. W. KOCH and H. HAUG
Treatmant of the Electron–Hole Droplet Nucleation in the Fokker-Planck Approximation
(b) Vol. **95**, No. 1, p. 155—161 (1979).
See also *13.5.1*, p. *188*; 20.1, p. 318; 20.3, p. 334.

G. MARTENS, P. RABE, G. TOLKIEHN, and A. WERNER
EXAFS in Photoelectron Yield Spectra at K Edges of Cu, Ni, and Ge
(a) Vol. **55**, No. 1, p. 105—108 (1979).
See also *17*, p. *258*; 20.1, p. 319; 21.1, p. 364.

K. B. TOLPYGO and S. M. ZUBKOVA
On the Indirect Exciton Energy Levels in Crystals of Diamond and Sphalerite Type with Spin–Orbit Interaction
(b) Vol. **95**, No. 2, p. 341—348 (1979).
See also 13.1, p. 152; *13.5.1*, p. *188*; 22.1.2, p. 424; 22.2, p. 431.

T. SOMA
Volume Effect on the Bulk Modulus of Si–Ge Solid Solutions
(b) Vol. **95**, No. 2, p. K117—K119 (1979).
See also *12.2*, p. *135*; 13, p. 141; 22.1.2, p. 424.

H.-J. MÖLLER, P. HAASEN, and H. EWALDT
Cross Slip of Single Dissociated Screw Dislocations in Silicon and Germanium
(a) Vol. **55**, No. 2, p. 469—478 (1979).
See also *10.2*, p. *114*; 22.1.2, p. 424.

N. PASHOV, S. SIMOV, M. KALITZOVA, M. G. MICHAILOV, and A. DJAKOV
Alignment of Defects in Ge Implanted with Te^+ Ions
(a) Vol. **55**, No. 2, p. 599—602 (1979).
See also *1.4*, p. *31*; *11*, p. *124*.

H. GOLLISCH
Density Functional Approach to Point Defect Propertis of Copper–Germanium Alloys
(b) Vol. **96**, No. 1, p. 225—231 (1979).
See also *10.1*, p. *99*; 21.1, p. 366.

D. R. Mašović and S. Zieković
Model Pseudopential for Elementary Semiconductors
(b) Vol. **96**, No. 1, p. 469–474 (1979).
See also *13.1*, p. *153*; 21.7, p. 392; 22.1.2, p. 425.

E. A. Bakulin, V. L. Hilarov, and V. V. Shcherbinina
Characteristic Energy Losses and Optical Properties of Intermetallic Compounds in the Fe–Ge System
(a) Vol. **56**, No. 1, p. 259–265 (1979).
See also *14.4*, p. *242*; 21.1.1, p. 377.

K. Seeger and H. Vana
Hot-Carrier Infrared Absorption in n-Type Germanium
(b) Vol. **96**, No. 2, p. 605–610 (1979).
See also *13*, p. *142*; 14.3.4, p. 237; *20.1*, p. *322*.

W. Pickin
The Kinetics of Electron Transitions for Donors with Excited States
(b) Vol. **96**, No. 2, p. 617–626 (1979).
See also *13.4*, p. *179*; *16*, p. *254*.

S. M. Darbinian and K. A. Ispirian
On High Energy Particle Channeling
(b) Vol. **96**, No. 2, p. 835–842 (1979).
See also *11*, p. *125*; 13, p. 142.

H. H. Boghossian and K. S. Dubey
Phonon Conductivity Correction Term and Electron–Phonon Scattering Relaxation Rate: Application to Doped and Undoped Ge
(b) Vol. **96**, No. 2, p. K107–K112 (1979).
See also *6*, p. *77*; *8*, p. *84*.

E. M. Trukhanov, S. I. Stenin, and A. G. Noskov
Dislocation Interactions in Germanium Covered with Dielectric Films
(a) Vol. **56**, No. 2, p. 443–450 (1979).
See also *1.5*, p. *37*; 2, p. 50; *10.2*, p. *117*; 22, p. 407; 22.6, p. 485.

B. D. Sawicka, J. A. Sawicki, J. Stanek, T. Tyliszczak, and J. Kowalski
Temperature-Dependet Conversion Electron Mössbauer Measurments of ^{57}Fe Implanted in Silicon and Germanium
(a) Vol. **56**, No. 2, p. 451–456 (1979).
See also *2*, p. *50*; *5*, p. *62*; 22.1.2, p. 426.

K. Zellama, J. F. Morhange, P. Germain, and J. C. Bourgoin
Growth Rate of Crystallization in Amorphous Germanium Produced by Ion Implantation. A Raman Spectroscopy Study
(a) Vol. **56**, No. 2, p. 717–723 (1979).
See also 1.2, p. 21; *2*, p. *50*; 6, p. 77; *11*, p. *126*; 20.1, p. 324.

R. R. Sharma and R. C. Sharma
Submillimeter Harmonic Generation in Semiconductor Waveguides
(a) Vol. **56**, No. 2, p. 743–747 (1979).
See also *14.3.4*, p. *237*; 22.2.3, p. 441.

22.1.2 Silicon

Original Papers and Short Notes

H. Köhler and M. Roos
Quantitative Determination of the Valley Splitting in n-Type Inverted Silicon (100) MOSFET Surfaces
(b) Vol. **91**, No. 1, p. 233–241 (1979).
See also *13.2*, p. *156*; *14.3.3*, p. *225*; 18, p. 261.

M. Nishida and T. Shimizu
Local Densities of States for the Overlayer of Al Chemisorbed on the Si (111) Surface
(b) Vol. **91**, No. 1, p. K5–K8 (1979).
See also *13.3*, p. *158*.

A. Makosa and T. Figielski
Spin-Dependent Reeombination at Dislocations in Heat Treated Silicon
(b) Vol. **91**, No. 1, p. K65–K67 (1979).
See also *10.2*, p. *103*; 13.4, p. 164; *16*, p. *252*.

M. Suezawa, K. Sumino, and I. Yonenaga
Dislocation Dynamics in the Plastic Deformation of Silicon Crystals (II): Theoretical Analysis of Experimental Results
(a) Vol. **51**, No. 1, p. 217–226 (1979).
See also *10.2*, p. *103*.

B. Joseph
The Time Dependence of the Flatband Voltage Shift of MOS Capacitors Caused by the Injection of Miniority Carries from Si into SiO_2
(a) Vol. **51**, No. 1, p. K9–K11 (1979).
See also *14.3.3*, p. *225*; *14.3.4*, p. *231*; 22.6, p. 477.

D. V. Morgan and J. Frey
Comments on the Modification of Schottky Barrier Height by Interfacial Oxides
(a) Vol. **51**, No. 1, p. K29–K33 (1979).
See also *14.3.2*, p. *223*; *14.3.4*, p. *231*; 22.2.1, p. 432; 22.2.2, p. 437.

W. Heywang, E. F. Krimmel, and H. Runge
Annealing Mechanism of Radiation Damage and Dopants in Pulsed Laser Light Irradiated Ion Implanted Layers
(a) Vol. **51**, No. 1, p. K79–K82 (1979).
See also **9**, p. **85**; *11*, p. *119*.

N. D. Stojadinović and S. D. Ristić
Effects of Emitter Diffusion-Induced Stresses on the Common-Emitter Current Gain of Silicon Planar Transistors
(a) Vol. **51**, No. 1, p. K83–K88 (1979).
See also *14.3.3*, p. *226*.

S. D. Ristić
Diffusion Coefficient in Heavily Doped Silicon
(a) Vol. **51**, No. 1, p. K89–K91 (1979).
See also *14.3*, p. *210*.

N. P. Morozov and D. I. Tetelbaum
Radiation Defect Formation at Ion Implantation of Semiconductors in the Presence of Force Fields
(a) Vol. **51**, No. 2, p. 629–640 (1979).
See also *11*, p. *120*.

M. Deb Roy
Effect of Random Variation in Impurity Concentration on Fermi Level in n-Type Silicon
(a) Vol. **51**, No. 2, p. K147–K149 (1979).
See also 13.1, p. 145; *14.3*, p. *210*.

V. G. Litovchenko, I. P. Lisovskii, R. O. Litvinov, H. Flietner, and W. Füssel
Specific Memory Effect in MOS Structures
(a) Vol. **51**, No. 2, p. K161–K164 (1979).
See also 11, p. 120; *13.3*, p. *158*; 14.3.2, p. 223.

M. A. Kumakhov and R. Wedell
Theory of Channelling of Relativistic Particles
(b) Vol. **92**, No. 1, p. 65–71 (1979).
See also *11*, p. *120*; 13, p. 137.

S. Marklund
Electron States Associated with Partial Dislocations in Silicon
(b) Vol. **92**, No. 1, p. 83–89 (1979).
See also *13.4*, p. *166*.

N. Lieske and R. Hezel
Surface States on Si(111) 7 × 7 Studied by Low-Energy Electron Loss Spectroscopy
(b) Vol. **92**, No. 1, p. 159–170 (1979).
See also *13.3*, p. *158*; *14.3.2*, p. *223*.

M. Jouanne, J. F. Morhange, and M. Balkanski
First and Second Order Raman Scattering by Pair Modes in Silicon
(b) Vol. **92**, No. 1, p. 255–258 (1979).
See also *6*, p. *66*; *20.1*, p. *307*.

A. Manoogian and A. Leclerc
Determination of the Dilation and Vibrational Contributions to the Energy Bands Gaps in Germanium and Silicon
(b) Vol. **92**, No. 1, K23–K27 (1979).
Erratum
(b) Vol. **94**, No. 2, p. K195 (1979).
See also 6, p. 66; *13.1*, p. *146*; 22.1.1, p. 410.

T. Soma
Equation ot State under High Pressure for Metallic Si and Ge
(b) Vol. **92**, No. 1, p. K51–K53 (1979).
See also *1.2*, p. *16*; 4, p. 55; 12.1, p. 129; 22.1.1, p. 410.

H.-J. Wünsche
On the Valley–Orbit Splitting of Multiple Bound Exciton States in Silicon
(b) Vol. **92**, No. 1, p. K57–K60 (1979).
See also *13.5.1*, p. *184*.

W. Schröter, H. G. Brion, and H. Siethoff
Self-Diffusion and Dynamical Recovery in Silicon at High Temperatures
(a) Vol. **52**, No. 2, p. K165–K168 (1979).
See also *9*, p. *86*; 10.2, p. 106.

E. P. Roth and A. C. Anderson
Phonon Scattering at a Low-Angle Grain Boundary in Silicon
(b) Vol. **93**, No. 1, p. 261–268 (1979).
See also *6*, p. *68*; *10.2*, p. *107*.

M. Asche and H. Kostial
Peculiarities of Current–Voltage Characteristics along ⟨110⟩ in n-Si
(b) Vol. **93**, No. 1, p. K89–K92 (1979).
See also *14.3.4*, p. *233*.

H. Pfleiderer
Magnetodiode Model with Gradation of Lifetime and Mobility
(a) Vol. **53**, No. 1, p. 187–194 (1979).
See also *14.3.3*, p. *227*; 14.3.4, p. 233.

I. P. Lisovskii, V. G. Litovchenko, and R. O. Litvinov
Effect of UV Illumination on the Electrical Properties of MOS Layer Structures
(a) Vol. **53**, No. 1, p. 253–262 (1979).
See also *13.3*, p. *159*; *14.3.1*, p. *219*.

J. Szuber and B. Salamon
Electron Spin Resonance Study on Low-Pressure-Baked Silicon Surfaces
(a) Vol. **53**, No. 1, p. 289–295 (1979).
See also 13.3, p. 159; *19*, p. *292*.

S. M. Lindsay
Laser Light Spot Mapping of Depletion in Power Semiconductor Devices
(a) Vol. **53**, No. 1, p. 311–320 (1979).
See also *14.3.2*, p. *223*; 14.3.3, p. 227.

A. Möschwitzer
A Simple Threshold Model for Depletion MOST
(a) Vol. **53**, No. 1, p. K43–K45 (1979).
See also *14.3.3*, p. *227*; *14.3.4*, p. *234*.

H. Feichtinger, A. Gschwandtner, and J. Waltl
Interstitial Iron and Thermal Defects in Silicon
(a) Vol. **53**, No. 1, p. K71–K74 (1979).
See also *10.2*, p. *108*; *13.4*, p. *170*.

P. E. van Camp, V. E. van Doren, and J. T. Devreese
A Phonon Dispersion Calculation of Silicon in the Dielectric Sceening Theory. A Closure Approximation
(b) Vol. **93**, No. 2, p. 483–491 (1979).
See also *6*, p. *68*; *13.1*, p. *148*.

D. Brown, S. Fairbairn, and G. J. Morgan
A Quasi-Polycrystalline Approximation for Electronic States in Liquid Transition Metals and Tetrahedrally Bonded Semiconductors
(b) Vol. **93**, No. 2, p. 617–627 (1979).
See also *2*, p. *44*; *13.1*, p. *148*; 21.1, p. 359; 22.1.1, p. 411.

J. Prechlet, J. Kalus, F. Lüscher, L. Pintschovius, and R. Ghosh
Measurement of Mode Grüneisen Parameters in Si
(b) Vol. **93**, No. 2, p. 653–659 (1979).
See also *6*, p. *69*.

G. Langouche, I. Dézsi, M. van Rossum, J. de Bruyn, and R. Coussement
Observation of a Strong Dose Dependence in ^{57}Co Implantations in Si and Ge
(b) Vol. **93**, No. 2, p. K107–K110 (1979).
See also *5*, p. *60*; *11*, p. *122*; 22.1.1, p. 411.

V. K Bashenov, I. K. Doicho, and A. G. Petukhov
Selection between Various Local Pseudopentials for Silicon
(b) Vol. **93**, No. 2, p. K131–K134 (1979).
See also *4*, p. *57*; *13.1*, p. *149*.

P. Köhler, W. Möhling, and M. Pasemann
Microdefects in Silicon Different from Swirls
(a) Vol. **53**, No. 2, p. 509–517 (1979).
Erratum
(a) Vol. **56**, No. 2, p. 769 (1979).
See also *1.4*, p. *28*; 10.2, p. 109.

A. George and G. Champier
Velocities of Screw and 60° Dislocations in n- and p-Type Silicon
(a) Vol. **53**, No. 2, p. 529–540 (1979).
See also *10.2*, p. *109*.

W. Wroński
On the Role of Chlorine in the Passivation of MOS Structures
(a) Vol. **53**, No. 2, p. 659–664 (1979).
See also *14.3.2*, p. *224*.

J. Wagner and R. Sauer
Electron–Hole Drops in Uniaxially Stressed Silicon
(b) Vol. **94**, No. 1, p. 69–78 (1979).
See also *13.5.1*, p. *186*; *20.3*, p. *332*.

M. Grünewald and P. Thomas
A Hopping Model for Activated Charge Transport in Amorphous Silicon
(b) Vol. **94**, No. 1, p. 125–133 (1979).
See also 2, p. 45; *14.3*, p. *213*; 15, p. 250.

F. Bechstedt and R. Enderlein
Electronic Polarization (Relaxation) Effects in the Core Level Spectra of Semiconductors (I): General Theory of Electronic Polarisation (Relaxation) in Semiconductors
(b) Vol. **94**, No. 1, p. 239–248 (1979).
See also *13.1*, p. *150*; 14.4, p. 241; 22.2, p. 430.

K. Hübner
Energy-Gap Variation at the Si–SiO_x–SiO_2 Interface
(b) Vol. **94**, No. 1, p. K35–K39 (1979).
See also *13.1*, p. *150*; 13.3, p. 160; 22.6, p. 481.

E. V. Suvorov, O. S. Gorelik, V. M. Kaganer, and V. L. Indenbom
Shape of Extinction Fringes and Determination of the Burgers Vector in Single-Crystal Interferometric Techniques
(a) Vol. **54**, No. 1, p. 29–35 (1979).
See also *10.2*, p. *109*.

M. G. Grimaldi, P. Baeri, S. U. Campisano, G. Foti, and E. Rimini
Laser Induced As Profile Broadening in Amorphous Silicon Layers
(a) Vol. **54**, No. 1, p. 55–59 (1979).
See also *2*, p. *45*; 9, p. 87; *11*, p. *122*.

M. Pasemann and P. Werner
Indentification of Small Defects in Silicon
(a) Vol. **54**, No. 1, p. 179–188 (1979).
See also *1.4*, p. *29*; *10.2*, p. *109*.

I. V. Gridneva, Yu. V. Milman, V. I. Trefilov, and S. I. Chugunova
Analysis of Dislocation Mobility under Concentrated Loads at Indentations of Single Crystals
(a) Vol. **54**, No. 1, p. 195–206 (1979).
See also *10.2*, p. *109*; 12.2, p. 135; 22.1.1, p. 411; 22.2.1, p. 434; 22.2.3, p. 440.

V. G. Kohn
On the Theory of the Bragg Reflection in the Case of Multiple X-Ray Diffraction
(a) Vol. **54**, No. 1, p. 375–384 (1979).
See also *1*, p. *9*.

N. D. Stojadinović
Conditions for the Creation of Dislocations by Diffusion of Phosphorus into Silicon
(a) Vol. **54**, No. 1, p. K5–K10 (1979).
See also *9*, p. *87*; *10.2*, p. *110*.

J. Salm, K. Steenbeck, and E. Steinbeiss
Current–Voltage Characteristic of Reactive Sputtering with Element Targets
(a) Vol. **54**, No. 1, p. K23–K26 (1979).
See also *14.3.1*, p. *220*.

D. Weaire and F. Wooten
Interpretation of Structural Anomalies in Hydrogenated Amorphous Silicon
(a) Vol. **54**, No. 1, p. K41–K43 (1979).
See also *2*, p. *45*.

V. A. Zhukov, V. V. Savyalov, B. V. Koba, V. L. Litvinov, V. P. Sadikov, and N. A. Ukhin
Electrical Properties of Silicon, Irradiated by Neutrons and γ-Quanta, in Strong Electric Fields
(a) Vol. **54**. No. 1, p. K49–K52 (1979).
See also *11*, p. *122*; *14.3.4*, p. *235*.

P. R. Vaya and D. Kakati
Improvement of Breakdown Characteristics of a Planar Microwave P–I–N Diode by the Bevelling Technique
(a) Vol. **54**, No. 1, p. K67–K69 (1979).
See also *14.3.3*, p. *228*.

A. Iida
Applications of X-Ray Triple Crystal Diffractometry to Studies on the Diffusion-Induced Defects in Silicon Crystals
(a) Vol. **54**, No. 2, p. 701–706 (1979).
See also *1.1*, p. *13*; *10.2*, p. *111*.

A. K. Owczarek
Potential and Carrier Concentration Distributions in MOS Transistors at Punch-Through
(a) Vol. **54**, No. 2, p. 761–771 (1979).
See also *14.3.3*, p. *229*; *14.3.4*, p. *235*; 22.6, p. 482.

K. E. Puttik and M. A. Shahid
Dislocation Movement at Annealed Scratches on Near (111) Silicon
(a) Vol. **54**, No. 2, p. K95–K98 (1979).
See also *10.2*, p. *112*.

H. Köhler and M. Roos
Evidence for a Landé-Factor $g = 2$ in n-Type Inverted Silicon MOSFET Surfaces
(b) Vol. **95**, No. 1, p. 107–115 (1979).
See also *13.2*, p. *157*; *14.3.3*, p. *229*; 18, p. 263.

F. Bechstedt and R. Enderlein
Electronic Polarization (Relaxation) Effects in the Core Level Spectra of Semiconductors (II): Application to Ga3d and Si2p Levels
(b) Vol. **95**, No. 1, p. 185–194 (1979).
See also *13.1*, p. *152*; 14.4, p. 242; 22.2, p. 430.

M. Nishida
Electronic States for the Al Overlayer on the Si (111) Surface
(b) Vol. **95**, No. 1, p. 263–272 (1979).
See also *13.3*, p. *160*.

L. Kubler, A. Jaéglé, and J. J. Koulmann
The Effect of Oxygen on the Properties of Evaporated Amorphous Silicon. ESR Studies
(b) Vol. **95**, No. 1, p. 307–315 (1979).
See also 2, p. 47; *13.4*, p. *175*; *19*, p. *295*.

M. D. Varentsov
Quantum Current Oscillations in γ-Irradiated n-Si with Layer Inhomogeneity
(b) Vol. **95**, No. 1, p. K47–K49 (1979).
See also 11, p. 123; *14.3*, p. *215*.

J. F. Graczyk
Structure of Glow Discharge Amorphous Silicon
(a) Vol. **55**, No. 1, p. 231–242 (1979).
See also *1.5*, p. *36*; *2*, p. *47*.

R. Erdmann, and H. Alexander
Photo-EPR of Discolations in Silicon
(a) Vol. **55**, No. 1, p. 251–259 (1979).
See also *10.2*, p. *113*; 13.4, p. 176; 19, p. 295.

M. Sato and K. Sumino
Motion of Extended Dislocations in Silicon Crystals Observed by HVEM
(a) Vol. **55**, No. 1, p. 297–306 (1979).
See also *10.2*, p. *113*.

P. GAWORZEWSKI and K. SCHMALZ
On the Electrical Activity of Oxygen in Silicon
(a) Vol. 55, No. 2, p. 699–707 (1979).
See also *13.4*, p. *177*; *14.3*, p. *216*.

A. OURMAZD and G. R. BOOKER
The Electrical Recombination Efficiency of Individual Edge Dislocations and Stacking Fault Defects in n-Type Silicon
(a) Vol. 55, No. 2, p. 771–784 (1979).
See also *10.2*, p. *115*; 14.3.2, p. 224.

G. BOSE, P. R. VAYA, and D. KAKATI
Silicon Beam Epitaxy for Fabrication of Devices
(a) Vol. 55, No. 2, p. K135–K138 (1979).
See also *3*, p. *54*; *14.3.3*, p. *230*.

D. E. POPOV, V. V. KAPLIN, and S. A. VOROBIEV
Flux Redistribution of Channeled Fast Electrons
(b) Vol. 96, No. 1, p. 263–269 (1979).
See also *11*, p. *125*; 13, p. 141.

K. HÜBNER and F. BECHSTEDT
Temperature Dependence of Atomic Core Levels in Solids (I): Separation of the Temperature Dependence of Conduction-Band and Core-Exciton Energies
(b) Vol. 96, No. 1, p. 351–357 (1979).
See also *13.1*, p. *153*; *13.5.1*, p. *189*; 22.2.2, p. 438; 22.6, p. 484.

D. R. MAŠOVIĆ and S. ZIEKOVIĆ
Model Pseudopotential for Elementary Semiconductors
(b) Vol. 96, No. 1, p. 469–474 (1979).
See also *13.1*, p. *153*; 21.7, p. 392; 22.1.1, p. 413.

B. N. MUKASHEV, K. H. NUSSUPOV, and M. F. TAMENDAROV
Divacancy in Silicon Irradiated by Protons
(b) Vol. 96, No. 1, p. K17–K19 (1979).
See also *10.2*, p. *115*; *11*, p. *125*; 20.1, p. 322.

J. SZUBER and B. SALAMON
Electron Spin Resonance Study on Low-Pressure-Baked (111) Si Surfaces Interacting with Adsorbed Gases
(a) Vol. 56, No. 1, p. 111–115 (1979).
See also 13.3, p. 161; *19*, p. *296*.

K. TEMPELHOFF, F. SPIEGELBERG, R. GLEICHMANN, and D. WRUCK
Precipitation of Oxygen in Dislocation-Free Silicon
(a) Vol. 56, No. 1, p. 213–223 (1979).
See also *1.4*, p. *32*; 6, p. 76; *10.2*, p. *116*; 20.1, p. 322.

Y. OHMURA, T. SHIGEMATSU, and R. NAKAYAMA
Effects of Irradiation Time and Subsequent Heat Treatment on Nd:YAG Laser Annealing of High-Dose As-Implanted Silicon
(a) Vol. 56, No. 1, p. 225–230 (1979).
See also *11*, p. *125*; 14.3, p. 217.

Y. OCHIAI and E. MATSUURA
Orbach Relaxation Process in Heavily Doped n-Type Silicon
(a) Vol. **56**, No. 2, p. K109–K112 (1979).
See also *19*, p. *297*.

22.1.3 Group VI Elements

Review Articles

O. A. Golikova
Boron and Boron-Based Semiconductors
(a) Vol. **51**, No. 1, p. 11—40 (1979).
See also *1*, p. *9*; 13.4, p. 163; *14.3*, p. *209*; 14.3.4, p. 231; 15, p. 249; 20.1, p. 304; 22, p. 393; 22.2, p. 430.

M. Averous
Symmetry-Induced Zero-Gap Semiconductors
(b) Vol. **95**, No. 1, p. 9—28 (1979).
See also *13.1*, p. *143*; 13.4, p. 163; 14.3, p. 209; 22.4, p. 448.

Original Papers and Short Notes

K. Schindler, K. Schmidt, and G. Staupendahl
Stimulated Scattering of CO_2-Laser Radiation in Tellurium
(b) Vol. **91**, No. 1, p. K17—K19 (1979).
See also *20.2*, p. *325*.

R. M. Mehra, Radhey Shyam, and P. C. Mathur
Negative Magnetoresistance of an Amorphous Se–Te–Ge Sample
(a) Vol. **51**, No. 2, p. K165—K169 (1979).
See also *14.3*, p. *211*; 22.1.1, p. 410.

Y. L. Rao and C. Mande
K X-Ray Absorption Discontinuity in Crystalline and Amorphous Selenium
(b) Vol. **92**, No. 1, p. K61—K64 (1979).
See also 2, p. 41; *20*, p. *299*.

H. Fiedler and H.-R. Beilich
Influence of the Substrate Temperature on the Dielectrie Relaxation of Amorphous Selenium Layers in the Glass Transition Range
(a) Vol. **52**, No. 2, p. K157—K159 (1979).
See also *2*, p. *43*; *14.4*, p. *239*.

G. Jones and R. A. Collins
Threshold and Memory Switching in Amorphous Selenium Thin Films
(a) Vol. **53**, No. 1, p. 339—350 (1979).
See also 2, p. 43; *14.3.1*, p. *220*; *14.3.3*, p. *227*.

H. Fiedler and H.-R. Beilich
Dielectric Relaxation in Amorphous Selenium Layers in the Glass Transition Range and Ageing
(a) Vol. **53**, No. 1, p. K9—K10 (1979).
See also *2*, p. *43*; *14.4*, p. *240*.

E. Gerstenhauer, G. Bauer, P. Grosse, and J. C. Doukhan
Influence of Dislocations on the Cyclotron Resonance Absorption in Tellurium
(b) Vol. **93**, No. 2, p. 629—640 (1979).
See also *10.2*, p. *108*; 20, p. 300.

U. Steigenberger and M. von Ortenberg
Cyclotron Resonance in Uniaxially Stressed Tellurium
(b) Vol. **93**, No. 2, p. K139–K141 (1979).
See also 12.2, p. 134; *13.1*, p. *149*; *19*, p. *293*.

Z. Dobrovolskis, W. Hoerstel, and A. Krotkus
Determination of the Impact Ionization Rate of Holes in Tellurium at 77 K
(a) Vol. **55**, No. 1, p. 197–202 (1979).
See also *14.3.4*, p. *236*.

J. A. Kaladė, E. A. Montrimas, and J. P. Varaneckas
Radiographic Contrast of Selenium Electroradiographic Layers
(a) Vol. **55**, No. 1, p. 261–269 (1979).
See also *16*, p. *254*.

J. P. Estienne and J. L. Farvacque
Experimental Evidence for Piezoelectrical Free-Carrier Scattering by Screw Dislocations in Tellurium
(b) Vol. **95**, No. 2, p. 593–600 (1979).
See also *10.2*, p. *113*; *14.3*, p. *215*.

S. Radhakrishnan and P. C. Mehendru
Electronic Conduction in Dye-Sensitized Selenium Films
(a) Vol. **55**, No. 2, p. 635–638 (1979).
See also *14.3.1*, p. *221*.

22.2 III–V Compounds

Review Article

Original Papers and Short Notes

K. B. Tolpygo and S. M. Zubkova
On the Indirect Exciton Energy Levels in Crystals of Diamond and Sphalerite Type with Spin–Orbit Interaction
(b) Vol. **95**, No. 2, p. 341–348 (1979).
See also 13.1, p. 152; *13.5.1*, p. *188*; 22.1.1, p. 412; 22.1.2, p. 424.

S. I. Stenin
Problems of the Formation of Dislocation Structures in Heteroepitaxial Layers
(a) Vol. **55**, No. 2, p. 519–527 (1979).
See also *1.5*, p. *36*; *10.2*, p. *114*; 22.1, p. 408.

22.2.1 Arsenides

Review Article

Original Papers and Short Notes

T. D. Dzhafarov, E. A. Skoryatina, E. S. Guds, and I. E. Moronchuk
The Diffusion and Electric Activity of Co in Epitaxial GaAs
(a) Vol. **51**, No. 2, p. K221—K225 (1979).
See also *9*, p. *85*; 14.3, p. 211.

K. Kurumada, T. Mizutani, and M. Fujimoto
Stationary Domains in Planar Gunn Elements
(a) Vol. **52**, No. 1, p. 259—268 (1979).
See also *14.3.3*, p. *226*.

W. Rühle, K. Lösch, and J. U. Fischbach
A New Optical Method to Investigate Deep Traps in GaAs and InP
(a) Vol. **52**, No. 1, p. K65—K67 (1979).
See also *13.4*, p. *167*; 20.1, p. 309; 22.2.2, p. 437.

W. H. Lamfried
Temperature-Dependent Density of States of n-GaAs
(b) Vol. **92**, No. 2, p. 433—439 (1979).
See also *13.1*, p. *146*; 13.4, p. 167.

J. H. Basson and C. A. B. Ball
Threading Dislocations in Single-Layer Heteroepitaxial Structures
(a) Vol. **52**, No. 2, p. 609—614 (1979).
See also *1.5*, p. *34*; *10.2*, p. *106*; 22.2.4, p. 442.

Y. Itoh, S. Adachi, and C. Hamaguchi
Analysis of Resonant Brillouin Scattering in GaAs
(b) Vol. **93**, No. 1, p. 381—389 (1979).
See also 13.5.1, p. 185; *20.1*, p. *311*.

U. Berg, R. Fuhrmann, and O. Brümmer
Angular-Dependent X-Ray Photoemission Study on the Surface of GaAs (111) and ($\bar{1}\bar{1}\bar{1}$) Single Crystals
(a) Vol. **53**, No. 1, p. K1—K4 (1979).
See also *1.6*, p. *38*; *17*, p. *256*.

N. S. Orlova
X-Ray Thermal Diffuse Scattering Measurements of the [100] and [111] Phonon Dispersion Curves of Indium Arsenide
(b) Vol. **93**, No. 2, p. 503—509 (1979).
See also *6*, p. *68*.

R. S. Turtelli and A. R. B. de Castro
LO Phonon Raman Lineshape in Highly Excited GaAs
(b) Vol. **93**, No. 2, p. 811—815 (1979).
See also *6*, p. *69*; *20.1*, p. *313*; 20.3, p. 331.

E. W. Kreutz and P. Schroll
Field Effect on Real GaAs Surfaces
(a) Vol. **53**, No. 2, p. 499—508 (1979).
See also *1.33*, p. *159*; *14.3.2*, p. *224*.

M. Büchel and H. Lüth
Defect-Induced Surface States on Cleaved GaAs (110) Surfaces Measured in UV Photoemission
(b) Vol. **94**, No. 2, p. K155–K158 (1979).
See also *13.3*, p. *160*; 17, p. 257.

N. L. Dmitruk and O. Yu. Borkovskaya
Photosensitivity Losses in Ideal and Real Schottky Barriers
(a) Vol. **54**, No. 2, p. 523–527 (1979).
See also *14.3.3*, p. *229*; *16*, p. *253*.

V. I. Osinskii and A. Ya. Peshko
Anomalous Temperature Dependence of Recombination Radiation of Gallium Arsenide
(a) Vol. **54**, No. 2, p. 577–584 (1979).
See also *13.4*, p. *174*; *20.3*, p. *334*.

T. Sawada and H. Hasegawa
Small-Signal Admittance Study of GaAs Anodic MOS System
(a) Vol. **54**, No. 2, p. 689–696 (1979).
See also *13.3*, p. *160*; *14.3.2*, p. *224*.

D. N. Bose, U. Sen Gupta, and H. N. Acharya
Inversion Layer Formation and Surface States at the GaAs-Electrolyte Interface
(a) Vol. **54**, No. 2, p. K155–K157 (1979).
See also *13.3*, p. *160*.

T. J. Magee, J. Peng, J. D. Hong, W. Katz, and C. A. Evans
Back Surface Gettering of Au in GaAs
(a) Vol. **55**, No. 1, p. 161–168 (1979).
See also 1.4, p. 30; *10.2*, p. *112*.

T. J. Magee, J. Peng, J. D. Hong, C. A. Evans, and V. R. Deline
Gettering of Cr in GaAs by Back Surface Mechanical Damage
(a) Vol. **55**, No. 1, p. 169–172 (1979).
See also 1.4, p. 30; *10.2*, p. *112*.

T. X. Hoai, L. V. Hong, and N. V. Vuong
On the Recombination Barrier in Highly Compensated GaAs
(a) Vol. **55**, No. 1, p. K99–K102 (1979).
See also *14.3.4*, p. *236*; *16*, p. *254*.

R. V. Konakova, G. D. Melnikov, Yu. A. Tkhorik, and M. Yu. Filatov
Influence of Structural Defects and Concentration Inhomogeneities in GaAs Epitaxial Layers on Reverse $I-U$ Characteristics of Schottky Barrier Diodes
(a) Vol. **55**, No. 2, p. K131–K133 (1979).
See also 1.5, p. 36; *14.3.1*, p. *221*; *14.3.3*, p. *230*.

T. I. Kolchenko, V. M. Lomako, and I. G. Tarutin
Introduction of Defect Clusters in GaAs by High-Energy γ-Rays
(a) Vol. **56**, No. 1, p. 387–393 (1979).
See also *11*, p. *125*; 14.3, p. 217.

22.2.2 Phosphides

Original Papers and Short Notes

P. P. Seregin, F. S. Nasredinov, and A. Sh. Bakhtiyarov
Study of the Charge Exchange Process of Iron Impurity Atoms in GaAs aud GaP by Mössbauer Spectroscopy
(b) Vol. **91**, No. 1, p. 35–41 (1979).
See also *5*, p. *59*; 22.2.1, p. 432.

O. Breitenstein, B. Rheinländer, and R. Bindemann
DLTS Capacitance Studies on Deep Level Centres in VPE GaP:N
(a) Vol. **51**, No. 1, p. 79–86 (1979).
See also *13.4*, p. *164*; *14.3.3*, p. *225*.

D. V. Morgan and J. Frey
Comments on the Modification of Schottky Barrier Height by Interfacial Oxides
(a) Vol. **51**, No. 1, p. K29–K33 (1979).
See also *14.3.2*, p. *223*; *14.3.4*, p. *231*; 22.1.2, p. 415; 22.2.1, p. 432.

L. A. Demberel, A. S. Popov, D. B. Kushev, and N. N. Zheleva
Deep Levels in Fe-Doped InP
(a) Vol. **52**, No. 1, p. 341–345 (1979).
See also *13.4*, p. *167*; *14.3.4*, p. *232*; 20.1, p. 308; 20.3, p. 328.

W. Rühle, K. Lösch, and J. U. Fischbach
A New Optical Method to Investigate Deep Traps in GaAs and InP
(a) Vol. **52**, No. 1, p. K65–K67 (1979).
See also *13.4*, p. *167*; 20.1, p. 309; 22.2.1, p. 433.

G. Weber and W. Rühle
Effect of Uniaxial Stress on Free and Bismuth-Bound Excitons in InP
(b) Vol. **92**, No. 2, p. 425–431 (1979).
See also 12.2, p. 134; *13.4*, p. *167*; *13.5.1*, p. *184*; 20.3, p. 329.

L. A. Demberel, A. S. Popov, and D. B. Kushev
Deep Levels in Fe-Doped GaP
(a) Vol. **52**, No. 2, p. 653–656 (1979).
See also *13.4*, p. *168*; *14.3.4*, p. *233*; 20.1, p. 310; 20.3, p. 329.

R. Bindemann, E. Hempel, and K. Kreher
Determination of the Dielectric Constant of GaP from S–C Donor–Acceptor Pair Spectra
(a) Vol. **52**, No. 2, p. K201–K203 (1979).
See also *13.4*, p. *169*; 14.4, p. 239; *20.3*, p. *329*.

G. Oelgart
Avalanche Muliplication of Electron-Beam Excited Carriers in GaP p–n Junction
(a) Vol. **52**, No. 2, p. K205–K208 (1979).
See also *14.3.4*, p. *233*.

J. S. Nkoma and S. K. Rwaboona
Frequency Dependence of Some Functions Associated with Surface Polaritons
(b) Vol. **93**, No. 1, p. 397–402 (1979).
See also *13.3*, p. *159*; *13.5*, p. *182*.

22.2.3 Antimonides

Review Article

R. S. Brazis, J. K. Furdyna, and J. K. Požela
Microwave Effects in Narrow-Gap Semiconductors (II)
(a) Vol. 54, No. 1, p. 11–27 (1979).
See also 13, p. 136; 18.1, p. 265; *19*, p. *289*; 20, p. 298; 22.2.1, p. 432; 22.2.4, p. 442; 22.4.3, p. 455; 22.4.4, p. 458.

Original Papers and Short Notes

A. J. C. Sampaio and R. Luzzi
Resonant Magneto-Raman Scattering in Polar Semiconductors
(b) Vol. 91, No. 1, p. 305–310 (1979).
See also *6*, p. *63*; *20.1*, p. *304*.

A. K. Walton and J. C. Dutt
Evidence of Surface Conduction on Magnetically Fronzen-out InSb
(b) Vol. 91, No. 1, p. K57–K60 (1979).
See also *14.3.2*, p. *223*.

V. K. Malyutenko and A. I. Liptuga
The Determination of the Intrinsic Concentration and Effective Mass of Carriers in Semiconductors by an Optical Method
(a) Vol. 51, No. 2, p. K137–K139 (1979).
See also 13.1, p. 145; 14.3, p. 210; *20.1*, p. *306*.

S. Guha and N. Apte
Effect of Nonparabolic Band Structure of Semiconductors on Acoustoelectric Instability
(b) Vol. 92, No. 1, p. K47–K50 (1979).
See also *7*, p. *78*; 14.3.4, p. 232.

S. S. Bolgov, A. G. Kollukh, V. K. Malyutenko, and I. I. Boiko
The Influence of an External Magnetic Field on the Recombination Radiation and Current–Voltage Characteristics of Indium Antimonide under Impact Ionization Conditions
(a) Vol. 52, No. 1, p. K7–K11 (1979).
See also *14.3.4*, p. *232*; *20.3*, p. *328*.

I. Gorczyca
Spin Dependent Wave Functions at Some Points of the Brillouin Zone of InSb
(b) Vol. 92, No. 2, p. K143–K146 (1979).
See also *13.1*, p. *147*.

L. F. Vereshchagin, L. G. Khvostantsev, and V. F. Skok
Thermoelectric Properties of Indium Antimonide under High Pressure up to 35 kbar
(a) Vol. 52, No. 2, p. 401–405 (1979).
See also 14.3, p. 212; *15*, p. *250*.

S. Guha, P. K. Sen, and S. Ghosh
Parametric Instability of Acoustic Waves in Transversely Magnetised Piezoelectric Semiconductors
(a) Vol. 52, No. 2, p. 407–414 (1979).
See also *7*, p. *78*; *14.3.4*, p. *233*.

Z. N. Chigogidze and N. P. Khuchua
Study of Electrical Breakdown Peculiarities in High-Resistivity Compensated p-InSb
(a) Vol. **54**, No. 2, p. 773–781 (1979).
See also *13.4*, p. *175*; *14.3.4*, p. *235*.

H. Y. Lam, R. E. Jones, Jr., W. J. Keeler, and V. V. Paranjape
Pressure Dependent Conductivity in InSb
(b) Vol. **95**, No. 1, p. 85–90 (1979).
See also *14.3*, p. *214*.

B. R. Nag and M. Deb Roy
Monte Carlo Calulation of the Low-Field Electron Mobility
(b) Vol. **95**, No. 1, p. K65–K67 (1979).
See also *14.3*, p. *215*.

A. L. Dawar, K. V. Krishna, O. P. Taneja, and P. C. Mathur
Negative Magnetoresistance of Two-Phase Polycrystalline p-Type InSb Films
(a) Vol. **55**, No. 1, p. K19–K22 (1979).
See also *14.3.1*, p. *221*.

S. Guha and P. Namjoshi
Parametric Decay of a Helicon in n-InSb
(b) Vol. **96**, No. 2, p. 713–716 (1979).
See also *13*, p. *142*; *20*, p. *302*.

R. R. Sharma and R. C. Sharma
Submillimeter Harmonic Generation in Semiconductor Waveguides
(a) Vol. **56**, No. 2, p. 743–747 (1979).
See also *14.3.4*, p. *237*; 22.1.1, p. 413.

S. Guha and P. K. Sen
Effect of Transverse Magnetic Field on the Electromagnetic Wave Propagation in Semiconductor Plasmas at High Electric Field
(a) Vol. **56**, No. 2, p. K117–K121 (1979).
See also *13*, p. *142*; 14.3, p. 218; *20*, p. *303*.

22.2.4 Mixed Crystals

Review Article

R. S. Brazis, J. K. Furdyna, and J. K. Požela
Microwave Effects in Narrow-Gap Semiconductors (II)
(a) Vol. **54**, No. 1, p. 11–27 (1979).
See also **13**, p. 136; 18.1, p. 265; *19*, p. *289*; 20, p. 298; 22.2.1, p. 432; 22.2.3, p. 439; 22.4.3, p. 455; 22.4.4, p. 458.

Original Papers and Short Notes

E. Jahne, W. Pilz, M. Giehler, and L. Hildisch
Non-Isodisplacement of P Atoms in Long-Wavelength Optical Phonons in $In_{1-x}Ga_xP$
(b) Vol. **91**, No. 1, p. 155–165 (1979).
See also *6*, p. *63*; *20.1*, p. *304*.

C. Ance
Variation of the Optical Transitions of $Ga_{1-x}Al_xAs$ and $Ga_{1-x}Al_xSb$
(b) Vol. **92**, No. 1, p. K33–K38 (1979).
See also *13.1*, p. *146*; *20.1*, p. *308*.

J. H. Basson and C. A. B. Ball
Threading Dislocations in Single-Layer Heteroepitaxial Structures
(a) Vol. **52**, No. 2, p. 609–614 (1979).
See also *1.5*, p. *34*; *10.2*, p. *106*; 22.2.1, p. 433.

Z. R. Żytkiewicz
Lattice Matching in $Ga_xIn_{1-x}As_yP_{1-y}$–GaAs and $Ga_xIn_{1-x}As_yP_{1-y}$–InP Structures
(a) Vol. **53**, No. 2, p. K165–K168 (1979).
See also *1.5*, p. *35*; *3*, p. *52*.

A. G. Petukhov, I. K. Doicho, and V. K. Bashenov
Electronic Density of States in Ternary Alloys Using the Cluster-Bethe-Lattice Method
(b) Vol. **94**, No. 1, p. K71–K75 (1979).
See also *13.1*, p. *150*.

C. Ance
Dispersion Energy in Semiconductors
(b) Vol. **94**, No. 2, p. 557–568 (1979).
See also *13.1*, p. *151*; *14.4*, p. *241*; 22, p. 402.

Y. Epelboin and P. Riglet
Boundary Conditions in the Numerical Integration of Takagi-Taupin Equations. Application to the Laue and Bragg Cases
(a) Vol. **54**, No. 2, p. 547–556 (1979).
See also *1*, p. *10*.

P. Kleinert
Cluster Bethe Lattice Treatment for Phonons in Pseudobinary Alloy Systems
(b) Vol. **95**, No. 1, p. 163–170 (1979).
See also *6*, p. *73*; *20.1*, p. *318*.

A. ZIĘBA, S. HANEDA, Y. YAMAGUCHI, and H. WATANABE
Latent Ferromagnetism in $MnAs_{0.96}P_{0.04}$
(a) Vol. 55, No. 1, p. K27–K29 (1979).
See also *18.2*, p. *278*.

A. K. SAXENA
Non-Shallow Levels and the Conduction Band Structure of $Ga_{1-x}Al_xAs$
(b) Vol. 96, No. 2, p. K77–K82 (1979).
See also *13.1*, p. *155*; 13.4, p. 180.

22.3 IV–IV, II–V, and III–VI Compounds

Original Papers and Short Notes

J. M. Pawlikowski, J. Misiewicz, B. Sujak-Cyrul, and J. Wróbel
Absorption Edge of Zn_3As_2 at 5, 80, and 300 K
(b) Vol. **92**, No. 2, p. K123 – K125 (1979).
See also *13.1*, p. *147*; *20.1*, p. *310*.

F. Lukeš, E. Schmidt, J. Humlíček, M. K. Kekoua, and E. Khoutsichvili
Determination of the Composition of GeSi Alloy by Ellipsometry
(a) Vol. **53**, No. 1, p. 321 – 325 (1979).
See also *1.3*, p. *23*; 20.1, p. 312.

L. Huldt, N. G. Nilsson, B. O. Sundström, and W. Żdanowicz
Photoluminescence of Zinc Phosphide — Zn_3P_2
(a) Vol. **53**, No. 1, p. K15 – K18 (1979).
See also *20.3*, p. *331*.

A. Rainò, A. M. Mancini, R. Murri, A. Rizzo, and L. Vasanelli
Cathodoluminescence of GaSe Single Crystals
(a) Vol. **53**, No. 2, p. 487 – 492 (1979).
See also *13.5.1*, p. *186*; *20.3*, p. *332*.

B. Dowgiałło-Plenkiewicz and P. Plenkiewicz
Inverted Band Structure of Cd_3As_2
(b) Vol. **94**, No. 1, p. K57 – K60 (1979).
See also *13.1*, p. *150*.

J. Weszka, I. Miotkowski, K. Kloc, and W. Żdanowicz
Raman Scattering in CdP_2 Single Crystals
(b) Vol. **94**, No. 1, p. K97 – K99 (1979).
See also *6*, p. *71*; *20.1*, p. *315*.

J. F. McGilp and A. W. Parke
Angle-Resolved UPS of GaS, GaSe, and $GaS_{0.6}Se_{0.4}$
(b) Vol. **94**, No. 2, p. 685 – 689 (1979).
See also *13.1*, p. *151*; *17*, p. *257*; 22.3.1, p. 447.

P. Plenkiewicz and B. Dowgiałło-Plenkiewicz
Energy Band Structure of Cd_3P_2 for Real D_{4h}^{15} Symmetry (II): Symmetry Properties
(b) Vol. **95**, No. 1, p. 29 – 37 (1979).
See also *13.1*, p. *151*.

N. M. Gasanly, N. F. Gakhramanov, B. M. Dzhavadov, V. I. Tagirov, and E. A. Vinogradov
Infrared Spectra of the Layer Compound InS
(b) Vol. **95**, No. 1, p. K89 – K92 (1979).
See also *6*, p. *73*; *20.1*, p. *319*.

C. de Blasi, S. Galassini, C. Manfredotti, G. Micocci, L. Ruggiero, and A. Tepore
Thermally Stimulated Currents in n-GaS
(a) Vol. **55**, No. 1, p. 291 – 296 (1979).
See also 13.4, p. 176; *14.3*, p. *215*.

T. Soma
The Electronic Theory of Si–Ge Solid Solutions
(b) Vol. **95**, No. 2, p. 427 – 431 (1979).
See also *4*, p. *57*; 13, p. 141.

22.3.1 Mixed Crystals

Original Papers and Short Notes

N. M. Gasanly, A. F. Goncharov, B. M. Dzhavadov, N. N. Melnik, V. I. Tagirov, and E. A. Vinogradov
Optical Phonons in Layer $TlInSe_2$ Single Crystals
(b) Vol. **92**, No. 2, p. K139–K142 (1979).
See also *6*, p. *67*; *20.1*, p. *310*.

K. R. Allakhverdiev, S. S. Babaev, and M. M. Tagiev
Two-Mode Behaviour of the Optical Phonons in GaS_xSe_{1-x} Solid Solutions
(b) Vol. **93**, No. 1, p. K67–K70 (1979).
See also *6*, p. *68*; 20.1, p. 311.

G. L. Belenkii, M. O. Godzhaev, N. T. Mamedov, E. Yu. Salaev, and R. A. Suleimanov
The Nature of Energy Bands of $A^{III}B^{VI}$ Layered Crystals near the Absorption Edge
(a) Vol. **53**, No. 1, p. 137–142 (1979).
See also *13.1*, p. *148*; 20.1, p. 312; 20.3, p. 330.

G. B. Abdullaev, K. R. Allakhverdiev, R. Kh. Nani, E. Yu. Salaev, and M. M. Tagyev
Neutron Diffraction, Infrared, and Raman Scattering Investigations of the Layered GaS_xSe_{1-x} System
(a) Vol. **53**, No. 2, p. 549–555 (1979).
See also *1.1*, p. *13* ; 6, p. 70; *20.1*, p. *314*.

J. F. McGilp and A. W. Parke
Angle-Resolved UPS of GaS, GaSe, and $GaS_{0.6}Se_{0.4}$
(b) Vol. **94**, No. 2, p. 685–689 (1979).
See also *13.1*, p. *151*; *17*, p. *257*; 22.3, p. 445.

E. A. Vinogradov, G. N. Zhizhin, N. N. Melnik, S. I. Subbotin, V. V. Panfilov, K. R. Allakhverdiev, E. Yu. Salaev, and R. Kh. Nani
Raman Scattering and Phase Transformations of $TlGaSe_2$ and $TlGaS_2$ under Pressure
(b) Vol. **95**, No. 2, p. 383–390 (1979).
See also 1.2, p. 20; *6*, p. *74*; *20.1*, p. *319*.

A. E. Bakhyshov, Sufuat Boules, F. E. Faradzhov, M. Sh. Mamedov, and V. I. Tagirov
The Fundamental Optical Absorption Edge of $TlGaS_{2x}Se_{2(1-x)}$ Solid Solutions
(b) Vol. **95**, No. 2, p. K121–K125 (1979).
See also *20.1*, p. *320*.

22.4 II–VI Compounds

Review Article

Original Papers and Short Notes

CH. UIHLEIN and S. FEIERABEND

Magnetoreflectance of Exciton-Polaritons in Cubic II–VI Compounds (II): Semiclassical Treatment

(b) Vol. **94**, No. 2, p. 421–428 (1979).

See also *13.5.1*, p. *187*; 20.1, p. 316.

22.4.1 Sulphides

Original Papers and Short Notes

J. Voigt, F. Spiegelberg, and M. Senoner
Band Parameters of CdS and CdSe Single Crystals Determined from Optical Exciton Spectra
(b) Vol. **91**, No. 1, p. 189–199 (1979).
See also *13.5.1*, p. *183*; *20.1*, p. *304*; 22.4.2, p. 453.

B. B. Laud and V. W. Kulkarni
Photo- and Electro-Luminescence of Polycrystalline BaS:Cu Phosphors
(a) Vol. **51**, No. 1, p. 269–274 (1979).
See also *20.3*, p. *327*.

J. Flohrer, E. Jahne, and M. Porsch
Energy Levels of A and B Excitons in Wurtzite-Type Semiconductors with Account of Electron–Hole Exchange Interaction Effects
(b) Vol. **91**, No. 2, p. 467–478 (1979).
See also *13.5.1*, p. *183*.

G. O. Müller, H. H. Weber, and I. Höricke
Linear Optical Response of Highly Excited Semiconductors. Electron–Hole Liquid in CdS
(b) Vol. **91**, No. 2, p. 531–539 (1979).
See also 13.5, p. 181; *20.1*, p. *305*.

P. R. Herczfeld and M. J. Fisher
Utilization of Noise Theory and Measurements in Modelling Electronic Process and Transport Phenomena in Photoconductors
(a) Vol. **52**, No. 1, p. 357–363 (1979).
See also *16*, p. *253*; 22.4.2, p. 453.

G. D. Sootha, G. K. Padam, and S. K. Gupta
ESR Study of Oxygen Radicals Formed in Cadium Sulphide
(a) Vol. **52**, No. 2, p. K125–K128 (1979).
See also *19*, p. *291*.

J. Röseler and K. Henneberger
Magnetopolaritons in Wurtzite-Type Crystals
(b) Vol. **93**, No. 1, p. 213–222 (1979).
See also *13.5.1*, p. *185*; 20, p. 300; 22.4.2, p. 453.

R. Mach, Yu. V. Bochkov, B. Selle, and A. N. Georgobiani
Physical Properties of Schottky Diodes on ZnS Single Crystals
(a) Vol. **53**, No. 1, p. 263–270 (1979).
See also *14.3.3*, p. *227*; *16*, p. *253*.

W. Busse, H.-E. Gumlich, A. Geoffroy, and R. Parrot
Spectral Distribution and Decay Times of the Luminescence of Mn^{2+} on Different Lattice Sites in ZnS
(b) Vol. **93**, No. 2, p. 591–596 (1979).
See also *13.4*, p. *170*; *20.3*, p. *331*.

B. Dietrich and J. Voigt
Spatial Dispersion Effects in the Transmission Spectra of Semiconductors in the Region of Exciton Resonances
(b) Vol. **93**, No. 2, p. 669–677 (1979).
See also *13.5.1*, p. *186*; 20.1, p. 313.

J. E. Ralph
Photoluminescence of CdS:Cu Doped by Diffusion
(a) Vol. **53**, No. 2, p. 611–615 (1979).
See also 13.4, p. 171; *20.3*, p. *332*.

J. Puls and J. Voigt
Excitation Spectroscopy of the I_2 Bound Exciton and the Excitonic Molecule in CdS (I): Excited States of the I_2 Bound Exciton
(b) Vol. **94**, No. 1, p. 199–204 (1979).
See also *13.5.1*, p. *186*; 20.3, p. 332.

M. A. S. Sweet and D. Urquhart
Shallow Traps in ZnS Single Crystals
(a) Vol. **54**, No. 1, p. K81–K84 (1979).
See also *13.4*, p. *173*; 20.3, p. 333.

I. Broser and M. Rosenzweig
A Quantitative Study of Excitonic Polariton Reflectance in CdS (II): B-Exciton and *k*-Linear Term
(b) Vol. **95**, No. 1, p. 141–154 (1979).
See also *13.5.1*, p. *188*; *20.1*, p. *318*.

R. N. Tripathi and S. Sivaraman
Fluorescence Spectra of CaS:(Tb, Eu) Phosphors under X-Ray Excitation
(a) Vol. **55**, No. 1, p. K123–K126 (1979).
See also 11, p. 124; *20.3*, p. *335*.

V. D. Negryi and Yu. A. Osipyan
Dislocation Emission in CdS
(a) Vol. **55**, No. 2, p. 583–588 (1979).
See also *10.2*, p. *114*; *20.3*, p. *335*.

W. Stössel and H. J. Wagner
The Influence of Spatial Dispersion and of the Angle of Incidence on Thermomodulation Spectra of the $A_{n=1}$Exciton in CdS
(b) Vol. **96**, No. 1, p. 369–378 (1979).
See also *13.5.1*, p. *189*; *20.1*, p. *321*.

A. Kuroiwa, H. Saito, and S. Shionoya
Exciton Reflection Spectra under High-Intensity Excitation in CdS and CdSe
(b) Vol. **96**, No. 1, p. 483–486 (1979).
See also *13.5.1*, p. *189*; *20.1*, p. *321*; 20.3, p. 336; 22.4.2, p. 454.

22.4.2 Selenides

Original Papers and Short Notes

J. Voigt, F. Spiegelberg, and M. Senoner
Band Parameters of CdS and CdSe Single Crystals Determined from Optical Exciton Spectra
(b) Vol. **91**, No. 1, p. 189–199 (1979).
See also *13.5.1*, p. *183*; *20.1*, p. *304*; 22.4.1, p. 450.

W. Szuszkiewicz
Interband Absorption and the Γ-Point Band Structure Parameters in HgSe
(b) Vol. **91**, No. 1, p. 361–369 (1979).
See also *13.1*, p. *144*; *20.1*, p. *304*.

J. Daunay, J. Baillou, J. Daunay and P. Bugnet
Application of Etch Figures to the Study of Growth Twins in Single-Crystalline ZnSe
(a) Vol. **51**, No. 1, p. K5–K8 (1979).
See also *3*, p. *52*; 10.2, p. 103.

P. R. Herczfeld and M. J. Fisher
Utilization of Noise Theory and Measurements in Modelling Electronic Processes and Transport Phenomena in Photoconductors
(a) Vol. **52**, No. 1, p. 357–363 (1979).
See also *16*, p. *253*; 22.4.1, p. 450.

R. Baubinas, F. Senulis, and J. Vaitkus
Temperature Dependence of the Optical Ionization Energy of Deep Local Levels in CdSe Single Crystals
(a) Vol. **52**, No. 2, p. K143–K146 (1979).
See also *13.4*, p. *169*; *16*, p. *253*.

J. Röseler and K. Henneberger
Magnetopolaritons in Wurtzite-Type Crystals
(b) Vol. **93**, No. 1, p. 213–222 (1979).
See also *13.5.1*, p. *185*; 20, p. 300; 22.4.1, p. 450.

E. A. Meneses and E. S. Braga
Photo-Excited Hot Electrons in ZnSe at 2 K
(b) Vol. **93**, No. 1, p. K21–K23 (1979).
See also *13*, p. *138*; *20.3*, p. *330*.

P. Lemasson and J. Gautron
Current Photoresponse of an Electrolyte–Zinc Selenide Junction
(a) Vol. **53**, No. 1, p. 303–309 (1979).
See also *14.3.3*, p. *227*; *16*, p. *253*.

A. I. Ryskin, A. V. Vasilev, and A. L. Natadze
Stress-Induced Alignment of Jahn-Teller Centres Cr^{2+} in ZnSe Crystals. Optical Study
(b) Vol. **93**, No. 2, p. 459–467 (1979).
See also 13.4, p. 170; *20.1*, p. *312*.

22.4.3 Tellurides

Review Articles

R. S. Brazis, J. K. Furdyna, and J. K. Požela
Microwave Effects in Narrow-Gap Semiconductors (I)
(a) Vol. **53**, No. 1, p. 11–41 (1979).
See also *13*, p. *136*; 13.1, p. 143; 14.3, p. 209; 18.1, p. 265; *20*, p. *298*; 22.4.4, p. 458; 22.7, p. 488.

R. S. Brazis, J. K. Furdyna, and J. K. Požela
Microwave Effects in Narrow-Gap Semiconductors (II)
(a) Vol. **54**, No. 1, p. 11–27 (1979).
See also 13, p. 136; 18.1, p. 265; *19*, p. *289*; 20, p. 298; 22.2.1, p. 432; 22.2.3, p. 439; 22.2.4, p. 442; 22.4.4, p. 458.

Original Papers and Short Notes

K. Szlenk, Z. Dziuba, and R. R. Gałązka
Contribution of Resonant States to the Conductivity of HgTe
(b) Vol. **91**, No. 1, p. 255–261 (1979).
See also 13.4, p. 163; *14.3*, p. *209*.

V. Kremenitskii, S. Odoulov, and M. Soskin
Dynamic Gratings in Cadmium Telluride
(a) Vol. **51**, No. 1, p. K63–K66 (1979).
See also *20.2*, p. *325*.

E. Y. Gutmanas, N. Travitzky, and P. Haasen
Negative and Positive Photoplastic Effect in CdTe
(a) Vol. **51**, No. 2, p. 435–444 (1979).
See also *10.2*, p. *104*.

M. Godlewski
Microwave Conductivity Measurements in CdTe
(a) Vol. **51**, No. 2, p. K141–K145 (1979).
See also 13.4, p. 165; *14.3*, p. *210*.

K. Takita, K. Onabe, and S. Tanaka
Anomalous Magnetoresistance and Band Crossing in Uniaxially Compressed HgTe
(b) Vol. **92**, No. 1, p. 297–306 (1979).
See also *13.1*, p. *145*; *14.3*, p. *211*.

N. Romeo, D. Seuret, and F. First
Properties of the ZnSe/ZnTe Heterojunction Prepared by a Multi-Source Evaporation of ZnTe:Sb on ZnSe Single Crystals
(a) Vol. **54**, No. 1, p. K27–K30 (1979).
See also *14.3.3*, p. *228*; 22.4.2, p. 454.

N. Magnea, D. Bensahel, J. L. Pautrat, and J. C. Pfister
Optical Indentification of Substitutional Acceptors in Refined ZnTe
(b) Vol. **94**, No. 2, p. 627–639 (1979).
See also *13.4*, p. *174*; 13.5.1, p. 187; *20.3*, p. *333*.

D. Valentovič, J. Červenák, S. Luby, M. L. Aldea, and T. Boţilă
Some Non-Equilibrium Phenomena in Sputtered CdTe Thin Films
(a) Vol. **56**, No. 1, p. 341–347 (1979).
See also *1.5*, p. *37*; 13.4, p. 179; 16, p. 254.

W. Jamroz and J. Karniewicz
Three-Dimensional Model of Nonlinear Magneto-Electric Susceptibilities
(b) Vol. **96**, No. 2, p. K87–K90 (1979).
See also *14.4*, p. *243*; *20*, p. *303*; 22.2.1, p. 436.

Z. Liliental, M. Pasemann, and P. Werner
Electron Microscopic Investigations of Precipitates in Cu-Doped ZnTe
(a) Vol. **56**, No. 2, p. 679–685 (1979).
See also *10.2*, p. *117*.

J. C. Pfister and D. Bensahel
Diffusion of Impurites Contained in Te Precipitates in ZnTe: Theoretical Model
(a) Vol. **56**, No. 2, p. K79–K82 (1979).
See also *9*, p. *88*.

22.4.4 Mixed Crystals

Review Articles

R. S. Brazis, J. K. Furdyna, and J. K. Požela
Microwave Effects in Narrow-Gap Semiconductors (I)
(a) Vol. **53**, No. 1, p. 11–41 (1979).
See also *13*, p. *136*; 13.1, p. 143; 14.3, p. 209; 18.1, p. 265; *20*, p. *298*; 22.4.3, p. 455; 22.7, p. 488.

R. S. Brazis, J. K. Furdyna, and J. K. Požela
Microwave Effects in Narrow-Gap Semiconductors (II)
(a) Vol. **54**, No. 1, p. 11–27 (1979).
See also 13, p. 136; 18.1, p. 265; *19*, p. *289*; 20, p. 298; 22.2.1, p. 432; 22.2.3, p. 439; 22.2.4, p. 442; 22.4.3, p. 455.

Original Papers and Short Notes

J. Niewodniczańska-Zawadzka and J. Grembowicz
Phonon-Assisted Tunneling in HgCdTe–Insulator–Metal and HgCdSe–Insulator–Metal Junctions
(b) Vol. **91**, No. 1, p. K13–K15 (1979).
See also *6*, p. *64*; *14.3.3*, p. *225*.

T. Grochulski, K. Leibler, A. Sienkiewicz, and R. R. Gałązka
EPR Investigations of Ordering Effects in $Hg_{1-x}Mn_xTe$
(b) Vol. **91**, No. 1, p. K73–K76 (1979).
See also *19*, p. *289*.

E. Shatkovskii, A. Tamaševičius, and E. Matulionis
Photomagnetoelectric Effect in Strongly Excited $Cd_xH_{1-x}Te$
(a) Vol. **51**, No. 1, p. K97–K99 (1979).
See also *15*, p. *249*; *16*, p. *252*.

M. Hoclet, P. Plumelle, M. Vandevyver, R. Triboulet, and Y. Marfaing
Lattice Dynamics of Cd-Rich $Cd_{1-x}Hg_xTe$ Alloys
(b) Vol. **91**, No. 2, p. 545–549 (1979).
See also *6*, p. *67*; *20.1*, p. *309*.

J. Stankiewicz
Transport Properties of $Cd_vHg_{1-v}Se$ Crystals
(b) Vol. **93**, No. 1, p. 113–118 (1979).
See also *13.1*, p. *147*; *14.3*, p. *212*.

G. Bastard, C. Rigaux, Y. Couder, H. Thome, and A. Mycielski
Effect of Exchange on the Impurity Magnetoabsorption in Zero-Gap $Hg_{1-k}Mn_kTe$ Mixed Crystals
(b) Vol. **94**, No. 1, p. 205–210 (1979).
See also 13.4, p. 172; *18*, p. *262*; *20.1*, p. *314*.

L. G. Suslina, A. G. Plyukhin, O. Goede, and D. Hennig
Compositional Fluctuation-Induced Broadening of Bound-Exciton Lines in II–VI Semiconductor Mixed Crystals
(b) Vol. **94**, No. 2, p. K185–K189 (1979).
See also *13.5.1*, p. *187*; *20.1*, p. *317*.

B. A. Orlowski
Photoemission Evidence of Mn(3d^5) Band in $Cd_{0.4}Mn_{0.6}Te$ Crystals
(b) Vol. **95**, No. 1, p. K31–K35 (1979).
See also *13.4*, p. *175*; *17*, p. *258*.

P. Byszewski, K. Szlenk, J. Kossut, and R. R. Gałązka
Energy Levels at Γ-Point in $Hg_{1-x}Mn_xTe$ in Intense Magnetic Fields
(b) Vol. **95**, No. 2, p. 359–367 (1979).
See also *13.4*, p. *176*; *14.3*, p. *215*.

V. K. Dugaev and P. P. Petrov
Impurity Band in $Hg_{1-x}Cd_xTe$ Alloys
(b) Vol. **96**, No. 1, p. 129–136 (1979).
See also *13.4*, p. *177*; 14.3, p. 216.

S. Takeyama and R. R. Gałązka
Band Structure of $Hg_{1-x}Mn_xSe$ from Anomalous Shubnikov-de Haas Effect
(b) Vol. **96**, No. 1, p. 413–423 (1979).
See also *13.1*, p. *153*; 14.3, p. 216; 18, p. 264.

A. Twardowski, M. Nawrocki, and J. Ginter
Excitonic Magnetoabsorption in $Cd_{1-x}Mn_xTe$ Mixed Crystals
(b) Vol. **96**, No. 2, p. 497–506 (1979).
See also *13.5.1*, p. *189*; 18, p. 264; 20.1, p. 322.

O. Goede, D. Hennig, and L. John
Disorder Effects on Free Excitons in $CdS_{1-x}Se_x$ Mixed Crystals
(b) Vol. **96**, No. 2, p. 671–681 (1979).
See also *13.5.1*, p. *189*; *20.1*, p. *323*.

J. Ginter, J. Kossut, and L. Świerkowski
Indirect Exchange Interaction via Electrons in Spin–Orbit Coupled Bands in Semiconductors
(b) Vol. **96**, No. 2, p. 735–744 (1979).
See also *13.1*, p. *154*.

22.5 Halides

Original Papers and Short Notes

M. A. Gusmão and C. Scherer
Reaction Field Approximation for the Ordered Phase of Heisenberg Systems
(b) Vol. **92**, No. 2, p. 595 – 600 (1979).
See also *18.2*, p. *273*; *18.4*, p. *286*.

F. Adduci, L. Baldassarre, G. Maggipinto, A. Minafra, and F. Levy
Defect Levels in PbI_2 by TSC Measurements
(a) Vol. **52**, No. 2, p. K173 – K178 (1979).
See also *13.4*, p. *169*; 14.3, p. 212.

P. Mialhe
Approximate Technique in ESR Transition Probability Treatments
(b) Vol. **93**, No. 1, p. 189 – 195 (1979).
See also *19*, p. *291*; 22.4, p. 448; 22.6, p. 479.

W. Ekardt and H.-J. Brocksch
Phonon-Assisted Two-Photon Absorption of the Excitonic Molecule in CuCl (I): Bare Exciton Approach
(b) Vol. **93**, No. 1, p. 203 – 211 (1979).
See also *13.5.1*, p. *185*; 20.1, p. 311.

G. Behnke and H. Büttner
Erratum to: The Exiton Dispersion in Polar Semiconductors
(b) Vol. **93**, No. 1, p. 447 (1979).
See also *13.5.1*, p. *185*; 13.5.3, p. 193.

K. K. Kim, J. Toulouse, and A. S. Nowick
Anelastic Relaxation of a Li^+-Li^+ Split Interstitial Defect in LiF-Doped MnF_2 Crystals
(b) Vol. **93**, No. 1, p. K1 – K4 (1979).
See also *10.2*, p. *107*.

I. Gregora, V. Vorlíček, and M. Závětová
Investigation of Phase Transition in Hg_2Cl_2 Crystals (I): Temperature Dependence of Refractive Index
(a) Vol. **53**, No. 1, p. 355—359 (1979).
See also *1.2*, p. *18*; 20.1, p. 312.

G. Boulon, J. C. Galon, D. Trottier, and A. Vedrine
Eu^{2+} Luminescence in KY_3F_{10}
(b) Vol. **93**, No. 2, p. 775—779 (1979).
See also *20.3*, p. *331*.

V. B. Sharan, D. N. S. Srivastwa, and S. C. Sen
On the Origin of the 375 nm Emission Band in KCl:Tl
(a) Vol. **54**, No. 1, p. 129—137 (1979).
See also *13.4*, p. *172*; 20.1, p. 316; *20.3*, p. *332*.

V. May, K. Hennenberger, and F. Hennenberger
Renormalization of Polaritons Due to Virtual Formation of Biexcitons at High Densities of Excitation
(b) Vol. **94**, No. 2, p. 611 – 620 (1979).
See also *13.5.1*, p. *187*; 20.1, p. 317; 20.2, p. 325.

A. I. Mitsek, D. I. Sirota, and N. P. Kolmakova
Phenomenological Theory of Electromagnetic and Sound Energy Attenuation in Many-Domain Magnetics
(b) Vol. **94**, No. 2, p. 739–747 (1979).
See also 7, p. 79; *18.2*, p. *277*; *18.4*, p. *287*; 20, p. 301; 21.1, p. 362; 22.6, p. 482.

A. V. Zalesskii, V. V. Vanchikov, V. G. Krivenko, and A. N. Ivashchenko
Spin Echo Spectra of ^{57}Fe and ^{19}F Nuclei in a Canted Antiferromagnet FeF_3
(a) Vol. **54**, No. 2, p. 471–476 (1979).
See also *18.4*, p. *288*; 19, p. 294.

J. Kötzler and W. Scheithe
Determination of Crystalline Field Interaction in $GdCl_3$
(b) Vol. **95**, No. 1, p. K1–K4 (1979).
See also *13*, p. *140*; *18*, p. *263*.

N. Shamir, M. Melamud, and H. Shaked
Crystalline Electric Field Studies of Uranium Compounds Using Neutron Inelastic Scattering at the NRCN
(a) Vol. **55**, No. 1, p. K49–K52 (1979).
See also *13*, p. *140*; 18, p. 263; 21.5, p. 386; 22.1.2, p. 424; 22.6, p. 483.

A. Kessler and B. Koreng
Spontaneous Polarization of NH_4Cl at the Order–Disorder Phase Transition
(b) Vol. **95**, No. 2, p. 517–526 (1979).
See also *1.2*, p. *20*; *14.4*, p. *242*.

B. Prevot and J. Biellmann
Raman Scattering in Red Mercury Iodide. New Improved Polarization Spectra
(b) Vol. **95**, No. 2, p. 601–606 (1979).
See also *6*, p. *74*; *20.1*, p. *320*.

M. Piotrowski, H. Ptasiewicz-Bąk, and A. Murasik
The Crystal Structures of HoF_3 and TbF_3 by Neutron Diffraction
(a) Vol. **55**, No. 2, p. K163–K166 (1979).
See also *1.1*, p. *14*.

H. J. Mattausch and Ch. Uihlein
Fine Structure of P Excitons in CuBr
(b) Vol. **96**, No. 1, p. 189–200 (1979).
See also *13.5.1*, p. *188*; 20.1, p. 321.

22.5.1 Silver Halides

Original Papers and Short Notes

S. E. Horan and L. M. Slifkin
An Internal Friction Study of Stress-Induced Migration and Aggregation of Solute Ions in Silver Bromide Single Crystals
(a) Vol. **51**, No. 2, p. 467–478 (1979).
See also *10.2*, p. *104*; *12.2*, p. *133*.

M. E. van Hulle and W. Maenhout-van der Vorst
Influence of Pb^{2+} Ions on the Ionic Conductivity and the Space Charge Characteristics of Cubic Silver Halide Microcrystals
(a) Vol. **52**, No. 1, p. 277–282 (1979).
See also *10.2*, p. *105*; *14.4*, p. *239*.

N. Koumvakalis and W. von der Osten
First-Order Resonant Raman Scattering in Mixed Silver Halides
(b) Vol. **92**, No. 2, p. 441–445 (1979).
See also *6*, p. *66*; 13.5.1, p. 184; 20.1, p. 309; *20.3*, p. *329*; 22.5.4, p. 475.

P. Varotsos and K. Alexopoulos
On the "Anomalous" Curvature of the Diffusion Plot of Bromine in AgBr
(b) Vol. **92**, No. 2, p. 633–637 (1979).
See also *9*, p. *86*.

S. C. Goyal and V. Kumar
Strain Derivatives of the Static Dielectric Constant of Silver Halides
(b) Vol. **92**, No. 2, p. K77–K80 (1979).
See also *12.2*, p. *134*; *14.4*, p. *239*.

S. E. Horan and L. M. Slifkin
Stress-Induced Anomalies in the Internal Friction of Strontium-Doped Silver Bromide
(a) Vol. **52**, No. 2, p. 657–662 (1979).
See also *10.2*, p. *106*; *12.2*, p. *134*.

A. P. Batra and L. M. Slifkin
Contributions from Vacancy Pairs and Single Vacancies to Chloride Impurity Diffusion in AgBr
(b) Vol. **93**, No. 1, p. K77–K80 (1979).
See also *9*, p. *86*; 10.2, p. 107.

A. V. Lelyakov
Absorption Band Shape of Free Excitons in Indirect Band Gap Semiconductors at High Excitation Density
(b) Vol. **93**, No. 2, p. 523–530 (1979).
See also *13.5.1*, p. *185*; 20.1, p. 313.

Y. Kayanuma
Anomalous Reduction of the Two Phonon Pinning Interaction in Magneto-Optics
(b) Vol. **94**, No. 1, p. K53–K56 (1979).
See also *13.5.3*, p. *193*; *20.1*, p. *315*.

P. N. Ram and M. D. Tiwari
Impurity-Induced Infrared Absorption in AgBr
(b) Vol. **94**, No. 2, p. 721 – 729 (1979).
See also *6*, p. *72*; *20.1*, p. *317*.

M. Ostrow and A. Goldmann
Density of Valence States of the Superionic Conductor AgI
(b) Vol. **95**, No. 2, p. 509 – 516 (1979).
See also *13.1*, p. *152*; 17, p. 258.

A. P. Marchetti and M. Scozzafava
The Optically Detected Magnetic Resonance of Nickel-Doped Silver Chloride
(b) Vol. **96**, No. 1, p. 441 – 445 (1979).
See also 13.4, p. 178; *19*, p. *296*; 20, p. 302.

F. Callens and W. Maenhout-van der Vorst
Theoretical Treatment of the Activation Energy ΔE for Ionic Conductivity in AgX Microcrystals
(a) Vol. **56**, No. 1, p. K71 – K74 (1979).
See also *10.2*, p. *116*; *14.4*, p. *243*.

22.5.2 Alkali Halides

Original Papers and Short Notes

G. G. Nilson
Phonon Frequencies in Off-Symmetry Directions of NaCl Determined by Neutron Inelastic Scattering
(b) Vol. **91**, No. 1, p. 83–86 (1979).
See also *6*, p. *63*.

J. Shanker, A. K. G. Lashkari, and V. P. Gupta
Effect of Three-Body Interactions on the Electronic Polarizabilities and Photoelastic Behaviour of Alkali Halides
(b) Vol. **91**, No. 1, p. 263–268 (1979).
See also *4*, p. *55*; *12.2*, p. *133*; 14.4, p. 238.

R. Sebastian, G. A. Savari Raj, and T. M. Haridasan
U Centres in Alkali Iodides
(b) Vol. **91**, No. 1, p. 371–376 (1979).
See also *13.4*, p. *163*; *20.1*, p. *304*.

L. Bosi, A. Longoni, and M. Nimis
Addendum to: Lifetime of Color Centers in NaBr
(b) Vol. **91**, No. 1, p. 377 (1979).
See also *10.2*, p. *103*; 13.4, p. 164; *20.3*, p. *327*.

R. K. Gupta and A. K. Singh
Modes of Vibrations Due to Single Ion Substitutional Impurities in Alkali Halides
(b) Vol. **91**, No. 2, p. 697–704 (1979).
See also *6*, p. *64*; 10.2, p. 104.

C. Ruíz Mejía, L. Cota Araiza, F. Broder, and R. R. Mijangos
The Charge of Dislocations in Alkali Halides for the Case of a Saturated Region
(b) Vol. **91**, No. 2, p. 705–711 (1979).
See also *10.2*, p. *104*; 13.4, p. 164.

S. Muramatso and N. Sakamoto
Calculation of the A Band Shape in KCl:Au^- in the D-Mode Model
(b) Vol. **91**, No. 2, p. K87–K90 (1979).
See also *13.4*, p. *165*; *20.1*, p. *306*.

S. Brissenden, J. W. Gardner, J. Illingworth, I. Kovaćević, and R. W. Whitworth
The Influence of an Electric Field on the Flow Stress of Crystals of NaCl
(a) Vol. **51**, No. 2, p. 521–526 (1979).
See also *10.2*, p. *104*.

H. Vogt and G. Neumann
Observation of Infrared Active and Silent Modes in Cubic Crystals by Hyper-Raman Scattering
(b) Vol. **92**, No. 1, p. 57–63 (1979).
See also *6*, p. *65*; *20.1*, p. *306*; 22.8.1, p. 499.

F. Lohse and J. M. Spaeth
Fluorescence of Interstitial Atomic Hydrogen in Alkali Halides Doped with I^-
(b) Vol. **93**, No. 1, p. 153–160 (1979).
See also *10.2*, p. *107*; 13.4, p. 169; *20.3*, p. *330*.

L. Bosi and M. Nimis
On the Interaction between F- and α-Centres in KBr
(b) Vol. **93**, No. 1, p. 183–187 (1979).
See also *10.2*, p. *107*; *20.3*, p. *330*.

A. A. Kiselev and K. Lüders
On the Symmetry Groups of Non-Rigid Impurtiy Centres in Crystals
(b) Vol. **93**, No. 1, p. 285–291 (1979).
See also 6, p. 68; *10.2*, p. *107*.

F. Dochy
Theoretical Calculation of the Absorption Energy of the F-Centre in Alkali Halides
(b) Vol. **93**, No. 1, p. 325–335 (1979).
See also *10.2*, p. *107*; 13.4, p. 169; *20.1*, p. *311*.

R. P. Singh and J. Shanker
Evaluation of Third-Order Elastic Constants and Pressure Dependence of Second-Order Elastic Constants of Ionic Crystals
(b) Vol. **93**, No. 1, p. 373–380 (1979).
See also *12.2*, p. *134*; 22.6, p. 480.

D. N. Popov, K. I. Terzijsky, and N. S. Backalova
Spectral and Temperature Dependence of the Initial Yield of the Z_3-Center Formation in Additively Coloured KCl:Ca
(b) Vol. **93**, No. 1, p. K35–K37 (1979).
See also *13.4*, p. *169*; *20.1*, p. *311*.

L. Mărculescu and S. Frunză
Optical Spectra of Lead Centres in NaBr Crystals
(b) Vol. **93**, No. 1, p. K59–K62 (1979).
See also *20.1*, p. *311*; *20.3*, p. *330*.

I. V. K. Bhagavan Raju and H. Strunk
Gilde on (111) Planes in NaCl
(a) Vol. **53**, No. 1, p. 211–216 (1979).
See also *10.2*, p. *108*.

B. S. Acharya and M. L. Mukherjee
Isothermal Decay of Luminescence and Photostimulated Decay of Luminescence in KCl and KCl:Sn Single Crystals
(a) Vol. **53**, No. 1, p. 373–378 (1979).
See also *20.3*, p. *331*.

Chitra Sankar, V. S. Sivasankar, and P. W. Whippey
F_x Centres in Cesium Halides
(a) Vol. **53**, No. 1, p. 383–390 (1979).
See also *10.2*, p. *108*; *20.1*, p. *312*.

D. L. Kirk
The Relationship between the Defect Structure and the Intrinsic Ionic Conductivity of Monocrystalline Sodium Chloride
(a) Vol. **54**, No. 1, p. 251 – 259 (1979).
See also *10.2*, p. *110*; *14.4*, p. *241*.

A. H. Edwards
F-Center Photoionization in the Semicontinuum Model
(b) Vol. **94**, No. 2, p. 367 – 370 (1979).
See also *13.1*, p. *151*; 20.1, p. 316.

Ch. Hoentzsch and J. M. Spaeth
Dynamical Superhyperfine Intercation of Atomic Hydrogen in KCl and RbCl
(b) Vol. **94**, No. 2, p. 497 – 503 (1979).
See also *13.4*, p. *173*; *19*, p. *294*.

S. Ghosh, A. N. Basu, and S. Sengupta
An Extendet Deformable Shell Model for Ionic Crystals Including Quadrupolar Deformation of the Charge Cloud
(b) Vol. **94**, No. 2, p. 673 – 683 (1979).
See also *6*, p. *72*; *12.2*, p. *135*.

D. M. G. Pinatti, S. C. Zílio, and M. de Souza
Point Defects in NaCl Crystals Doped with Divalent Ions. The NaCl–NaF:Pb^{++} Crystals
(b) Vol. **94**, No. 2, p. 749 – 755 (1979).
See also *10.2*, p. *111*; *20.1*, p. *317*.

S. B. S. Sastry and S. Sapru
Thermoluminescence Glow and Emission Studies on Suprapure Rubidium Bromide Irradiated at Room Temperature
(b) Vol. **94**, No. 2, p. K149 – K154 (1979).
See also *10.2*, p. *111*; *20.3*, p. *333*.

M. D. Tiwari
Low Temperature Specific Heat of KCl:Pb^{2+}
(b) Vol. **94**, No. 2, p. K173 – K178 (1979).
See also *8*, p. *82*.

B. Koiller and H. S. Brandi
A Self-Consistent Model for the Optical Excitations of the U_2 and U_1 Centers in Alkali-Halides
(b) Vol. **94**, No. 2, p. K179 – K183 (1979).
See also *13.4*, p. *174*; *20.1*, p. *317*.

J. Prakash
Photo-Induced Polarization and Photo-Induced Depolarization of S^{2-}-Anion Vacany Dipoles in Potassium Iodide
(a) Vol. **54**, No. 2, p. 681 – 688 (1979).
See also *10.2*, p. *111*; 14.4, p. 242; 20.1, p. 317.

J. Z. Kristapson, V. I. Ozersyii, and V. A. Chadaev
Annealing of NaCl Crystals Irradiated by High Electron Doses at High Temperatures
(a) Vol. **54**, No. 2, p. K119 – K121 (1979).
See also *11*, p. *123*.

A. P. Gupta and Jai Shanker
Evaluation of the Anderson-Grüneisen Parameter for Alkali Halides
(b) Vol. 95, No. 2, p. K103–K106 (1979).
See also *8*, p. *83*.

D. Kashchiev
On the Saturation Cluster Density in Deposition of Thin Films from Vapours
(a) Vol. 55, No. 2, p. 369–374 (1979).
See also *1.5*, p. *36*; *3*, p. *53*; 10.2, p. 114; 21.6, p. 389.

M. Krohn, G. Gerth, and H. Stenzel
Conditions of Step Decoration on Crystal Surfaces
(a) Vol. 55, No. 2, p. 375–383 (1979).
See also *1.6*, p. *39*.

L. Malicskó and H. Stenzel
On the Gold Decoration of KCl:Pb Crystal Surfaces
(a) Vol. 55, No. 2, p. 395–401 (1979).
See also *1.6*, p. *39*.

P. Bennema, J. P. van der Eerden, W. J. P. van Enckevort, B. van der Hoek, and K. Tsukamoto
Observation and Interpretation of Spirals for Spiral Growth from the Liquid State
(a) Vol. 55, No. 2, p. 403–413 (1979).
See also 1.6, p. 39; *3* p. *53*; 22.8.1, p. 503.

P. Grau and F. Fröhlich
Aggregation and Precipitation Hardening in NaCl:Ca Crystals
(a) Vol. 55, No. 2, p. 479–487 (1979).
See also *10.2*, p. *114*.

D. Nenow and S. Gueleva
Formation of Radiation-Induced Voids on Dislocations in LiF
(a) Vol. 55, No. 2, p. 489–496 (1979).
See also 10.2, p. 114; *11*, p. *124*.

S. V. Moharil, V. S. Kamavisdar, and B. T. Deshmukh
Thermoluminescence of Z_1 Centres in NaCl:Ca
(a) Vol. 55, No. 2, p. K167–K172 (1979).
See also 10.2, p. 115; *20.3*, p. *336*.

A. Shluger, E. Kotomin, and Yu. Dzelme
Quantum-Chemical Calculations of Electronic and Hole Centres and Surface of NaCl Crystals (I)
(b) Vol. 96, No. 1, p. 91–99 (1979).
See also *10.2*, p. *115*; 13.3, p. 161; *13.4*, p. *177*.

G. Reuter, F. Lohse, and J. M. Spaeth
Generation of Dichroic H Centres in KCl by Polarized Optical Bleaching of Interstitial Atomic Hydrogen
(b) Vol. 96, No. 1, p. 243–250 (1979).
See also *10.2*, p. *115*; 13.4, p. 178; 19, p. 296; *20.1*, p. *321*.

22.5.3 Alkali-Earth Halides

Original Papers and Short Notes

J. M. DE SIEBENTHAL and H. BILL
Investigation of Sulfur Molecular Paramagnetic Centers in $SrCl_2$
(b) Vol. **91**, No. 2, p. 479–486 (1979).
See also *10.2*, p. *103*; 19, p. 290; 20.1, p. 305.

R. BIRSOV and L. E. MURR
Coloring of Natural Fluorites
(a) Vol. **52**, No. 1, p. 77–84 (1979).
See also *10.2*, p. *105*; *20.1*, p. *308*.

D. SCHROEER, CH. S. KUO, and R. L. LAMBE
Isomer Shift of Eu^{2+} in Fluorites
(b) Vol. **92**, No. 2, p. 565–572 (1979).
See also *5*, p. *60*; 10.2, p. 106.

V. TEODORESCU, L. C. NISTOR, and S. V. NISTOR
Electron Microscopy Study of Pure and Doped Synthetically Grown CaF_2 Crystals
(a) Vol. **52**, No. 2, p. 711–717 (1979).
See also *1.4*, p. *28*; 10.2, p. 106.

C. M. SUNTA
Mechanism of Phototransfer of Thermoluminescence Peaks in Natural CaF_2
(a) Vol. **53**, No. 1, p. 127–135 (1979).
See also *13.4*, p. *170*; *20.3*, p. *330*.

E. NICKLAUS
Optical Properties of Some Alkaline Earth Halides
(a) Vol. **53**, No. 1, p. 217–224 (1979).
See also 13.1, p. 148; *20.1*, p. *312*; 22.5.4, p. 475.

A. G. MAKHANEK and G. A. SKRIPKO
Application of Two-Photon Spectroscopy in the Study of Trivalent Rare-Earth Ions in Crystals
(a) Vol. **53**, No. 1, p. 243–252 (1979).
See also 13.4, p. 170; *20.1*, p. *312*; *20.3*, p. *331*.

A. WOLKENBERG, S. MĄCZYŃSKI, and B. WASILEWSKA
Electrochemical Method to Determine Physical Surface Properties of Rare-Earth and Manganese Fluoride Doped CdF_2 Crystals
(a) Vol. **53**, No. 2, p. 447–453 (1979).
See also *13.3*, p. *159*; 14.4, p. 240; 20.3, p. 332.

S. S. BEDI and M. P. VERMA
On the Long-Wavelength Behaviour of Solids with Fluorite Structure
(b) Vol. **94**, No. 1, p. 297–300 (1979).
See also 6, p. 70; *12.2*, p. *135*.

J. KOSHY and C. C. OGBONNA
Etching of Synthetic Barium Fluoride Single Crystals
(a) Vol. **54**, No. 1, p. K11–K12 (1979).
See also *1.6*, p. *38*.

22.5.4 Mixed Crystals

Original Papers and Short Notes

M. F. de Jesus Filho, J. P. Sanchez, and J. M. Friedt
^{121}Sb Mössbauer Spectroscopy in Some Mixed-Valence Antimony Hexachlorides
(b) Vol. **91**, No. 2, p. 443–446 (1979).
See also *5*, p. *59*.

M. Bacci
Calculation of First-Order and Second-Order Jahn-Teller Coupling Constants in $MgO:V^{2+}$ and $KMgF_3:V^{2+}$
(b) Vol. **92**, No. 1, p. 193–198 (1979).
See also *13.4*, p. *166*; *20.1*, p. *307*; 22.6, p. 478.

M. Siu Li and M. de Souza
Off-Center Cu^+ in Mixed Crystals
(b) Vol. **92**, No. 1, p. 287–291 (1979).
See also *10.2*, p. *105*; *20.1*, p. *307*.

Zh. Egemberdiev, V. Seeman, S. Zazubovich, and N. Jaanson
Polarized Luminescence and EPR Study of Sn^+ Centres in $KCl:SnCl_2$ Crystals
(b) Vol. **92**, No. 1, p. 335–345 (1979).
See also *10.2*, p. *105*; 19, p. 291; 20.1, p. 308; *20.3*, p. *328*.

N. Koumvakalis and W. von der Osten
First-Order Resonant Raman Scattering in Mixed Silver Halides
(b) Vol. **92**, No. 2, p. 441–445 (1979).
See also *6*, p. *66*; 13.5.1, p. 184; 20.1, p. 309; *20.3*, p. *329*; 22.5.1, p. 463.

S. Chandra and V. K. Mohabey
Superionic Solid Films (I): Formation and Electrical Conductivity of KAg_4I_5
(a) Vol. **53**, No. 1, p. 63–73 (1979).
See also *1.5*, p. *34*; 10.2, p. 108; 14.4, p. 240.

E. Nicklaus
Optical Properties of Some Alkaline Earth Halides
(a) Vol. **53**, No. 1, p. 217–224 (1979).
See also 13.1, p. 148; *20.1*, p. *312*; 22.5.3, p. 473.

V. B. Sharan, D. N. S. Srivastwa, and S. C. Sen
Absorption Spectra of RbCl:Tl
(a) Vol. **53**, No. 2, p. 469–472 (1979).
See also *13.4*, p. *171*; *20.1*, p. *314*.

H. R. Anand
Mössbauer Study of Iron Precipitates in LiF and NaF
(a) Vol. **53**, No. 2, p. 671–679 (1979).
See also 1.3, p. 23; *5*, p. *61*; 22.5.2, p. 468.

M. M. Beg, J. Aslam, N. Ahmad, Q. H. Khan, and N. M. Butt
Concentration Dependence of the Debye Temperature for the $K_xRb_{1-x}I$ System
(b) Vol. **94**, No. 1, p. K45–K49 (1979).
See also *8*, p. *82*.

22.6 Simple Oxides (CdO, ZnO See 22.4)

Original Papers and Short Notes

F. Bechstedt
Chemical Shift of Si 2p Core Level in SiO_x. Calculation of Relaxation Contribution
(b) Vol. **91**, No. 1, p. 167–176 (1979).
See also *13*, p. *136*.

G. Müller, M. Trapp, R. Schimko, and C. E. Richter
Measurement of Range Distributions of Zinc and Nitrogen Ions in Mutiple-Layer Substrates with the Secondary Ion Microprobe
(a) Vol. **51**, No. 1, p. 87–92 (1979).
See also *11*, p. *119*.

V. Siegel, W. Rasp, and H.-H. Kirchner
Adsorption-Induced Exoelectron Maximum of BeO Ceramics
(a) Vol. **51**, No. 1, p. 123–130 (1979).
See also *17*, p. *256*.

J. Schoenes, E. Kaldis, W. Thöni, and P. Wachter
Optical, Magnetic, and Magnetooptical Properties of the Europium Silicate Glass $Eu_{0.14}Si_{0.31}O_{0.55}$
(a) Vol. **51**, No. 1, p. 173–181 (1979).
See also 2, p. 40; *18*, p. *261*; *20.1*, p. *305*.

H. Nanto and M. Kawanishi
Effect of Reanodization on Exoelectron Emission of Anodic Ta_2O_5 Films
(a) Vol. **51**, No. 1, p. 191–196 (1979).
See also *17*, p. *256*.

B. Joseph
The Time Dependence of the Flatband Voltage Shift of MOS Capacitors Caused by the Injection of Minority Carriers from Si into SiO_2
(a) Vol. **51**, No. 1, p. K9–K11 (1979).
See also *14.3.3*, p. *225*; *14.3.4*, p. *231*; 22.1.2, p. 415.

S. P. S. Arya and H. P. Singh
I–U Characteristics of Al–Al_2O_3–Hg Systems
(a) Vol. **51**, No. 1, p. K67–K70 (1979).
See also *14.3.3*, p. *226*; *14.3.4*, p. *231*.

D. P. Madacsi, L. A. Kappers, and J. F. Houlihan
Crystal Field Splitting in Rutile-Structure Oxides from Correlation of ESR and Optical Results: V^{4+} in Tetragonal GeO_2
(b) Vol. **91**, No. 2, p. K105–K109 (1979).
See also *13.4*, p. *165*; 19, p. 290; 20.1, p. 306.

E. R. Vance and W. C. Mallard
Infrared Studies of Hydrogen in Crystalline CaO and MgO
(b) Vol. **91**, No. 2, p. K155–K160 (1979).
See also *10.2*, p. *104*; *20.1*, p. *306*.

A. GOLTZENÉ and C. SCHWAB
Impurity Scattering Effect on the Cyclotron Resonance of Carriers in Cu_2O
(b) Vol. 92, No. 2, p. 483–487 (1979).
See also *13.1*, p. *146*; 19, p. 291.

K. HÜBNER
Frequency- and Wavenumber-Dependent Dielectric Function in Terms of Bond Parameters
(b) Vol. 92, No. 2, p. K73–K76 (1979).
See also *4*, p. *56*; *13.1*, p. *146*; 14.4, p. 239.

H. J. EICHLER, J. EICHLER, J. KNOF, and CH. NOACK
Lifetimes of Laser-Induced Population Density Gratings in Ruby
(a) Vol. 52, No. 2, p. 481–486 (1979).
See also *13.4*, p. *168*; 20.2, p. 325; *20.3*, p. *329*.

K. HÜBNER
Chemical Bond and Related Properties of SiO_2 (VI): Electronic Structure of SiO_x
(a) Vol. 52, No. 2, p. 541–547 (1979).
See also *4*, p. *56*; *13.1*, p. *147*; 14.4, p. 239.

H. KRAUSE
Tunnel Hopping Current and Trap Filling in Insulating Layers
(a) Vol. 52, No. 2, p. 565–575 (1979).
See also 13.4, p. 168; *14.3.3*, p. *227*.

G. RUDLOF, J. BECHERER, and H. GLAEFEKE
A Remark on Determining Activation Energies by Fractional Glow Technique (FGT) in Case of Complex Glow Spectra – A Procedure of Improving This Method
(a) Vol. 52, No. 2, p. K137–K141 (1979).
See also *13.4*, p. *169*; *17*, p. *256*.

R. MUCCILLO and L. L. CAMPOS
Thermally Stimulated Depolarization Currents in Ceramic ThO_2
(a) Vol. 52, No. 2, p. K183–K187 (1979).
See also *10.2*, p. *106*; *14.4*, p. *239*.

J. ISOYA and J. A. WEIL
Uncompensated Titanium (3+) Center in α-Quartz
(a) Vol. 52, No. 2, p. K193–K196 (1979).
See also *10.2*, p. *107*; 11, p. 121; *19*, p. *291*.

R. A. BARTELS, J. C. KOO, and M. L. THOMAS
Temperature ad Pressure Dependence of the Dielectric Constants of CaO and SrO
(a) Vol. 52, No. 2, p. K213–K216 (1979).
See also *14.4*, p. *239*.

Z. A. RACHKO and J. A. VALBIS
Luminescence of Free and Relaxed Excitons im MgO
(b) Vol. 93, No. 1, p. 161–166 (1979).
See also *13.5.1*, p. *185*; 20.3, p. 330.

P. MIALHE
Approximate Technique in ESR Transition Probability Treatments
(b) Vol. 93, No. 1, p. 189–195 (1979).
See also *19*, p. *291*; 22.4, p. 448; 22.5, p. 461.

I. V. Abarenkov and I. M. Antonova
The Electronic Structure of MgO (II): Results of Calculations
(b) Vol. **93**, No. 1, p. 315–323 (1979).
See also *13*, p. *138*.

R. P. Singh and J. Shanker
Evaluation of Third-Order Elastic Constants and Pressure Dependence of Second-Order Elastic Constants of Ionic Crystals
(b) Vol. **93**, No. 1, p. 373–380 (1979).
See also *12.2*, p. *134*; 22.5.2, p. 467.

V. D. Korolev, S. A. Peskovatskii, and V. M. Shulga
Spin–Lattice Relaxation in Fast Neutron Irradiated Quartz at Helium Temperatures and below 1 K
(b) Vol. **93**, No. 1, p. K45–K48 (1979).
See also 11, p. 121; *19*, p. *292*.

H. Haberlandt, H.-J. Weckner, G. Gobsch, and J. Reinhold
The g-Tensor of the Interstitial Oxygen and the Broken Si–O Bond in Silicon Dioxide–A MO Model Calculation
(b) Vol. **93**, No. 1, p. K71–K75 (1979).
See also *13.4*, p. *169*; *19*, p. *292*.

P. A. Kulis, M. J. Springis, I. A. Tale, and J. A. Valbis
Recombination Luminescence in Single Crystal Al_2O_3
(a) Vol. **53**, No. 1, p. 113–119 (1979).
See also *10.2*, p. *108*; *20.3*, p. *330*.

J. L. Janson and Z. A. Rachko
Nature of Impurity-Induced UV Luminescence of MgO Crystals
(a) Vol. **53**, No. 1, p. 121–126 (1979).
See also *13.4*, p. *170*; *20.3*, p. *330*.

K. Hübner and R. Engelke
Statistical Investigations of the Structure of SiO_x
(a) Vol. **53**, No. 1, p. K79–K83 (1979).
See also *2*, p. *44*; *4*, p. *56*.

R. Enderlein, P. Renner, and M. Scheele
Electro-Optic Effects Due to the Stark Splitting of Energy Bands
(b) Vol. **93**, No. 2, p. 539–547 (1979).
See also 13.1, p. 148; 13.5.1, p. 185; *14.3.4*, p. *234*; *20.1*, p. *313*.

A. Bianconi, S. Stizza, R. Bernardini, and S. Nannarone
Electronic Structure of Ti_2O_3 from Single Crystal Thermoreflectance Spectroscopy
(b) Vol. **93**, No. 2, p. 767–774 (1979).
See also *13.1*, p. *149*; *20.1*, p. *313*.

F. Walz, H. Deusch, and H. Kronmüller
Investigation of Magnetic After-Effects in Magnetite near the Verwey Temperature
(a) Vol. **53**, No. 2, p. 519–528 (1979).
See also 10.2, p. 109; *18.3*, p. *284*.

F. DETTMANN and G. ALBRECHT
The Influence of Vanadium Oxide on the Properties of Josephson Tunnel Junctions
(a) Vol. **53**, No. 2, p. 541–548 (1979).
See also *14.2*, p. *206*; 14.3.1, p. 220; 14.3.3, p. 228.

J. HÖNIGSCHMID and G. WILL
A Neutron Diffraction Investigation of the H–T Phase Diagram of α-Fe_2O_3
(a) Vol. **53**, No. 2, p. 557–563 (1979).
See also *18.4*, p. *287*.

H. ZOGHEIB, B. PARDO and D. RENARD
Conditions of Appearance and Extinction of Kinematically Forbidden Electronic Diffraction Spots in the Dynamical Theory
(a) Vol. **53**, No. 2, p. 577–587 (1979).
See also *1.1*, p. *13*; 1.4, p. 29.

J. P. SINGH, S. K. SINGH, R. S. SRIVASTAVA, and K. SINGH
Current Transport in MOS System on the Basis of T-D Emission Theory
(a) Vol. **53**, No. 2, p. 627–630 (1979).
See also *14.3.3*, p. *228*.

L. BOUSSE, R. VOUNCKX, and O. STEENHAUT
The Influence of Silicon Surface Potential on the Fowler-Nordheim Tunnel Injection into SiO_2
(a) Vol. **53**, No. 2, p. K215–K218 (1979).
See also *13.3*, p. *160*; *14.3.4*, p. *234*.

J. ZAREMBOWITCH, J. GOUTERON, and A. M. LEJUS
Raman Spectra of Lanthanide Sesquioxide Single Crystals with A-Type Structure
(b) Vol. **94**, No. 1, p. 249–256 (1979).
See also *6*, p. *70*; *20.1*, p. *315*.

Y. HARA and M. NICOL
Raman Spectra and the Structure of Rutile at High Pressures
(b) Vol. **94**, No. 1, p. 317–322 (1979).
See also *6*, p. *70*; *20.1*, p. *315*.

K. HÜBNER
Energy-Gap Variation at the Si–SiO_x–SiO_2 Interface
(b) Vol. **94**, No. 1, p. K35–K39 (1979).
See also *13.1*, p. *150*; 13.3, p. 160; 22.1.2, p. 420.

R. P. GOYAL and M. K. SINGHAL
Dispersion of Phonons in Transition Metal Oxides (FeO)
(b) Vol. **94**, No. 1, p. K51–K52 (1979).
See also *6*, p. *71*.

Š. LÁNYI
Contact-Limited Conduction in V_2O_5 Single Crystals
(a) Vol. **54**, No. 1, p. 37–43 (1979).
See also *14.3.4*, p. *234*.

S. P. S. BADWAL and H. J. DE BRUIN
Electrode Kinetics at the Pt/Yttria-Stabilized Zirconia Interface by Complex Impedance Dispersion Analysis
(a) Vol. **54**, No. 1, p. 261–270 (1979).
See also *13.3*, p. *160*; *14.3.4*, p. *235*; 21.6, p. 388.

S. Singh, S. L. Pandey, and R. N. Saxena
Dielectric Measurements of Thin Films of Bi O
(a) Vol. 54, No. 2, p. K91 – K93 (1979).
See also *14.4*, p. *242*.

H.-G. Neumann and Th. Roy
Diffraction Investigation on the Structure of SiO_x
(a) Vol. 54, No. 2, p. K107 – K111 (1979).
See also *1*, p. *10*; *2*, p. *46*

A. S. Rao and R. J. Kearney
Logarithmic Derivative Reflectance Spectra of BaO and SrO
(b) Vol. 95, No. 1, p. 243 – 250 (1979).
See also 13.5.1, p. 188; *20.1*, p. *319*.

I. E. Lacis and J. A. Valbis
Vacuum-Ultraviolett Luminescence of Be-Doped MgO Epitaxial Layers
(b) Vol. 95, No. 1, p. K21 – K26 (1979).
See also *20.3*, p. *335*.

Yu. D. Chistyakov, S. V. Potapov, D. N. Gulidov, and H. G. Schneider
The Effect of Residual Mechanical Stresses and Strains on the Superconductivity of Heteroepitaxial Layers of Nb on α-Al_2O_3
(a) Vol. 55, No. 1, p. 147 – 152 (1979).
See also 1.5, p. 36; *14.2*, p. *207*.

R. D. Gould and C. A. Hogarth
The Time Dependence of Circulating and Emission Currents in Evaporated Thin Film Sandwich Structures of Au–SiO_x–Au
(a) Vol. 55, No. 1, p. 325 – 332 (1979).
See also *14.3.3*, p. *229*; 17, p. 258.

S. P. S. Arya and H. P. Singh
Negative Resistance Characteristics of Metal–Antimony Oxide–Metal Thin Films
(a) Vol. 55, No. 1, p. K11 – K12 (1979).
See also *14.3.4*, p. *236*.

N. Shamir, M. Melamud, and H. Shaked
Crystalline Electric Field Studies of Uranium Compounds Using Neutron Inelastic Scattering at the NRCN
(a) Vol. 55, No. 1, p. K49 – K52 (1979).
See also *13*, p. *140*; 18, p. 263; 21.5, p. 386; 22.1.2, p. 424; 22.5, p. 462.

V. Siegel and W. Rasp
The Exoelectron Emission from Hydrated and Dehydrated Surfaces of BeO
(a) Vol. 55, No. 1, p. K95 – K98 (1979).
See also *17*, p. *258*.

A. Scharmann and U. Wiessler
The Contribution of Internal Electron Sources to TSEE Trap Population
(a) Vol. 55, No. 1, p. K117 – K118 (1979).
See also 14.3.2, p. 224; *17*, p. *258*.

B. Lalevic, M. Gvishi, and M. Shoga
Schottky Barrier Effects on the Electrical Properties of the Bi–NbO_2–Bi System
(a) Vol. **56**, No. 1, p. 379–386 (1979).
See also *14.3.3*, p. *230*; 14.4, p. 243.

L. N. Skuja and A. R. Silin
Optical Properties and Energetic Structure of Non-Bridging Oxygen Centers in Vitreous SiO_2
(a) Vol. **56**, No. 1, p. K11–K13 (1979).
See also *2*, p. *49*; *20.1*, p. *322*; 20.3, p. 336.

J. Purāns, J. Kliava, and I. Millere
Computer Simulation of Mn^{2+} EPR Spectra in Glasses with $g_{ef} = 4.29$
(a) Vol. **56**, No. 1, p. K25–K28 (1979).
See also *2*, p. *49*; *19*, p. *297*.

K. Hübner and F. Bechstedt
Temperature Dependence of Atomic Core Levels in Solids (II): Chemical Shift and Relaxation Energy of Core Levels and Their Dependence on Temperature
(b) Vol. **96**, No. 2, p. 595–603 (1979).
See also *13.1*, p. *154*; *13.5.1*, p. *189*; 22.1.2, p. 426; 22.2.2, p. 438.

B. Lorenz and D. Ihle
Electron–Phonon versus Coulomb Interaction Effects at the Verwey Transition of Fe_3O_4
(b) Vol. **96**, No. 2, p. 659–669 (1979).
See also *1.2*, p. *21*; 4, p. 58; 6, p. 76; 13, p. 142.

C. Lacour
Sintering and Random Structures
(b) Vol. **96**, No. 2, p. 785–796 (1979).
See also *1.4*, p. *32*.

E. M. Trukhanov, S. I. Stenin, and A. G. Noskov
Dislocation Interactions in Germanium Covered with Dielectric Films
(a) Vol. **56**, No. 2, p. 443–450 (1979).
See also *1.5*, p. *37*; 2, p. 50; *10.2*, p. *117*; 22, p. 407; 22.1.1, p. 413.

A. Coret and A. Haydar
Photoconductivity in the IS Excitonic Line of Cu_2O. Shape of the Response
(a) Vol. **56**, No. 2, p. 531–534 (1979).
See also *13.5.1*, p. *190*; *16*, p. *255*.

D. Haack and H. G. Wagemann
Electronic Conductivity in Highly Doped SiO_2 by Means of Ion Implantation
(a) Vol. **56**, No. 2, p. 581–591 (1979).
See also *13.4*, p. *180*; *14.3.4*, p. *237*.

V. A. Krylov, W. Ulrici, and L. S. Sochava
Optical Spectra of Off-Centre Ni^{2+} Ions in SrO
(a) Vol. **56**, No. 2, p. 615–621 (1979).
See also *10.2*, p. *117*; *20.1*, p. *324*.

G. Dhalenne, A. Revcolevschi, and C. Monty
Grain Boundaries in NiO (II): Determination of Mass Transport Mechanisms by Thermal Grooving
(a) Vol. **56**, No. 2, p. 623–636 (1979).
See also *1.4*, p. *32*; *10.2*, p. *117*.

J. W. Laughner, R. E. Newnham, and L. E. Cross
Ferroelastoelectricity in Quartz
(a) Vol. **56**, No. 2, p. K83 – K86 (1979).
See also *14.4.1*, p. *248*.

V. F. Antyushin, B. I. Sysoev, and V. F. Synorov
Identification of Alkali Metal Ions in Silicon Dioxide Films
(a) Vol. **56**, No. 2, p. K91 – K95 (1979).
See also *10.2*, p. *117*; *14.4*, p. *243*.

R. P. Khare
Electroluminescence of CaO:Ag and CaO:Au
(a) Vol. **56**, No. 2, p. K97 – K99 (1979).
See also *20.3*, p. *337*.

M. Kalliomäki, V. Meisalo, and A. Laisaar
High Pressure Transformations in Cuprous Oxide
(a) Vol. **56**, No. 2, p. K127 – K131 (1979).
See also *1.2*, p. *21*.

22.6.1 Mixed Crystals

Original Papers and Short Notes

I. Ardelean and W. Mieskes
Diffuse Reflectance Spectra of $x\mathrm{Fe_2O_3} \cdot (1 - x)$ [3 $B_2O_3 \cdot PbO$] Glasses
(a) Vol. **51**, No. 2, p. K129–K132 (1979).
See also *2*, p. ***41***; *20.1*, p. ***306***.

G. Frank, H. Köstlin, and A. Rabenau
X-Ray and Optical Measurements in the In_2O_3–SnO_2 System
(a) Vol. **52**, No. 1, p. 231–238 (1979).
See also *1.3*, p. *22*; *20.1*, p. ***308***.

J. K. Srivastava and B. W. Dale
Effect of Annealing Temperature on Cross Relaxation in the Cr^{3+}-Doped Magnetically Dilute Paramagnetic Al_2O_3–Fe_2O_3 System
(b) Vol. **96**, No. 1, p. K47–K50 (1979).
See also *5*, p. *61*; 13.4, p. 178.

J. Pebler und K. Schmidt
X-Ray Diffraction and Mössbauer Measurements of $V_{1-x}Fe_xO_{2-x}F_x$ in the Concentration Range $0 < x < 0.20$
(a) Vol. **56**, No. 1, p. 75–85 (1979).
See also *1.2*, p. *20*; *5*, p. *62*.

D. Gourier, D. Vivien, and J. Livage
ESR of Paramagnetic Defects in the Conduction Planes of Na^+, K^+, and H_3O^+ β-Alumina
(a) Vol. **56**, No. 1, p. 247–257 (1979).
See also *19*, p. *297*.

C. A. Hogarth and M. Y. Nadeem
The Effect of Electroforming on the Alternating Current Behaviour of Thin Amorphous Insulating Films
(a) Vol. **56**, No. 1, p. K37–K40 (1979).
See also *2*, p. *49*; *14.3.1*, p. *221*.

J. Pebler, K. Schmidt, and G. Weiser
Magnetic Susceptibility and Electrical Conductivity Measurements of $V_{1-x}Fe_xO_{2-x}F_x$ in the Concentration Range $0 < x < 0.20$
(a) Vol. **56**, No. 2, p. 457–465 (1979).
See also *1.2*, p. *21*; 13.4, p. 180; 14.3, p. 218; 18.1, p. 269.

22.7 Semiconducting Intermetallic Compounds

Review Article

Original Papers and Short Notes

K. H. J. Buschow, U. Goebel, and E. Dormann
Valence Instabilities in $CeSn_3$ and $YbAl_3$
(b) Vol. **93**, No. 2, p. 607–615 (1979).
See also *13*, p. *139*; 18, p. 262; 19, p. 292.

H. Krüger, R. Herrmann, B. Fellmuth, and W. Kraak
Measurement of the Cyclotron Masses in n-Conducting BiSb under High Hydrostatic Pressure
(b) Vol. **93**, No. 2, p. 661–667 (1979).
See also *13.1*, p. *148*; 19, p. 292.

D. V. Gitsu, F. M. Muntyanu, M. I. Onu, and A. I. Russu
Magnetoresistance Anisotropy of $Bi_{0.91}Sb_{0.09}$ Alloy in Strong Magnetic Field
(b) Vol. **93**, No. 2, p. K127–K129 (1979).
See also *14.3*, p. *213*; 18, p. 262.

E. V. Kapitanov and E. N. Yakovlev
Mössbauer Study of Phase Transitions under High Hydrostatic Pressures (I): The Phase Transition of Mg_2Sn
(a) Vol. **53**, No. 2, p. 473–480 (1979).
See also *1.2*, p. *18*; *5*, p. *61*.

E. V. Kapitanov and E. N. Yakovlev
Mössbauer Study of Phase Transitions under High Hydrostatic Pressures (II): The Phase Transition of SnTe
(a) Vol. **54**, No. 1, p. 139–143 (1979).
See also *1.2*, p. *19*; *5*, p. *61*.

22.7.1 Mixed Crystals

Original Papers and Short Notes

H. Süssmann und W. Heilinger
On the Anisotropy of the Thermoelectric Properties of p-$Sb_xBi_{2-x}Te_3$
(a) Vol. 51, No. 2, p. K201 – K206 (1979).
See also *15*, p. *249*.

H. Köhler and R. Weidner
Electronic Properties of n-Type $(Bi_{1-x}Sb_x)_2Te_3$ from Magneto-Quantum Oscillations
(b) Vol. 95, No. 2, p. 635 – 641 (1979).
See also *13.1*, p. *152*; 13.2, p. 157; 14.3, p. 216.

22.8 Tertiary and Higher Compounds

Original Papers and Short Notes

N. A. Bakhyshov, N. M. Gasanly, B. M. Yavadov, V. I. Tagirov, and Sh. M. Efendiev
Mixed One- and Two-Mode Behaviour of Optical Phonons in $TlGaS_{2x}Se_{2(1-x)}$ and $TlInS_{2x}Se_{2(1-x)}$ Layer Solid Solutions
(b) Vol. **91**, No. 1, p. K1–K3 (1979).
See also *6*, p. *63*; *20.1*, p. *304*.

A. E. Bakhyshov, M. F. Agaeva, and A. M. Darvish
Electrical and Optical Properties of $TlInSe_2$ Single Crystals
(b) Vol. **91**, No. 1, p. K31–K34 (1979).
See also *14.3*, p. *210*; 16, p. 252; *20.1*, p. *305*.

W. Hörig, H. Neumann, E. Reccius, H. Weinert, G. Kühn, and B. Schumann
Temperature Dependence of the Absorption Edge in $CuGaS_2$
(a) Vol. **51**, No. 1, p. 57–62 (1979).
See also 13.1, p. 144; *20.1*, p. *305*.

R. Hrabański, P. B. Sczaniecki, and J. Stankowski
Phase Transition Effects in the EPR Spectrum of Mn^{2+} in Magnesium Fluosilicate
(a) Vol. **51**, No. 1, p. 243–247 (1979).
See also *1.2*, p. *15*; *19*, p. *290*.

F. Studer and B. Raveau
Erratum to: Electron Microscopic Study of Intergrowth Phenomena in Barium Siliconiobates (II): Multiple Intergrowths and Microdomains
(a) Vol. **51**, No. 1, p. 313 (1979).
See also *1.1*, p. *11*; *1.4*, p. *26*.

J. Przedmojski and B. Pałosz
X-Ray Investigation of Needle-Like Crystals of $CdIn_2Se_4$
(a) Vol. **51**, No. 1, p. K1–K3 (1979).
See also *1.1*, p. *11*.

S. Niziol, R. Fruchart, and J. P. Senateur
Magnetic Properties of CoMnSi and $CoMnSi_{0.85}Ge_{0.15}$
(a) Vol. **51**, No. 1, p. K23–K27 (1979).
See also *18.2*, p. *271*; *18.4*, p. *286*.

S. M. Wasim
Thermal Conductivity of Ternary Compounds
(a) Vol. **51**, No. 1, p. K35–K40 (1979).
See also *8*, p. *80*.

J. Leciejewicz, H. Ptasiewicz-Bąk, and A. Zygmunt
Magnetic Phase Transition in UPd_2Si_2
(a) Vol. **51**, No. 1, p. K71–K73 (1979).
See also *1.1*, p. *11*; *18.2*, p. *271*; 18.4, p. 286.

N. V. Fedoseeva, T. A. Velikanova, and A. G. Zvegintsev
High-Pressure Cubic Phase of $RbMnCl_3$ — Magnetic Properties
(a) Vol. **51**, No. 1, p. K93–K96 (1979).
See also *18.1*, p. *265*; *18.4*, p. *286*.

E. Dynowska and J. Stankowski
X-Ray Study of the Structural Phase Transition in Nickel Hexammine Perchlorate
(a) Vol. 52, No. 2, p. 381 – 386 (1979).
See also *1.2*, p. *17*.

M. F. Vereshchak, A. K. Zhetbaev, and Sh. Sh. Ibragimov
A Mössbauer Study of the Effect of γ-Rays and Neutron Irradiation on Iron Oxalates
(a) Vol. 52, No. 2, p. 415 – 419 (1979).
See also *5*, p. *60*; *11*, p. *121*.

H. Neumann, D. Peters, B. Schuhmann, and G. Kühn
Electrical Properties of $CuGaTe_2$ Epitaxial Layers
(a) Vol. 52, No. 2, p. 559 – 564 (1979).
See also 13.4, p. 168; *14.3.1*, p. *219*.

J. Novotný and Z. Spurný
Absorption Shift in Aluminophosphate Glasses
(a) Vol. 52, No. 2, p. 603 – 608 (1979).
See also 2, p. 43; 13.4, p. 168; *20.1*, p. *310*.

S. B. Patil, H. V. Keer, and D. K. Chakrabarty
Structural, Electrical, and Magnetic Properties in the System $Ba_xLa_{1-x}CoO_3$
(a) Vol. 52, No. 2, p. 681 – 686 (1979).
See also 1.1, p. 12; *14*, p. *197*; *18.2*, p. *273*.

H. B. Tripathi, H. C. Kandpal, and A. K. Agarwal
Non-Radiative Energy Transfer from $Dy^{3+} \rightarrow Ho^{3+}$ in Calibo Glass
(a) Vol. 52, No. 2, p. 697 – 701 (1979).
See also *13.4*, p. *168*; *20.3*, p. *329*.

B. Chevalier, G. Demazeau, J. Etourneau, and P. Hagenmuller
Pressure Dependence of the Resistivity of the Rare Earth Oxide Nitrides $Eu^{2+}_{1-x}Ln^{3+}_xN_xO_{1-x}$ (Ln = Nd, En)
(b) Vol. 93, No. 1, p. K63 – K66 (1979).
See also 12.2, p. 134; 14.1, p. 200; *14.3*, p. *212*.

J. Hrdý, E. Krouský, and O. Renner
A Search for Faraday Rotation in X-Ray Region
(a) Vol. 53, No. 1, p. 143 – 146 (1979).
See also *18.2*, p. *274*; *20*, p. *300*; 21.1, p. 358.

H. Pykacz
Investigation of a Phase Transition in $NaH_3(SeO_3)_2$ by Measurement of the Pyroelectric Current
(a) Vol. 53, No. 1, p. 367 – 372 (1979).
See also *14.4.1*, p. *246*.

A. Witek, J. Wosik, and W. Zbieranowski
The Elastic Constants and Debye Temperature of the Monocrystal $NaMgAl(C_2O_4)_3 \cdot 9\ H_2O$
(a) Vol. 53, No. 1, p. K27 – K29 (1979).
See also *7*, p. *78*; *12.2*, p. *134*.

B. Etlinger
Dielectric Properties of $Al_2In_{12}S_{21}$
(a) Vol. 53, No. 1, p. K35 – K38 (1979).
See also *14.4*, p. *240*.

A. A. Kaminskii, V. A. Fedorov, S. E. Sarkisov, J. Bohm, P. Reiche, and D. Schultze
Stimulated Emission of Ho^{3+} and Er^{3+} Ions in $Gd_3Ga_5O_{12}$ Crystals and Cascade Laser Action of Ho^{3+} Ions over the $^5S_2 \rightarrow {^5I_5} \rightarrow {^5I_6} \rightarrow {^5I_8}$ Scheme
(a) Vol. **53**, No. 2, p. K219—K222 (1979).
See also *20.2*, p. *325*.

M. J. Thwaites, R. D. Tomlinson, and M. J. Hampshire
The Observation of Optical Transitions from Valence Band States in $CuInTe_2$
(b) Vol. **94**, No. 1, p. 211—214 (1979).
See also *13.1*, p. *150*; *20.1*, p. *315*.

K. R. Allakhverdiev, A. N. Abbasov, T. R. Mekhtiev, and R. Kh. Nani
Far-Infrared Reflection Spectra of $CdInGaS_4$ Single Crystals
(b) Vol. **94**, No. 1, p. K31—K34 (1979).
See also *6*, p. *71*; *20.1*, p. *315*.

V. M. Fridkin, R. Nitsche, N. Korchagina, N. A. Kosonogov, R. Magomadov, A. I. Rodin, and K. A. Verkhovskaya
Photoferroelastic Phenomena in Sb_5O_7I Crystals
(a) Vol. **54**, No. 1, p. 231—237 (1979).
See also 12.2, p. 135; *16*, p. *253*.

G. Kühnel, W. Siegel, and E. Ziegler
Deep Levels in $ZnSiP_2$ Determined by Schottky TSC Measurements
(a) Vol. **54**, No. 1, p. 315—320 (1979).
See also *13.4*, p. *173*; *14.3*, p. *214*.

W. Wojdowski, T. Łukasiewicz, W. Nazarewicz, and J. Żmija
Infrared Studies of Lattice Vibrations in $Bi_{12}GeO_{20}$ and $Bi_{12}SiO_{20}$ Crystals
(b) Vol. **94**, No. 2, p. 649—658 (1979).
See also *6*, p. *72*; *20.1*, p. *317*.

R. Rai, R. Morehouse, and E. Dibaj
Dynamic Shift of the ESR Field in the One-Dimensional Magnetic Copper Dipyridine Dichloride
(b) Vol. **94**, No. 2, p. K159—K164 (1979).
See also 18.1, p. 268; 18.4, p. 287; *19*, p. *294*.

A. Ratuszna, A. Pietraszko, A. Chełkowski, and K. Łukaszewicz
The Temperature Dependence of Lattice Parameters of $KMeF_3$ and $KMn_{0.9}Me_{0.1}F_3$ Compounds (Me = Mn^{2+}, Co^{2+}, and Ni^{2+})
(a) Vol. **54**, No. 2, p. 739—743 (1979).
See also *1.1*, p. *13*; *1.2*, p. *19*.

O. Calzadilla, J. Fuentes, and L. Hernández
$ZnIn_2S_4$III(a) Absorption Spectrum Tail Changes Due to Microstresses
(a) Vol. **54**, No. 2, p. K99—K101 (1979).
See also *20.1*, p. *318*.

H. Tichá, L. Tichý, and M. Frumar
Temperature Dependence of the Exponential Absorption Edge in $Ge_{20}Sb_{25}S_{55}$ Chalcogenide Glass
(a) Vol. **54**, No. 2, p. K163—K166 (1979).
See also *20.1*, p. *318*.

I. Savatinova, E. Anachkova, J. Rohleder, and B. Jakubovski
C ≡ N Vibrations in Potassium Ferrocyanide Trihydrate
(b) Vol. **96**, No. 2, p. K97–K101 (1979).
See also *6*, p. *77*; 14.4.1, p. 247; *20.1*, p. *323*.

A. A. Kaminskii, S. E. Sarkisov, T. I. Butaeva, G. A. Denisenko, B. Hermoneit, J. Bohm, W. Grosskreutz, and D. Schultze
Growth, Spectroscopy, and Stimulated Emission of Cubic $Bi_4Ge_3O_{12}$ Crystals Doped with Dy^{3+}, Ho^{3+}, Er^{3+}, Tm^{3+}, or Yb^{3+} Ions
(a) Vol. **56**, No. 2, p. 725–736 (1979).
See also 20.1, p. 324; *20.2*, p. *326*; *20.3*, p. *337*.

H. Neumann, R. D. Tomlinson, E. Nowak, and N. Avgerinos
Electrical Properties of p-Type $CuInSe_2$ Single Crystals
(a) Vol. **56**, No. 2, p. K137–K140 (1979).
See also *13.4*, p. *180*; *14.3*, p. *218*.

J. Schneider, K. Zaveta, A. Handstein, F. Zounova, and V. Nekvasil
Temperature Dependence of the Magnetization in Amorphous $(Fe_{0.04}Ni_{0.96})_{80}P_{10}B_{10}$
(a) Vol. **56**, No. 2, p. K141–K144 (1979).
See also *2*, p. *51*; *18*, p. *264*.

22.8.1 Salts of Oxy-Acids (Carbonates, Phosphates, Silicates Including Mica etc.)

Review Article

J. Fousek and J. Petzelt
Changes of Refractive Indices of Crystals Induced by Structural Phase Transitions
(a) Vol. **55**, No. 1, p. 11–40 (1979).
See also *1.2*, p. *15*; 12.2, p. 133; ***14.4.1***, p. ***244***; 20, p. 298.

Original Papers and Short Notes

S. A. Marshall, T. Marshall, S. A. Zager, and D. N. Zimmerman
Electron Spin Resonance Absorption Spectrum of Divalent Manganese in Single Crystal Calcite
(b) Vol. **91**, No. 1, p. 275–282 (1979).
See also *13.4*, p. *163*; *19*, p. *289*.

M. Scrocco
A Raman Spectroscopic Study of K_2SO_4 in the D_{2h}^{16} and C_{2v}^{9} Phases
(b) Vol. **91**, No. 1, p. K21–K25 (1979).
See also *6*, p. *64*; *20.1*, p. *305*.

V. K. Jain
Electron Spin Resonance of VO^{2+} Ions in Antiferroelectric ADP Crystals
(b) Vol. **91**, No. 1, p. K35–K37 (1979).
See also 14.4.1, p. 244; *19*, p. *289*.

D. Chakravorty, A. R. Haranahalli, and D. Kumar
Electrical Properties of V_2O_5–P_2O_5 Glasses Containing Metallic Particles
(a) Vol. **51**, No. 1, p. 275–280 (1979).
See also *14*, p. *197*.

N. M. Nesterenko and V. I. Fomin
Soft Modes and the Raman Spectrum in $KSc(MoO_4)_2$ at Ferroelastic Phase Transitions
(a) Vol. **51**, No. 1, p. K101–K104 (1979).
See also 6, p. 64; *14.4.1*, p. *244*; *20.1*, p. *305*.

I. V. Stasyuk, R. R. Levitskii, and N. A. Korinevskii
Collective Vibrations of Protons in Compounds of the KH_2PO_4-Type. The Cluster Approximation
(b) Vol. **91**, No. 2, p. 541–550 (1979).
See also *13.5*, p. *181*; *14.4.1*, p. *244*.

A. M. Balagurov, I. D. Dutt, Z. Gheorghiu, B. N. Savenko, and L. A. Shuvalov
A Method for Observing the Ferroelectric Domain Structure by Using a Time-of-Flight Neutron Diffractometer
(a) Vol. **51**, No. 2, p. 367–374 (1979).
See also *14.4.1*, p. *245*.

H. Hess and E. Krautz
Photoconductivity of Densely Sintered Zincsilicate
(a) Vol. **51**, No. 2, p. 391–398 (1979).
See also *16*, p. *252*.

Z. Czapla, T. Lis, and L. Sobczyk
Ferroelectric Properties of NH_4HSeO_4 Crystals
(a) Vol. 51, No. 2, p. 609–612 (1979).
See also *1.2*, p. *16*; *14.4.1*, p. *245*.

H. Shimizu, H. Yamada, N. Yasuda, and S. Fujimoto
Effect of Hydrostatic Pressure on the Ferroelectric Phase Transition in NH_4IO_3
(a) Vol. 51, No. 2, p. K121–K124 (1979).
See also 1.2, p. 16; *14.4.1*, p. *245*.

H. Vogt and G. Neumann
Observation of Infrared Active and Silent Modes in Cubic Crystals by Hyper-Raman Scattering
(b) Vol. 92, No. 1, p. 57–63 (1979).
See also *6*, p. *65*; *20.1*, p. *306*; 22.5.2, p. 465.

K. Hermann and S. Kobe
Thermodynamical Behaviour of Finite Amorphous Ising Chains with Antiferromagnetic Interaction
(b) Vol. 92, No. 1, p. K69–K72 (1979).
See also *2*, p. *41*; *18.4*, p. *286*.

Hj. Bernhardt
Defect Electron Centres in Lead Molybdate Crystals
(a) Vol. 52, No. 1, p. 223–229 (1979).
See also 10.2, p. 105; 13.5.3, p. 193; *20.1*, p. *308*; *20.3*, p. *328*.

Z. Jirák, S. Vratislav, and J. Zajiček
The Magnetic Structure of $Pr_{0.9}Ca_{0.1}MnO_3$
(a) Vol. 52, No. 1, p. K39–K43 (1979).
See also *18.3*, p. *283*.

Sh. Sh. Bashkirov and V. N. Lebedev
Isomer Shift and Quadrupolar Splitting of the Mössbauer Spectrum of Ferroelectrics
(b) Vol. 92, No. 2, p. 397–402 (1979).
See also *5*, p. *60*; 14.4.1, p. 245.

I. H. Ismailzade, A. Yu. Kudzin, and L. Ya. Sadovskaya
X-Ray Diffractional and Optical Investigations of Ferroelectric $SrTeO_3$
(a) Vol. 52, No. 2, p. K105–K109 (1979).
See also 1.1, p. 12; *14.4.1*, p. *245*.

J. Arsene, J. Lopitaux, M. Drifford and M. Lenglet
Evidence of Tetrahedral Coordination of Cr^{3+} in a Lithium Chromium Gallate from Electronic Spectra
(a) Vol. 52, No. 2, p. K111–K114 (1979).
See also *1.1*, p. *12*; *20.1*, p. *310*.

H. Pykacz
Notes on the First-Order Ferroelectric Phase Transitions in an Electric Field
(a) Vol. 52, No. 2, p. K179–K182 (1979).
See also *14.4.1*, p. *245*.

W. Meyer, P. Würfel, R. Munser, and G. Müller-Vogt
Kinetics of Fixation of Phase Holograms in $LiNbO_3$
(a) Vol. **53**, No. 1, p. 171–180 (1979).
See also 14.4.1, p. 245; *16*, p. *253*; 20.1, p. 312.

U. Knauer
Distribution of the Iron Dopant in Barium Titanate Ceramic, Determined by the Scanning Transmission Electron Microscope
(a) Vol. **53**, No. 1, p. 207–210 (1979).
See also *1.4*, p. *28*; 10.2, p. 108.

M. K. Kostov, N. P. Ivanchev, and E. F. Dudnik
Electrically Induced Optical Activity in the Centrosymmetric Crystal $NaBi(MoO_4)_2$
(a) Vol. **53**, No. 1, p. K47–K49 (1979).
See also *20*, p. *300*.

Y. A. Badr, F. El-Kabbany, and M. Tosson
DTA and Dielectric Study of the High Temperature Phase of Na_2SO_4
(a) Vol. **53**, No. 1, p. K51–K55 (1979).
See also *1.2*, p. *18*; *14.4*, p. *240*.

W. Wegener, J. Vanderhaeghen, D. Scheerlinck, S. Hautecler, E. Legrand, and G. Will
Magnetic Interactions in α-$MnSO_4$ and β-$MnSeO_4$
(b) Vol. **93**, No. 2, p. 597–605 (1979).
See also *18.1*, p. *267*; 18.4, p. 287.

Y. A. Badr and R. Kamel
Low-Temperature Structure Changes in Nitrate Crystals
(a) Vol. **53**, No. 2, p. K161–K164 (1979).
See also *1.2*, p. *18*; *14.4*, p. *241*.

M. P. Ozolinsh
Pulse Response of the Electrooptic Effect in PMN Single Crystals
(a) Vol. **53**, No. 2, p. K169–K171 (1979).
See also *14.1*, p. *246*; *20.1*, p. *314*.

J. G. Kliava, Z. A. Konstants, J. J. Purāns, and A. I. Dimante
EPR of Mn^{2+} in Polycrystals of Low Symmetry. Magnesium Metaphosphate
(a) Vol. **54**, No. 1, p. 61–66 (1979).
See also *19*, p. *293*.

Č. Koňák
Changes of Optical Properties at an Incommensurate–Commensurate Phase Transition in $(NH_4)_2BeF_2$ Crystals
(a) Vol. **54**, No. 1, p. 99–104 (1979).
See also *1.2*, p. *18*; 14.4.1, p. 246; 20.1, p. 316.

J. A. Groenink, C. Hakfoort, and G. Blasse
The Luminescence of Calcium Molybdate
(a) Vol. **54**, No. 1, p. 329–336 (1979).
See also *13.4*, p. *173*; *20.3*, p. *333*.

V. S. Sivasankar and P. W. Whippey
Thermoluminescence and Optical Absorption of X-Ray Irradiated $NaBrO_3$
(a) Vol. 54, No. 1, p. 369–373 (1979).
See also *11*, p. *122*; 20.1, p. 316; *20.3*, p. *333*.

B. V. R. Chowdari and Y. Ravi Sekhar
Electrical Conductivity and Thermal Depolarisation Studies of CrO_4^{2-} and $Cr_2O_7^{2-}$ Doped KDP and ADP Crystals
(a) Vol. 54, No. 1, p. 413–419 (1979).
See also *13.4*, p. *173*; *14.3*, p. *214*.

M. Friedrich and W. Karthe
Structure and Kinetics of Fe^{3+}-Doped $PbMoO_4$
(b) Vol. 94, No. 2, p. 451–456 (1979).
See also *13.4*, p. *173*; *19*, p. *294*.

A. V. Turik, N. B. Shevchenko, V. G. Gavrilyachenko, and E. G. Fesenko
On the Nature of Dielectric Permittivity of $PbTiO_3$ Single Crystals
(b) Vol. 94, No. 2, p. 525–528 (1979).
See also 6, p. 72; *14.4*, p. *241*.

D. Lapraz, A. Baumer, and P. Iacconi
On the Thermoluminescent Properties of Hydroxyapatite $Ca_5(PO_4)_3OH$
(a) Vol. 54, No. 2, p. 605–613 (1979).
See also *10.2*, p. *111*; *20.3*, p. *334*.

M. Krystek
Lattice Parameters of $(Ba_xSr_{100-x})SO_4$ Doped with Europium
(a) Vol. 54, No. 2, p. K133–K135 (1979).
See also *1.1*, p. *13*.

M. M. Nadoliisky, S. D. Toshev, and T. K. Vasileva
Electric Conductivity and Dielectric Properties of Ferroelectric PZT Ceramics Doped with Cr_2O_3
(a) Vol. 54, No. 2, p. K145–K150 (1979).
See also *14.4.1*, p. *246*.

V. N. Bapat
Thermoluminescence Process in $CaSO_4$:RE
(a) Vol. 54, No. 2, p. K171–K173 (1979).
See also *10.2*, p. *112*; *20.3*, p. *334*.

M. Razeghi, J. P. Buisson, and B. Houlier
EPR Investigation of Gd^{3+} and Eu^{2+} in the α- and β-Phase of Lead Phosphate
(b) Vol. 95, No. 1, p. 283–289 (1979).
See also *1.2*, p. *19*; *19*, p. *295*.

V. K. Jain
EPR of Cr^{3+} in $(NH_4)_xRb_{1-x}Al(SO_4)_2 \cdot 12\ H_2O$ Single Crystals
(b) Vol. 95, No. 1, p. K79–K82 (1979).
See also *19*, p. *295*.

P. Bennema, J. P. van der Eerden, W. J. P. van Enckevort, B. van der Hoek, and K. Tsukamoto
Observation and Interpretation of Spirals for Spirals Growth from the Liquid State
(a) Vol. **55**, No. 2, p. 403–413 (1979).
See also 1.6, p. 39; *3*, p. *53*; 22.5.2, p. 471.

J. B. Bilde-Sørensen
TEM Method for Identification of Domains in Materials Which Undergo a Pm3m → R3c or R$\bar{3}$c Phase Transition
(a) Vol. **55**, No. 2, p. 747–750 (1979).
See also *1.2*, p. *20*; *1.4*, p. *31*.

A. E. Krumins and P. Guenter
Photovoltaic Effect and Photoconductivity in Reduced Potassium Niobate Crystals
(a) Vol. **55**, No. 2, p. K185–K189 (1979).
See also 14.4.1, p. 247; *16*, p. *254*.

G. D. Bagratishvili, R. B. Dzhanelidze, D. A. Jishiashvili, L. V. Piskanovskii, and Z. N. Shiolashvili
Boron Diffusion from a Reactively Sputtered Glass Source in Si and SiO_2
(a) Vol. **56**, No. 1, p. 27–35 (1979).
See also 1.5, p. 36; *9*, p. *88*; 20.1, p. 322.

L. Wolber
Simultaneous Measurement of Thermally Stimulated Exoelectron Emission and Surface Charging of $BaSO_4$
(a) Vol. **56**, No. 1, p. 207–212 (1979).
See also 13.3, p. 162; *17*, p. *259*.

M. Połomska, M. Malinowski, and H. H. Otto
Dielectric and Electric Study of $Pb_5(Ge, Si)_3O_{11}$ Single Cystals
(a) Vol. **56**, No. 1, p. 335–339 (1979).
See also *14.4*, p. *243*.

S. L. Chaplot and V. C. Sahni
Lattice Dynamics of $(NH_4)_2CuCl_4 \cdot 2\ H_2O$
(b) Vol. **96**, No. 2, p. 575–580 (1979).
See also *6*, p. *76*.

S. K. Misra and B. Mikolajczak
EPR of Mn^{2+} Doped Single Crystals of $Ni(NH_4)_2(SO_4)_2 \cdot 6\ H_2O$ and $NiSO_4 \cdot 7\ H_2O$
(b) Vol. **96**, No. 2, p. 807–815 (1979).
See also 13.4, p. 197; *19*, p. *297*.

P. E. Tomaszewski and A. Pietraszko
Thermal Expansion of Lithium Ammonium Sulphate
(a) Vol. **56**, No. 2, p. 467–472 (1979).
See also *1.2*, p. *21*; 14.4.1, p. 247.

W. Kubitzki and B. Hoffmann
Application of an AC Thermopower Measurement to Electronic Ceramics
(a) Vol. **56**, No. 2, p. 573–579 (1979).
See also *14.3*, p. *218*; *15*, p. *251*.

22.8.2 Spinels, Garnets, Ferrites

Original Papers and Short Notes

M. Bałanda and S. Nizioł
Spin Reorientation in HoIG Investigated by Means of Faraday Effect
(b) Vol. **91**, No. 1, p. 291–298 (1979).
See also *18.3*, p. *283*.

Z. M. Stadnick and W. Zarek
On the Exchange Integrals in Nickel Ferrite
(b) Vol. **91**, No. 1, p. K83–K85 (1979).
See also *18.3*, p. *283*.

M. A. Sigal
Ferromagnetic Resonance Absorption in a Thin Uniaxial Platelet with Stripe Domain Structure Magnetized along the Easy Axis
(a) Vol. **51**, No. 1, p. 151–161 (1979).
See also *18.2*, p. *270*; *19*, p. *289*.

S. Prasad, S. B. Singh, R. M. Mehra, and P. K. Singh
A New Expression for Diagonal Components of the Permeability Tensor and Resonance in Partially Magnetised Ferrites
(a) Vol. **51**, No. 1, p. K75–K78 (1979).
See also *18.3*, p. *283*.

G. A. Jones and I. B. Puchalska
The Birefringent Effects of Magnetic Colloid Applied to the Study of Magnetic Domain Structures
(a) Vol. **51**, No. 2, p. 549–558 (1979).
See also *18.2*, p. *271*.

T. Sakata and A. Enomura
Studies of the Rare Earth–Iron Interactions in the Orthoferrites $GdFeO_3$ and $HoFeO_3$
(a) Vol. **52**, No. 1, p. 311–314 (1979).
See also *18.1*, p. *266*; 18.2, p. 272.

A. Stępniak and J. Skwarcz
Optical and Magnetic-Optical Properties of $Sm_{0.55}Tb_{0.45}FeO_3$
(a) Vol. **52**, No. 1, p. K77–K79 (1979).
See also *18.3*, p. *283*; *20.1*, p. *309*.

Yu. E. Perlin, A. A. Kaminskii, V. N. Enakii, and D. N. Vylegzhanin
Nonradiative Multiphonon Relaxation in $Y_3Al_5O_{12}:Nd^{3+}$
(b) Vol. **92**, No. 2, p. 403–410 (1979).
See also 6, p. 66; *13.4*, p. *167*; *20.3*, p. *329*.

M. Popescu and C. Ghizdeanu
Cation Distribution in Cobalt Ferrite–Aluminates
(a) Vol. **52**, No. 2, p. K169–K172 (1979).
See also *1.1*, p. *12*; *18.3*, p. *283*.

K. Fischer, D. Linzen, and E. Sinn
On the Structure of Oxidic High Temperature Solutions Used for the Growth of LPE Garnet Films
(a) Vol. **55**, No. 1, p. K75—K77 (1979).
See also *3*, p. *53*.

Y. Kawai, M. Tanabe, and T. Ogawa
Electrical Conductivity of a Manganese–Zinc Ferrite Single Crystal
(a) Vol. **55**, No. 1, p. K119—K122 (1979).
See also *14*, p. *198*.

J. Jasenek
Solution of a Transient Phenomenon in the Spin System Using the Eigen-Coordinate System of the Magnetization Vector ***M***
(b) Vol. **95**, No. 2, p. K131—K134 (1979).
See also *18.2*, p. *278*.

N. D. Zakharov, W. Neumann, R. Hillebrand, and V. N. Rozhanskii
Analysis of Many-Beam Lattice Images of Y–Al Garnet Crystals
(a) Vol. **55**, No. 2, p. 573—582 (1979).
See also *1.1*, p. *14*; 1.4, p. 31.

G. Srinivasan
Domain Wall Resonance in Zn and Cu Substituted Magnetite
(a) Vol. **55**, No. 2, p. K149—K152 (1979).
See also *18.3*, p. *285*.

H. B. von Basel and K. A. Hempel
High Orientation of Coprecipitated Barium Ferrite by Uniaxial Hot-Pressing
(a) Vol. **55**, No. 2, p. K183—K184 (1979).
See also *18.3*, p. *285*.

Z. Šimša, P. Široký, F. Lukeš, and E. Schmidt
Optical Properties of Manganese Ferrites
(b) Vol. **96**, No. 1, p. 137—144 (1979).
See also *20.1*, p. *320*.

R. M. Nayak and B. D. Padalia
Chemical Shifts of the K-Absorption Edge of Mn in Some Mixed Oxidic Spinels
(b) Vol. **96**, No. 1, p. 259—262 (1979).
See also *4*, p. *57*; *20*, p. *302*.

G. N. Falileev
Calculation of the Magnetization–Temperature Dependence of Rare-Earth and Yttrium Iron Garnet Solid Solutions
(b) Vol. **96**, No. 1, p. 429—435 (1979).
See also *18.3*, p. *285*.

A. I. Merkulov, Yu. M. Yakovlev, S. I. Radautsan, V. E. Tezlevan, and T. N. Lyanguzova
The Influence of Acceptor States on the Exchange Interaction in $CdCr_2Se_4$ Single Crystals
(b) Vol. **96**, No. 1, p. K57—K59 (1979).
See also *18.2*, p. *279*; *18.4*, p. *288*.

22.9 Organic Semiconductors

Original Papers and Short Notes

C. Benoit, M. Galtier, and A. Montaner
A Model for TTF-TCNQ Conductivity
(b) Vol. **91**, No. 1, p. 269 – 274 (1979).
See also *14*, p. *197*; *20*, p. *298*.

U. Schmelzer, B. Dorner, J. Kalus, I. Natkaniec, R. Ostertag, G. S. Pawley, G. A. Mackenzie
E. F. Sheka, and E. L. Bokhenkov
Pressure Dependence of Lattice Frequencies of Deuterated Naphthalene at 100 K
(b) Vol. **91**, No. 1, p. K27 – K29 (1979).
See also *6*, p. *64*.

K. Kojima
Slip Systems in Deformed Anthracene Crystals
(a) Vol. **51**, No. 1, p. 71 – 78 (1979).
See also *10.2*, p. *103*.

G. Sorge and U. Straube
Dielectric Behaviour of Ferroelastic Monodomain TSCC Crystals
(a) Vol. **51**, No. 1, p. 117 – 121 (1979).
Erratum
(a) Vol. **53**, No. 2, p. 681 (1979).
See also *14.4.1*, p. *244*.

A. Brau, J. P. Farges, and H. Grassi
Anomalies of the Electrical Conductivity in the TCNQ Complex Salt Formed from the Cation (Methyl-1-N-Ethyl Benzimidazolium)
(a) Vol. **51**, No. 1, p. K45 – K48 (1979).
See also *14.3*, p. *210*.

D. Hauschildt, W. Fuhs, and H.-J. Hesse
Photoconduction in Single Crystals of a Pentamethinium Cyanine Dye
(b) Vol. **91**, No. 2, p. 447 – 453 (1979).
See also *16*, p. *252*.

Cz. Pawlaczyk, G. Luther, and H. E. Müser
Influence of Crystal Lattice Defects on the Ferroelectric Dispersion of Triglycine Sulphate
(b) Vol. **91**, No. 2, p. 627 – 631 (1979).
See also 10.2, p. 103; *14.4.1*, p. *244*.

D. D. Kolendritskii, M. V. Kurik, and Yu. P. Piryatinskii
Exciton Reflection Spectra of Tetracene Single Crystals
(b) Vol. **91**, No. 2, p. 741 – 751 (1979).
See also *13.5.1*, p. *183*; *20.1*, p. *306*.

A. Szyszkowski
Polarized IR Reflection Spectra of Crystals of Derivatives of Pyridine-Tetracyanoquinodimethane. An Evidence for Electron–Vibration Coupling
(b) Vol. **91**, No. 2, p. K95 – K98 (1979).
See also 6, p. 65; *20.1*, p. *306*.

T. V. Rao and K. L. Chopra
Electron Transport Properties of Cu-Doped PVC Films
(a) Vol. **53**, No. 1, p. 43–50 (1979).
See also *14.3.1*, p. *219*.

J. Godlewski and J. Kalinowski
Photo-Enhanced Currents in Organic Insulators
(a) Vol. **53**, No. 1, p. 161–170 (1979).
See also *14.3.4*, p. *233*; *16*, p. *253*.

R. C. Thompson, V. K. Gujral, H. J. Wagner and C. F. Schwerdtfeger
Magnetic and Electric Properties of $Mn(TCNQ)_2$ and Its Hydrates
(a) Vol. **53**, No. 1, p. 181–186 (1979).
See also *14.3*, p. *212*; *18*, p. *262*; 19, p. 292.

N. I. Ostapenko, V. I. Sugakov M. P. Chernomorets, and M. T. Shpak
Spectra of Directionally Deformed Naphthalene Crystals and Exciton–Phonon Interaction Constants
(b) Vol. **93**, No. 2, p. 493–501 (1979).
See also 6, p. 68; 8, p. 81; *20.1*, p. *312*.

K. N. Subramanyam and L. R. Khare
The X-Band Microwave Dielectric Spectra of a Certain Polystyrene Foam
(a) Vol. **53**, No. 2, p. K139–K141 (1979).
See also *14.4*, p. *240*.

S. Kania, A. Lipiński, and W. Mycielski
Field-Dependent Mobility in Oriented p-Quaterphenyl Layers
(a) Vol. **53**, No. 2, p. K223–K224 (1979).
See also *14.3.4*, p. *234*.

S. Arnold, M. Pope, and T. K. T. Hsieh
Double Quantum External Photoelectric Effect in Crystalline Tetracene
(b) Vol. **94**, No. 1, p. 263–272 (1979).
See also 13.4, p. 172; *17*, p. *257*.

G. Mihály, A. Jánossy, J. Kürti, L. Forró, and G. Grüner
Nonlinear Transport in $Qn(TCNQ)_2$
(b) Vol. **94**, No. 1, p. 287–296 (1979).
See also *14*, p. *197*.

K. Rozwadowska
Preparation and DC Conductivity of the Low-Temperature Glassy Carbon
(a) Vol. **54**, No. 1, p. 93–97 (1979).
See also *2*, p. *45*; *14.3.1*, p. *220*.

P. Petelenz
Mixing of Frenkel Excitons and Ionic Excited States in a Linear Crystal of Donor–Acceptor Complex
(b) Vol. **94**, No. 2, p. 371–379 (1979).
See also *13.4*, p. *173*; *13.5.1*, p. *186*.

Ch. Kleint
Stationary Stochastic Adparticle Flip-Flop as a Precursor to Surface Diffusion
(a) Vol. 55, No. 2, p. 447–456 (1979).
See also *1.6*, p. *39*; *9*, p. *88*; 17.1, p. 260.

M. Godlewska, J. Ściesiński, and E. Ściesińska
Spectroscopic Study of Phase Polymorphism and Vibrational Assignments for 2,2-Dinitropropane
(a) Vol. 55, No. 2, p. 723–734 (1979).
See also 1.2, p. 20; *6*, p. *75*; *20.1*, p. *320*.

E. Pink, H. Bäck, and B. Ortner
The Stress Relaxation of Polyvinyl Chloride in the Macrodeformation Range
(a) Vol. 55, No. 2, p. 751–761 (1979).
See also *10.2*, p. *115*.

H. Grassi, A. Brau, and J. P. Farges
Observation of Conductivity Anomalies in Small Needle-Like Crystals of $TEA(TCNQ)_2$
(a) Vol. 55, No. 2, p. K179–K182 (1979).
See also *14.3*, p. *216*.

Y. Kaahwa
Electronic Mobility in Polyethylene Terephthalate
(a) Vol. 55, No. 2, p. K197–K199 (1979).
See also 14.3.4, p. 236; *14.4*, p. *242*.

W. J. Gunning, C. K. Chiang, A. J. Heeger, and A. J. Epstein
Effects of Controlled Disorder on the Phase Transitions and Charge Density Wave State in Tetrathiofulvalene–Tetracyanoquinodimethane
(b) Vol. 96, No. 1, p. 145–151 (1979).
See also *1.2*, p. *20*; *11*, p. *124*; 14.3, p. 216; 18, p. 264.

M. A. Shaikh
Electrical and Thermoelectric Properties of Dibenzotetrathiofulvalinium-Tetracyanoquinodimethane
(b) Vol. 96, No. 1, p. K7–K9 (1979).
See also *14.3*, p. *216*; *15*, p. *251*.

J. Godlewski and J. Kalinowski
Trapping Effects on Emission-Limited Current Flow in Narrow-Band Insulators
(a) Vol. 56, No. 1, p. 293–303 (1979).
See also *14.3.4*, p. *237*.

J.-L. Goudard, Amir A. Lakhani, and N. K. Hota
Dielectric Constant of High and Low Temperature Phases of K-TCNQ
(a) Vol. 56, No. 1, p. K53–K56 (1979).
See also *14.4*, p. *243*.

J. Godlewski and J. Kalinowski
On Action Spectra of Photocurrent in Organic Insulators
(a) Vol. 56, No. 1, p. K57–K60 (1979).
See also *14.3.4*, p. *237*; *16*, p. *254*.

23. Solidified Gases

Review Article

V. K. PERSHIN and VL. K. PERSHIN
Conditions of the Existence of a Molecular Crystal as a Mesogenous System
(b) Vol. 92, No. 1, p. 9–22 (1979).
See also 1.2, p. 15; 2, p. 40; *6*, p. *63*.

Original Papers and Short Notes

M. S. KUSHWAHA and S. S. KUSHWAHA
The Frequency Spectra of Solid Neon and Xenon
(b) Vol. 91, No. 2, p. 689–692 (1979).
See also *6*, p. *64*.

D. Y. CHUNG and V. KONECY
The Determination of the Orientation of Cubic Crystals from Measured Sound Velocities
(a) Vol. 52, No. 1, p. 29–33 (1979).
See also *1*, p. *9*.

S. LEHRI and M. P. VERMA
Phonon Dispersion in Rare Gas Solids
(b) Vol. 92, No. 2, p. 363–370 (1979).
See also 4, p. 56; *6*, p. *66*; 12.2, p. 133.

M. S. KUSHWAHA and S. S. KUSHWAHA
The Lattice Dynamics of Quantum Crystals of F.C.C. ^{4}He
(b) Vol. 93, No. 1, p. 237–241 (1979).
See also *6*, p. *68*; 12.2, p. 134.

V. RAMAMURTHY and S. B. RAJENDRAPRASAD
On the Homology of Lattice Vibrations of Rare-Gas Solids
(b) Vol. 93, No. 2, p. 793–797 (1979).
See also *6*, p. *69*; 21.2, p. 379.

H. J. JODL, G. THEYSOHN, and R. BRUNO
Frequency Shift Analysis of Matrix Isolated Impurities
(b) Vol. 94, No. 1, p. 161–169 (1979).
See also *13.4*, p. *172*.

N. E. MASSA, S. S. MITRA, and J. F. VETELINO
Volume Dependent Lattice Dynamics of Solid Argon. A Pseudo-Harmonic Approach
(b) Vol. 94, No. 2, p. 387–396 (1979).
See also *6*, p. *71*; 8, p. 82.

V. I. ZUBOV
On the Theory of Phase Transition between a Strongly Anharmonic Crystal and Gas
(b) Vol. 94, No. 2, p. K105–K109 (1979).
See also *1.2*, p. *19*.

Author Index

For comments on the use of the classification schema see p. 6.

32913/32914

Manuscripts and letters for physica status solidi (b) — basic research and physica status solidi (a) — applied research

should be sent to

DDR-1020 Berlin, Neue Schönhauser Str. 20

or

Internationaler Buchversand GmbH, 1000 Berlin 19, Eschenallee 21.

The object of "physica status solidi (b)" and "physica status solidi (a)" is to publish rapidly important and topical results concerning fundamental research in the field of solid state physics and aspects of application and experimental techniques of fundamental solid state research, respectively. The issues of "physica status solidi (b)" and "physica status solidi (a)" will be published on the 1st and the 16th day of each month, respectively. Either series will contain the following categories:

1. Review articles written by arragement with the editors. They will be remunerated.
2. Original papers selected from those which have not been submitted for publication elsewhere, after being reviewed. (The latest date for acceptance of contributions is the 8th and the 23rd day of each month for "physica status solidi (b)" and "physica status solidi (a)", respectively, 50 days before the delivery of the journal.)
3. Short notes published in the off-set part of the journal. They are, if necessary after being reviewed, selected by the editorial staff. (The latest date for acceptance of manuscripts for this part is the 12th and the 27th day of each month for "physica status solidi (b)" and "physica status solidi (a)", respectively, 16 days before the delivery of the journal.)
4. Pre-printed titles of papers to be published in the next issue of both series.

Editorial Information

Manuscripts should be submitted preferentially in English. Besides, papers in German, French, and Russian are admitted. The final decision on the incorporation of the papers in the series (b) or (a) is taken by the editorial staff after consultation of the referee. Authors are asked to observe the following rules:

1. Manuscripts have to be written as concisely as possible. As most, 20 and 4 pages (28 lines per page; abstracts, tables, references, figures, and captions of figures included, where 3 figures are counted as about one page) are admitted for original papers and short notes, respectively. All manuscripts should be type-written on one side only. Two copies are required. Badly arranged manuscripts are returned.
2. The author's name, initials, and institution should be given.
3. One abstract of about ten lines in English and one in German, French, or Russian are required for original papers.
4. Because of lack of space conclusions (if any) must not have the character of a summary (i. e. substantially a repetition of the contents of the abstract).
5. Formulae should not be written too small and not with pencil. Separate lines for formulae are desirable. Indistinct writing will lead to a return of the manuscript.
6. Drawings should be enclosed on separate sheets of paper, not exceeding the maximum and minimum sizes of 20×30 and $10 \times 15\ \text{cm}^2$, respectively. Photographs should be submitted, on white paper, only for review articles and original papers. Duplicates are required.
7. Captions for the figures should be submitted on separate sheets of papers, and special instructions with regard to the size of illustrations should be given on these.
8. References should be numbered in order of appearance and should be given, on a separate sheet, in the form used in physica status solidi.
9. SI-units should be used.
10. Each paper should contain a proposal for the subject classification according to the classification scheme appearing on the second and third cover of each issue.

The authors will be sent galley proofs of review articles and original papers. Proof reading of short notes and pre-printed titles will be done by the editorial staff.

For each original paper, and short note, a total of 75 reprints is available free of charge. The number of reprints free of charge per review article is 25. Requests for additional reprints (against payment) should be given together with the submitted manuscript. **The deadline for orders of additional reprints is the 15th and the 30th day of each month** for "physica status solidi (b)" and "physica status solidi (a)", respectively, two weeks before the delivery of the journal.

Bei Fragen zur Produktsicherheit wenden Sie sich bitte an:
If you have any questions regarding product safety,
please contact:

Walter de Gruyter GmbH
Genthiner Straße 13
10785 Berlin
productsafety@degruyterbrill.com